Philosophy

This series contains both philosophical texts and critical essays about philosophy, concentrating especially on works originally published in the eighteenth and nineteenth centuries. It covers a broad range of topics including ethics, logic, metaphysics, aesthetics, utilitarianism, positivism, scientific method and political thought. It also includes biographies and accounts of the history of philosophy, as well as collections of papers by leading figures. In addition to this series, primary texts by ancient philosophers, and works with particular relevance to philosophy of science, politics or theology, may be found elsewhere in the Cambridge Library Collection.

The Philosophy of the Inductive Sciences

First published in 1840, this two-volume treatise by Cambridge polymath William Whewell (1794–1886) remains significant in the philosophy of science. The work was intended as the 'moral' to his three-volume *History of the Inductive Sciences* (1837), which is also reissued in this series. Building on philosophical foundations laid by Immanuel Kant and Francis Bacon, Whewell opens with the aphorism 'Man is the Interpreter of Nature, Science the right interpretation'. Volume 2 contains the final sections of Part 1, addressing namely the philosophy of biology and palaetiology. Part 2, 'Of Knowledge', includes a selective review of opinions on the nature of knowledge and the means of seeking it, beginning with Plato. Whewell's work upholds throughout his belief that the mind was active and not merely a passive receiver of knowledge from the world. A key text in Victorian epistemological debates, notably challenged by John Stuart Mill and his *System of Logic*, Whewell's treatise merits continued study and discussion in the present day.

Cambridge University Press has long been a pioneer in the reissuing of out-of-print titles from its own backlist, producing digital reprints of books that are still sought after by scholars and students but could not be reprinted economically using traditional technology. The Cambridge Library Collection extends this activity to a wider range of books which are still of importance to researchers and professionals, either for the source material they contain, or as landmarks in the history of their academic discipline.

Drawing from the world-renowned collections in the Cambridge University Library and other partner libraries, and guided by the advice of experts in each subject area, Cambridge University Press is using state-of-the-art scanning machines in its own Printing House to capture the content of each book selected for inclusion. The files are processed to give a consistently clear, crisp image, and the books finished to the high quality standard for which the Press is recognised around the world. The latest print-on-demand technology ensures that the books will remain available indefinitely, and that orders for single or multiple copies can quickly be supplied.

The Cambridge Library Collection brings back to life books of enduring scholarly value (including out-of-copyright works originally issued by other publishers) across a wide range of disciplines in the humanities and social sciences and in science and technology.

The Philosophy of the Inductive Sciences

Founded Upon their History

VOLUME 2

WILLIAM WHEWELL

CAMBRIDGE
UNIVERSITY PRESS

University Printing House, Cambridge, CB2 8BS, United Kingdom

Published in the United States of America by Cambridge University Press, New York

Cambridge University Press is part of the University of Cambridge.
It furthers the University's mission by disseminating knowledge in the pursuit of
education, learning and research at the highest international levels of excellence.

www.cambridge.org
Information on this title: www.cambridge.org/9781108064033

This edition first published 1840
This digitally printed version 2014

ISBN 978-1-108-06403-3 Paperback

THE

PHILOSOPHY

OF THE

INDUCTIVE SCIENCES,

FOUNDED UPON THEIR HISTORY.

BY THE

REV. WILLIAM WHEWELL, B.D.,

FELLOW OF TRINITY COLLEGE, AND PROFESSOR OF MORAL PHILOSOPHY IN THE UNIVERSITY
OF CAMBRIDGE, VICE PRESIDENT OF THE GEOLOGICAL SOCIETY
OF LONDON.

IN TWO VOLUMES.

Λαμπάδια ἔχοντες διαδώσουσιν ἀλλήλοις.

VOLUME THE SECOND.

LONDON:

JOHN W. PARKER, WEST STRAND.

CAMBRIDGE: J. AND J. J. DEIGHTON.

M.DCCC.XL.

CONTENTS

OF

THE SECOND VOLUME.

THE

PHILOSOPHY

OF THE

INDUCTIVE SCIENCES.

PART I.

OF IDEAS.

Τῶν καλῶν καὶ τιμίων τὴν εἴδησιν ὑπολαμβάνοντες, μᾶλλον δ'
ἑτέραν ἑτέρας ἢ κατ' ἀκρίβειαν ἢ τῷ βελτιόνων τε καὶ θαυμασιω-
τέρων εἶναι, δι' ἀμφότερα ταῦτα τὴν ΤΗΣ ΨΥΧΗΣ ΊΣΤΟΡΙΑΝ
εὐλόγως ἂν ἐν πρώτοις τιθείημεν· δοκεῖ δε καὶ προς ἀλήθειαν
ἅπασαν ἡ γνῶτις αὐτῆς μεγάλα συμβάλλεσθαι, μάλιστα δε πρὸς
τὴν φυσιν· ἔστι γαρ οἷον ΑΡΧΗ ΤΩΝ ΖΩΩΝ.

ARISTOT. Πέρι Ψυχῆς. 1.

BOOK IX.

THE PHILOSOPHY OF BIOLOGY.

CHAPTER I.

ANALOGY OF BIOLOGY WITH OTHER SCIENCES.

1. IN the History of the Sciences, after treating of the Sciences of Classification, we proceeded to what are there termed the Organical Sciences, including in this term Physiology and Comparative Anatomy. A peculiar feature in this group of sciences is that they involve the notion of *living* things. The notion of *Life*, however vague and obscure it may be in men's minds, is apprehended as a peculiar Idea, not resolvable into any other Ideas, such, for instance, as Matter and Motion. The separation between living creatures and inert matter, between organized and unorganized beings, is conceived as a positive and insurmountable barrier. The two classes of objects are considered as of a distinct kind, produced and preserved by different forces. Whether the Idea of Life is really thus original and fundamental, and whether, if so, it be one Idea only, or involve several, it must be the province of true philosophy to determine. What we shall here offer may be considered as an attempt to contribute something to the determination of these questions; but we shall perhaps be able to make it appear that science is at present only in the course of its progress towards a complete solution of such problems.

Since the main feature of those sciences of which

we have now to examine the philosophy is, that they
involve the idea of life, it would be desirable to have
them designated by a name expressive of that circumstance.   The word *Physiology*, by which they have most
commonly been described, means *the Science of Nature;*
and though it would be easy to explain, by reference to
history, the train of thought by which the word was
latterly restricted to living nature, it is plain that the
name is, etymologically speaking, loose and improper.
The term *Biology*, which means exactly what we wish to
express, *the Science of Life*, has often been used, and has
of late become not uncommon among good writers.   I
shall therefore venture to employ it, in most cases, rather
than the word *Physiology*.

2. As I have already intimated, one main inquiry
belonging to the Philosophy of Biology, is concerning the
Fundamental Idea or Ideas which the science involves.
If we look back at the course and the results of our disquisitions respecting other sciences in this work, and
assume, as we may philosophically do, that there will be
some general analogy between those sciences and this,
in their developement and progress, we shall be enabled
to anticipate in some measure the nature of the view
which we shall now have to take.   We have seen that
in other subjects the Fundamental Ideas on which science depended, and the Conceptions derived from these,
were at first vague, obscure, and confused;—that by
gradual steps, by a constant union of thought and observation, these conceptions become more and more clear,
more and more definite;—and that when they approached
complete distinctness and precision, there were made
great positive discoveries into which these conceptions
entered, and thus the new precision of thought was
fixed and perpetuated in some conspicuous and lasting
truths.   Thus we have seen how the first confused

mechanical conceptions (Force and the like,) were from time to time growing clearer down to the epoch of Newton; —how true conceptions of Genera and of wider classes, gradually unfolded themselves among the botanists of the sixteenth and seventeenth centuries;—how the idea of Substance became steady enough to govern the theories of chemists only at the epoch of Lavoisier;—how the Idea of Polarity, although often used by physicists and chemists, is even now somewhat vague and indistinct in the minds of the greater part of speculators.   In like manner we may expect to find that the Idea of Life, if indeed that be the governing Idea of the science which treats of living things, will be found to have been gradually approaching towards a distinct and definite form among the physiologists of all ages up to the present day.   And if this be the case, it may not be considered superfluous, with reference to so interesting a subject, if we employ some space in tracing historically the steps · of this progress;—the changes by which the originally loose idea of Life, or of Vital Powers, became more nearly suited to the purposes of science.

3. But we may safely carry this analogy between Biology and other sciences somewhat further.   We have seen, in other sciences, that while men in their speculations were thus tending towards a certain peculiar Idea, but before they as yet saw it clearly to be peculiar and independent, they naturally and inevitably clothed their speculations in conceptions borrowed from some other extraneous idea.   And the unsatisfactoriness of all such attempts, and the necessary consequence of this, a constant alteration and succession of such inappropriate hypotheses, were indications and aids of the progress which was going on towards a more genuine form of the science.   For instance, we have seen that in chemistry, so long as men refused to recognise a peculiar and distinct kind of power

in the *Affinity* which binds together the elements of bodies, they framed to themselves a series of hypotheses, each constructed according to the prevalent ideas of the time, by which they tried to represent the relation of the compound to the ingredients:—first supposing that the elements bestowed upon the whole qualities *resembling* their own:—then giving up this supposition, and imagining that the properties of the body depended upon the *shape* of the component particles;—then, as their view expanded, assuming that it was not the shape, but the mechanical *forces* of the particles which gave the body its attributes;—and finally acquiescing in, or rather reluctantly admitting, the idea of *Affinity*, conceived as a peculiar power, different not only from material contact, but from any mechanical or dynamical attraction.

Now we cannot but think it very natural to find that the history of Biology offers a series of occurrences of the same nature. The notions of Life in general, or of any Vital Functions or Vital Forces in particular, are obviously very loose and vague as they exist in the minds of most men. The discrepancies and controversies respecting the definitions of all such terms, which are found in all works on physiology, afford us abundant evidence that these notions are not, at least not generally, apprehended with complete clearness and steadiness. We shall therefore find approaches and advances, intermediate steps, gradually leading up to the greatest degree of distinctness which has yet been attained. And in those stages of imperfect apprehension in which the notions of life and of vital powers are still too loose and unformed to be applied independently, we may expect to find them supported and embodied by means of hypotheses borrowed from other subjects, and thus made so distinct and substantial as to supply at least a temporary possibility of scientific reasoning upon the laws of life.

4. For example, if we suppose that men begin to speculate upon the properties of living things without acknowledging a peculiar Vital Power, but making use successively of the knowledge supplied by the study of other subjects, we may easily imagine a series of hypotheses along which they would pass.

They would probably, first, in this as in other sciences, have their thoughts occupied by vague and *mystical* notions in which material and spiritual agency, natural and supernatural events, were mixed together without discrimination, and without any clear notion at all. But as they acquired a more genuine perception of the nature of knowledge, they would naturally try to explain vital motions and processes by means of such forces as they had learnt the existence of from other sciences. They might first have a *mechanical* hypothesis, in which the mechanical forces of the solids and fluids which compose organized bodies should be referred to, as the most important influences in the process of life. They might then attend to the actions which the fluids exercise in virtue of their affinity, and might thus form a *chemical* theory. When they had proved the insufficience of these hypotheses, borrowed from the powers which matter exhibits in other cases, they might think themselves authorized to assume some peculiar power or agency, still material, and thus they would have the hypothesis of a *vital fluid*. And if they were driven to reject this, they might think that there was no resource but to assume an immaterial principle of life, and thus they would arrive at the doctrine of an animal *soul*.

Now, through the cycle of hypotheses which we have thus supposed, physiology has actually passed. The conclusions to which the most philosophical minds have been led by a survey of this progress is, that by the failure of all these theories, men have exhausted this path of inquiry,

and shown that scientific truth is to be sought in some
other manner.   But before I proceed further to illustrate
this result, it will be proper, as I have already stated, to
exhibit historically the various hypotheses which I have
described.   In doing this I shall principally follow the
History of Medicine of Sprengel.   It is only by taking
for my guide a physiologist of acknowledged science and
judgment, that I can hope, on such a subject, to avoid
errors of detail.   I proceed now to give in succession an
account of the Mystical, the Iatrochemical, the Iatromathematical, and the Vital-Fluid Schools ; and finally of
the Psychical School who hold the Vital Powers to be
derived from the Soul (*Psychè*).

Chapter II.
SUCCESSIVE BIOLOGICAL HYPOTHESES.

Sect. I. *Mystical School.*—In order to abbreviate as
much as can conveniently be done the historical view
which I have now to take, I shall altogether pass over
the physiological speculations of the ancients, and begin
my sur, with the general revival of science in modern
times.

We need not dwell long on the fantastical and unsubstantial doctrines concerning physiology which prevailed
in the sixteenth century, and which flowed in a great
measure from the fertile but ill-regulated imaginations of
the cultivators of Alchemy and Magic.   One of the prominent doctors of this school is the celebrated Paracelsus,
whose doctrines contained a combination of biblical interpretations, visionary religious notions, fanciful analogies, and bold experiments in practical medicine.   The
opinion of a close but mystical resemblance of parts
between the universe and the human body, the *Macrocosm*

and the *Microcosm*, as these two things, thus compared, were termed, had probably come down from the Neoplatonists; it was adopted by the Paracelsists*, and connected with various astrological dreams and cabbalistic riddles. A succession of later Paracelsists†, Rosicrucians, and other fanatics of the same kind, continued into the seventeenth century. Upon their notions was founded the pretension of curing wounds by a sympathetic powder, which Sir Kenelm Digby, among others, asserted; while animal magnetism, and the transfer of diseases from one person to another‡, were maintained by others of this school. They held, too, the doctrines of *astral bodies* corresponding to each terrestrial body; and of the *signatures* of plants, that is, certain features in their external form by which their virtues might be known. How little advantage or progress real physiology could derive from speculations of this kind may be seen from this, that their tendency was to obliterate the distinction between living and lifeless things: according to Paracelsus, all things are alive, eat, drink, and excrete; even minerals and fluids§. According to him and his school, besides material and immaterial beings, there are *elementary spirits* which hold an intermediate place, *sylvans, nymphs, gnomes, salama·ʒ ·ers*, &c. by whose agency various processes of enchantment may be achieved, and things apparently supernatural explained. Thus this spiritualist scheme dealt with a world of its own by means of fanciful inventions and mystical visions, instead of making any step in the study of nature.

Perhaps, however, one of the most fantastical of the inventions of Paracelsus may be considered as indicating a perception of a peculiar character in the vital powers. According to him, the business of digestion is performed by a certain demon whom he calls *Archæus*, who has his

---

* Spr., iii., 456.   † *Ib.*, iv., 270.   ‡ *Ib.*, iv., 276.

§ *Ib.*, iii. 458.   Parac., *De Vita Rerum Naturàlium*, p. 889.

abode in the stomach, and who, by means of his alchemical processes, separates the nutritive from the harmful
part of our food, and makes it capable of assimilation*.
This fanciful notion was afterwards adopted and expanded
by Van Helmont†. According to him the stomach and
spleen are both under the direction of this master-spirit,
and these two organs form a sort of *Duumvirate* in the body.

But though we may see in such writers occasional
gleams of physiological thought, the absence of definite
physical relations in the speculations thus promulgated
was naturally intolerable to men of sound understanding
and scientific tendencies. Such men naturally took hold of
that part of the phenomena of life which could be most distinctly conceived, and which could be apparently explained
by means of the sciences then cultivated; and this was the
part which appeared to be reducible to chemical conceptions and doctrines. It will readily be supposed that the
processes of chemistry have a considerable bearing upon
physiological processes, and might, till their range was
limited by a sound investigation, be supposed to have
still more than they really had; and thus a physiology
was formed which depended mainly upon chemistry, and
the school which held this doctrine has been called the
iatrochemical school.

Sect. II. *The Iatrochemical School.*—That all physical
properties, and therefore chemical relations, have a material influence on physiological results, was already recognised, though dimly, in the Galenic doctrine of the four
elementary qualities. But at the time of Paracelsus,
chemical action was more distinctly than before separated
from other kinds of physical action; and therefore a
physiological doctrine, founded upon chemistry, and freed
from the extravagance and mysticism of the Paracelsists,
was a very promising path of speculation. Andrew Liba-

---

* Spr., iii.; 468.   † *Ib.*, iv., 302.

vius* of Halle, in Saxony, Physician and Teacher in the Gymnasium at Koberg, is pointed out by Sprengel as the person who began to cultivate chemistry, as distinct from the theosophic fantasies of his predecessors ; and Angelus Sala of Vienna†, as his successor. The latter has the laudable distinction of having rejected the prevalent conceits about potable gold, a universal medicine, and the like‡. In Germany already at the beginning of the seventeenth century a peculiar chair of *Chymiatria* was already created at Marpurg : and many in various places pursued the same studies, till, in the middle of the seventeenth century, we come to Lemery§, the principal reformer of pharmaceutical chemistry. But we are not here so much concerned with the practical as with the theoretical parts of iatrochemistry; and hence we pass on to Sylvius‖ and his system.

The opinion that chemistry had an important bearing upon physiology did not, however, begin with Sylvius. Paracelsus, among his extravagant absurdities, did some service to medicine by drawing attention to this important truth. He used¶ chemical principles for the explanation of particular diseases; most or all of them according to him, arise from the effervescence of salts, from the combustion of sulphur, or from the coagulation of mercury. His medicines were chemical preparations ; and it was**, an undeniable advantage of the Paracelsian doctrine that chemistry thus became indispensable to the physician. We still retain a remnant of the chemical nomenclature of Paracelsus in the term *tartar*, denoting the stony concretion which forms on the teeth††. According to him there is a certain substance, the basis of all diseases which arise from a thickening of the juices and a collection of

---

* Spr., iii., 550.   † *Ib.*, iv., 281.   ‡ *Ib.*, iv., 2J3.
§ *Ib.*, iv., 291.   ‖ *Ib.*, iv., 336.
¶ *Ib.*, iii., 472.   ** *Ib.*, iii., 482.   †† *Ib.*, iii., 475.

earthy matter; and this substance he calls *Tartarus,* because it "burns like the fire of hell." Helmont, the successor of Paracelsus in many absurdities, also followed him in the attempt to give a chemical account, however loose and wild, of the functions of the human body; and is by Sprengel considered, with all his extravagancies, as a meritorious and important discoverer. The notion of the fermentation of fluids*, and of the aërial product thence resulting, to which he gave the name of *Gas,* forms an important part of his doctrines; and of the six digestions which he assumes, the *first* prepares an acid, which is neutralized by the gall when it reaches the duodenum, and this constitutes the *second* digestion.

I have already, in the History of Chemistry†, stated, that the doctrine of the opposition of acid and alkali, the great step which theoretical chemistry owes to Sylvius, was first brought into view as a physiological tenet, although we had then to trace its consequences in another science. The explanation of all the functions of the animal system, both healthy and morbid, by means of this and other chemical doctrines, and the prescription of methods of cure founded upon such explanations, form the scheme of the *iatrochemical* school; a school which almost engrossed the favour of European physicians during the greater part of the seventeenth century.

Sylvius taught medicine at Leyden, from the year 1658, with so much success, that Boerhaave alone surpassed him‡. His notions, although he piqued himself on their originality, were manifestly suggested in no small degree (as all such supposed novelties are) by the speculations of his predecessors, and the spirit of the times. Like Helmont§, he considers digestion as consisting in a fermentation; but he states it more definitely

* Vol. v., 315.
† *Hist. Ind. Sci.,* iii., 108.
‡ Spr., iv., 336.
§ *Ib.,* 338.

as the effervescence of an acid, supplied by the saliva and the pancreatic juice, with the alkali of the gall. By various other hypothetical processes, all of a chemical nature, the blood becomes a collection of various juices, which are the subjects of the speculations of the iatro-chemists, to the entire neglect of the solid parts of the body. Diseases were accounted for by a supposed prevalence of one or the other of the acrid principles, the acid or the alkaline: and Sylvius* was bold enough to found upon these hypotheses practical methods of cure, which were in the highest degree mischievous.

The Sylvian doctrine was often combined with some of the notions of the Cartesian system of philosophy; but this mixture I shall not notice, since my present object is to trace the history of a mere chemical physio-logy as one of the unsuccessful attempts at a philosophy of life. With various modifications, this doctrine was diffused over Europe. It gave rise to several contro-versies, which turned upon the questions of the novelty of the doctrine, and the use of chemical remedies to which it pointed, as well as upon its theoretical truth. We need not dwell long upon these controversies, although they were carried on with no small vehemence in their time. Thus the school of Paris opposed all innovation, remained true to the Galenic dogmatism, and declared itself earnestly against all combination of che-mistry with medicine; and even against the chemical preparation of medicaments. Guy Patin, a celebrated and learned professor of that day, declares† that the chemists are no better than forgers, and ought to be pun-ished as such. The use of antimonial medicines was a main point of dispute between the iatrochemists and their opponents; Patin maintained that more men had been destroyed by antimony than by the thirty years war

---

* Spr., iv., 345. † Ib., 349.

of Germany; and endeavoured to substantiate this assertion by collecting all such cases in his *Martyrologium
Antimonii*. It must have been a severe blow to Patin
when*, in 1666, the doctors of the faculty of Paris,
assembled by command of the parliament, declared, by a
majority of ninety-two voices, that the use of antimonial
medicines was allowable and laudable, and when all
attempts to set aside this decision failed.

Florentius Schuyl of Leyden sought to recommend
the iatrochemical doctrines, by maintaining that they
were to be found in the Hippocratic writings; nor was
it difficult to give a chemical interpretation of the
humoral pathology of the ancients. The Italian† physicians also, for the most part, took this line, and
attempted to show the agreement of the principles of
the ancient school of medicine with the new chemical
notions. This, indeed, is the usual manner in which the
diffusion of new theoretical ideas becomes universal.

The progress of the chemical school of medicine in
England‡ requires our more especial notice. Willis was
the most celebrated champion of this sect. He assumed,
but with modifications of his own, the three Paracelsian
principles, salt, sulphur, and mercury; considered digestion as the effect of an acid, and explained other parts of
the animal economy by distillation, fermentation, and the
like. All diseases arise from the want of the requisite
ferment; and the physician, he says§, may be compared to
a vintner, since both the one and the other have to take
care that the necessary fermentations go on, that no
foreign matter mixes itself with the wine of life, to inter-
rupt or derange those operations. In the middle of the
seventeenth century, medicine had reached a point in
which the life of the animal body was considered as

* Spr., iv., 350.   † *Ib.*, 368.
‡ *Ib.*, 353.   § *Ib.*, 354.

merely a chemical process; the wish to explain every-thing on known principles left no recognised difference between organized and unorganized bodies, and diseases were treated according to this delusive notion. The condition of chemistry itself during this period, though not one of brilliant progress, was sufficiently stable and flourishing to give a plausibility to any speculation which was founded on chemical principles, and the real influ-ence of these principles in the animal frame could not be denied.

The iatrochemists were at first resisted, as we have seen, by the adherents of the ancient schools; they were attacked on various grounds, and finally deposed from their ascendency by another sect, which we have to speak of, as the iatromathematical, or mechanical school. This sect was no less unsatisfactory and erroneous in its posi-tive doctrines than the chemists had been; for the animal frame is no more a mere machine than a mere laboratory: but it promoted the cause of truth, by detecting and exposing the insufficient explanations and unproved assertions of the reigning theory.

Boyle was one of the persons who first raised doubts against the current chemical doctrines of his time, as we have elsewhere noted; but his objections had no peculiar physiological import. Hermann Conring*, the most learned physician of his time, a contemporary with Syl-vius, took a view more pertinent to our present object; for he not only rejected the alchemical and hermetical medicines, but taught expressly that chemistry, in its then existing condition, was better fitted to be of use in the practice of pharmacy, than in the theories of physio-logy and pathology. He made the important assertion, also, that chemical principles do not pre-exist as such in the animal body; and that there are higher powers which

* SPR., iv., 361.

operate in the organic world, and which do not depend on the form and mixture of matter.

Attempts were made to prove the acid and alkaline nature of the fluids of the human body by means of experiments, as by John Viridet of Geneva*, and by Raimond Vieussens†, the latter of whom maintained that he had extracted an acid from the blood, and detected a ferment in the stomach. In opposition to him, Hecquet, a disciple of the iatromathematical school, endeavoured to prove that digestion was performed, not by means of fermentation, but by trituration. Hecquet's own opinions cannot be defended; but his objections to the chemical doctrines, and his assertion of the difference of chemical and organical processes, are evidences of just thought‡.

The most important opponents of the iatrochemical school were Pitcairn in England, Bohn and Hoffman in Germany, and Boerhaave in Holland. These eminent physicians, about the end of the seventeenth century, argued on the same grounds of observation, that digestion is not fermentation, and that the Sylvian accounts of the origin of diseases by means of acid and alkali are false. The arguments and authority of these and other persons finally gained an ascendency in the medical world, and soon after this period we may consider the reign of the chemical school of physiology as past. In fact, the attempts to prove its assertions experimentally were of the feeblest kind, and it had no solid basis on which it could rest, so as to resist the shock of the next hypothesis which the progress of the physical sciences might impel against it. We may, therefore, now consider the opinion of the mere chemical nature of the vital processes as disproved, and we proceed next to notice the history of another unsuccessful essay to reduce vital actions to known actions of another kind.

* Spn., iv., 329.   † Ib., 350, (1715.)   ‡ Ib., 401.

SECT. III. *The Iatromathematical School.*—In the first Section of this chapter, we enumerated the biological hypotheses which at first present themselves, as the mystical, the mechanical, the chemical. We might have expected that they should occur to men's minds in the order thus stated: and in fact they did so; for the physiology of the ancient materialists, as Democritus and Lucretius, is mechanical so far as it is at all distinct in its views, and thus the mechanical preceded the chemical doctrine. But in modern times, the fluid or chemical physiology was developed before the solid or mechanical: of which the reason appears to have been this;—that mechanics and chemistry began to assume a scientific character about the same time; and that of the two, chemistry not only appeared at first sight more applicable to the functions of the body, because all the more rapid changes appear to be connected with modifications of the fluids of the animal system, but also, by its wider range of facts and more indefinite principles, afforded a better temporary refuge for the mind when perplexed by the difficulties and mysteries which spring out of the speculations concerning life. But if chemistry was thus at first a more inviting field for the physiologist, mechanics soon became more attractive in virtue of the splendid results obtained by the schools of Galileo and Newton. And when the insufficiency of chemical physiology was discovered by trial, as we have seen it was, the hope naturally arose, that the mechanical principles which had explained so many of the phenomena of the external universe might also be found applicable to the smaller world of material life; that the *microcosm* as well as the *macrocosm* might have its mechanical principles. From this hope sprung the iatromathematical school, or school of mechanical physiologists.

We may, however, divide this school into two parts,

the Italian, and the Cartesio-Newtonian sect. The former employed themselves in calculating and analysing a number of the properties of the animal frame which are undoubtedly mechanical; the latter, somewhat intoxicated by the supposed triumphs of the corpuscular philosophy, endeavoured to extend these to physiology, and for this purpose introduced into the subject many arbitrary and baseless hypotheses. I will very briefly mention some of the writers of both these sects.

The main points to which the Italian or genuine mechanical physiologists attended, were the application of mechanical calculations to the force of the muscles, and of hydraulical reasonings to the motion of the fluids of the animal system. The success with which Galileo and his disciples had pursued these branches of mechanical philosophy, and the ascendency which they had obtained, first in Italy, and then in other lands, made such speculations highly interesting. Borelli may be considered as the first great name in his line, and his book, *De Motu Animalium*, (*Opus Posthumum*, Romæ, 1680,) is even now a very instructive treatise on the force and action of the bones and muscles. This, certainly one of the most valuable portions of mechanical physiology, has not even yet been so fully developed as it deserves, although John Bernoulli[*] and his son Daniel[†] applied to it the resources of analysis, and Pemberton[‡], in England, pursued the same subject. Other of these mechanico-physiological problems consisted in referring the pressure of the blood and of the breath to hydrostatical principles. In this manner Borelli was led to assert that the muscles of the heart exert a force of 180,000 pounds[§]. But a little later, Keill reduced this force to a few ounces[||]. Keill and others attempted to determine, on

---

[*] *De Motu Musculorum.*   [†] *Act. Acad. Petrop.* i., 170.
[‡] *Course of Physiology*, 1773.   [§] SPR., iv. 110.   [||] *Ib.* 443.

similar principles, the velocity of the blood; we need not notice the controversies which thus arose, since there is not involved in them any peculiar physiological principle.

The peculiar character of the iatromathematical school, as an attempt at physiological theory, is more manifest in its other section, which we have called the Cartesio-Newtonian. The Cartesian system pretended to account for the appearances and changes of bodies by means of the size, figure, and motion of their minute particles. And though this system in its progress toward the intellectual empire of Europe was suddenly overturned by the rise of the Newtonian philosophy, these corpuscular doctrines rather gained than lost by the revotion; for the Newtonian philosophy enlarged the powers of the corpuscular hypothesis, by adding the effects of the attractive and repulsive forces of particles to those of their form and motion. By this means, although Newton's discoveries did not in fact augment the probability of the corpuscular hypothesis, they so far increased its plausibility, that this hypothesis found favour both with Newton himself and his contemporaries, no less than it had done with the Cartesians.

The attempt to apply this corpuscular hypothesis to physiology was made by Des Cartes himself. The general character of such speculations may easily be guessed *. The secretions are effected by the organs operating after the manner of sieves. Round particles pass through cylindrical tubes, pyramidal ones through triangular pores, cubical particles through square apertures, and thus different kinds of matter are separated. Similar speculations were pursued by other mathematicians; the various diameter of the vessels†, their curvatures, folds, and angles, were made subjects of calculation. Bellini, Donzellini, Gulielmini, in Italy; Perrault, Dodart, in

ει \ * SPR., iv., 329.   †Ib., 432.

France; Cole, Keill, Jurin, in England, were the principal cultivators of such studies. In the earlier part of the eighteenth century, physiological theorists considered it as almost self-evident that their science required them to reason concerning the size and shape of the particles of the fluids, the diameter and form of the invisible vessels. Such was, for instance, the opinion of Cheyne[*], who held that acute fevers arise from the obstruction of the glands, which occasions a more vehement motion of the blood. Mead, the physician of the King, and the friend of Newton, in like manner explained the effects of poisons by hypotheses concerning the form of their particles[†], as we have already seen in speaking of chemistry.

It is not necessary for us to dwell longer on this subject, or to point out the total insufficiency of the mere mechanical physiology. The iatrochemists had neglected the effect of the solids of the living frame; the iatro-mathematicians attended only' to these [‡]. And even these were considered only as canals, as cords, as levers, as lifeless machines. These reasoners never looked for any powers of a higher order than the cohesion, the resistance, the gravity, the attraction, which operate in inert matter. If the chemical school assimilated the physician to a vintner or brewer, the mechanical physiologists made him an hydraulic engineer; and, in fact, several of the iatromathematicians were at the same time teachers of engineering and of medicine.

Several of the reasoners of this school combined chemical with their mechanical principles; but it would throw no additional light upon the subject to give any account of these, and I shall therefore go on to speak of the next form of the attempt to explain the processes of life.

---

[*] SPR., v., 223.　　　[†] *Mechanical Account of Poisons.* 1702.
[‡] SPR., iv., 419.

SECT. IV. *The Vital-Fluid School.*—I speak here, not of that opinion which assumes some kind of fluid or ether as the means of communication along the *nerves* in particular, but, of the hypothesis that *all* the peculiar functions of life depend upon some subtle ethereal substance diffused through the frame;—not of a *nervous* fluid, but of a *vital* fluid. Again, I distinguish this opinion from the doctrine of an *immaterial* vital power or principle, an animal soul, which will be the subject of the next Section: nor is this distinction insignificant; for a material element, however subtle, however much spiritualized, must still act everywhere according to the same laws; whereas we do not conceive an immaterial spirit or soul to be subject to this necessity.

The iatromathematical school could explain to their own satisfaction how motions once begun, were transferred and modified; but in many organs of the living frame there seemed to be a power of beginning motion, which is beyond all mere mechanical action. This led to the assumption of a principle of a higher kind, though still material. Such a principle was asserted by Frederick Hoffman, who was born at Halle, in 1660[*], and became Professor of Medicine at the newly-established University there in 1694. According to him[†], the reason of the greater activity of organized bodies lies in the influence of a material substance of extreme subtilty, volatility, and energy. This is, he holds, no other than the Ether, which, diffused through all nature, produces in plants the bud, the secretion and motion of the juices, and is separated from the blood and lodged in the brain of animals[‡]. From this, acting through the nerves, must be derived all the actions of the organs in the animal frame; for when the influence of the nerve upon the muscle ceases, muscular motion ceases also.

---

[*] SPR., v. 254. [†] *Ib.*, v. 257.
[‡] *De Differentiâ Organismi et Mechanismi*, pp. 48. 67.

The mode of operation of this vital fluid was, however, by no means steadily apprehended by Hoffmann and his followers. Its operations are so far mechanical* that all effects are reduced to motion, yet they cannot be explained according to known mechanical laws. At one time the effects are said to take place according to laws of a Higher Mechanics which are still to be discovered†. At another time, in complete contradiction of the general spirit of the system, metaphysical conceptions are introduced: each particle of the vital fluid is said to have a determined *idea* of the whole mechanism and organism‡, and according to this, it forms the body and preserves it by its motion. By means of this fluid the soul operates upon the body, and the instincts and the passions have their source in this material sensitive soul. This attribution of ideas to the particles of the fluid is less unaccountable when we recollect that something of the same kind is admitted into Leibnitz's system, whose monads have also ideas.

Notwithstanding its inconsistencies, Hoffman's system was received with very general favour both in Germany and in the rest of Europe; the more so, inasmuch as it fell in very well with the philosophy both of Leibnitz and of Newton. The Newtonians were generally inclined to identify the vital fluid with the Ether, of which their master was so strongly disposed to assume the existence: and indeed he himself suggested this identification.

When the discoveries made respecting Electricity in the course of the eighteenth century had familiarized men with the notion of a pervading subtle agent, invisible, intangible, yet producing very powerful effects in every part of nature, physiologists also caught at the suggestion of such an agent, and tried, by borrowing or imitating it, to aid the imperfection of their notions of

* Spr., v. 262, 3.   † *Opp.*, vol. v. p. 123.
‡ *De Diff. Organ. et Mechan.*, p. 81.

the vital powers.   The Vital Principle* was imagined to be a substance of the same kind, by some to be the same substance, with the electric fluid.   By its agency all these processes in organized bodies were accounted for which cannot be explained by mechanical or chemical laws, as the secretion of various matters (tears, milk, bile, &c.) from an homogeneous fluid, the blood; the production of animal heat, digestion, and the like. According to John Hunter, this attenuated substance pervaded the blood itself, as well as the solid organic frame; and the changes which take place in the blood which has flowed out of the veins into a basin are explained by saying that it is, for a time, till this vital fluid evaporates, truly alive.

The notion of a vital fluid appears also to be favourably looked upon by Cuvier; although with him this doctrine is mainly put forwards in the form of a nervous fluid.   Yet in the following passage he extends the operation of such an agent to all the vital functions†. " We have only to suppose that all the medullary and nervous parts produce the nervous agent, and that they alone conduct it; that is, that it can only be transmitted by them, and that it is changed or consumed by their actions.   Then everything appears simple.   A detached portion of muscle preserves for some time its irritability; on account of the portion of nerve which always adheres to it. The sensibility and the irritability reciprocally exhaust each other by their exercise, because they change or consume the same agent.   All the interior motions of digestion, secretion, excretion, participate in this exhaustion, or may produce it.   All local excitation of the nerves brings thither more blood by augmenting the irritability of the arteries, and the afflux of blood augments the real sensi-

---

* PRICHARD, *on the Doctrine of a Vital Principle*, p. 12.
† *Hist. Sci. Nat. depuis* 1789, i. 214.

bility by augmenting the production of the nervous agent. Hence the pleasures of titillations, the pains of inflammation. The particular sensations increase in the same manner and by the same causes; and the imagination exercises, (still by means of the nerves,) upon the internal fibres of the arteries or other parts, and through them on the sensations, an action analogous to that of the will upon the voluntary motions. As each exterior sense is exclusively disposed to admit the substances which it is to perceive, so each interior organ, secretory or other, is also more excitable by some one agent than by another: and hence arises what has been called the *proper sensibility* or *proper life of the organs;* and the influence of specifics which introduced into the general circulation affect only certain parts. In fine, if the nervous agent cannot become sensible to us, the reason is that all sensation requires that this agent should be altered in some way or other; and it cannot alter itself.

"Such is the summary idea which we may at present form of the mutual and general working of the vital powers in animals."

Against the doctrine of a vital fluid as one uniform material agent pervading the organic frame, an argument has been stated which points out extremely well the philosophical objection to such an hypothesis*. If the vital principle be the *same* in all parts of the body, how does it happen, it is asked, that the secretions are so *different?* How do the particles in the blood, separated from their old compounds and united into new ones, under the same influence, give origin to all the different fluids which are produced by the glands? The liver secretes bile, the lacrymal gland tears, and so on. Is the vital principle different in all these organs? To assert this, is to multiply nominal principles without limit, and

* Prichard, *on a Vital Principle,* p. 98.

without any advance in the explanation of facts. Is the vital principle the same, but its operation modified by the structure of the organ? We have then two unknown causes, the vital principle and the organic structure, to account for the effect. By such a multiplication of hypotheses nothing is gained. We may as well say at once, that the structure of the organ, acting by laws yet unknown, is the cause of the peculiar secretion. It is as easy to imagine this structure acting to produce the whole effect, as it is to imagine it modifying the activity of another agent. Thus the hypothesis of the vital fluid in this form explains nothing, and does not in any way help onwards the progress of real biological knowledge.

The hypothesis of an immaterial vital principle must now be considered.

SECT. V. *The Psychical School.*—The doctrine of an animal soul as the principle which makes the operations of organic different from those of inorganic matter, is quite distinct from, and we may say independent ₊of, the doctrine of the soul as the intelligent, moral, responsible part of man's nature. It is the former doctrine alone of which we have here to speak, and those who thus hold the existence of an immaterial agent as the cause of the phenomena of life, I term the *Psychical School.*

Such a view of the constitution of living things is very ancient. For instance, Aristotle's Treatise "*on the Soul,*" goes entirely upon the supposition that the soul is the cause of motion, and he arrives at the conclusion that there are different *parts* in the soul; the *nutritive* or *vegetative*, the *sensitive*, and the *rational**.

But 'this doctrine is more instructive to us, when it appears as the antagonist of other opinions concerning the nature of life. In this form it comes before us as promulgated by Stahl, whom we have already noticed as

* ARIST. Περὶ Ψυχῆς. ii. 2.

one of the great discoverers in chemistry. Born in the same year as Hoffman, and appointed at his suggestion professor at the same time in the same new university of Halle, he soon published a rival physiological theory. In a letter to Lucas Schröck, the president of the Academy of Naturalists, he describes the manner in which he was led to form a system for himself*. Educated in the tenets of Sylvius and Willis, according to which all diseases are derived from the acidity of the fluids, Stahl, when a young student, often wondered how these fluids, so liable to be polluted and corrupted, are so wonderfully preserved through innumerable external influences, and seem to be far less affected by these than by age, constitution, passion. No material cause, could, he thought, produce such effects. No attention to mechanism or chemistry alone could teach us the true nature and laws of organization.

So far as Stahl recognised the influence, in living bodies, of something beyond the range of mechanics and chemistry, there can be no doubt of the sound philosophy of his views; but when he proceeds to found a positive system of physiology, his tenets become more precarious. The basis of his theory is this†: the body has, as body, no power to move itself, and must always be put in motion by immaterial substances. All motion is a spiritual act‡. The source of all activity in the organic body, from which its preservation, the permanency of its composition, and all its other functions proceed, is an immaterial being, which Stahl calls the *Soul;* because, as he says, when the effects are so similar, he will not multiply powers without necessity. Of this principle, he says, as the Hippocratians said of Nature, that "it does without teaching what it ought to do§," and does it " without consideration‖." These ancient tenets Stahl

* SPR., v. 303.　　　† *Ib.*, v. 308.　　　‡ *Ib.*, v. 314.
§ STAHL, περι φυσεως ἀπάιδευτου.　　　‖ ὀυκ ἐκ διανοίης.

interprets in such a manner that even the involuntary motions proceed from the soul, though without reflection or clear consciousness.   It is indeed evident, that there are many customary motions and sensations which are perfectly rational, yet not the objects of distinct consciousness: and thus instinctive motions, and those of which we are quite unconscious, may still be connected with reason.   The questions which in this view offer themselves, as, how the soul passes from the mother to the child, he dismisses as unprofitable*   He considers nutrition and secretion as the work of the soul.   The corpuscular theory and the doctrine of animal spirits are, he rightly observes, mere hypotheses, which are arbitrary in their character, and only shift the difficulty.   For, if the animal spirits are not matter, how can they explain the action of an immaterial substance on the body; and if they are matter, how are they themselves acted on?

This doctrine of the action of the soul on the body, was accepted by many persons, especially by the iatro-mathematicians, who could not but feel the insufficiency of their system without some such supplement: such were Cheyne and Mead.   In Germany, Stahl's disciples in physiology were for the most part inconsiderable persons†.   Several Englishmen who speculated concerning the metaphysics as well as the physiology of Sensation and Motion, inclined to this psychical view, as Porterfield and Whytt.   Among the French, Boissier de Sauvages was the most zealous defender of the Stahlian system. Actions, he says‡, which belong to the preservation of life are determined by a moral not a mechanical necessity.   They proceed from the soul, but cannot be con-

---

* This was of course an obvious problem.   HARVEY, *On Generation,* Exercise 27, p. 148, teaches, "That the egg is not the production of the womb, but of the soul."

† SPR., v. 339, &c.   ‡ *Ib.,* 358.

trolled by it, as the starting from fear, or the trembling at danger. Unzer, a physician at Altona*, was also a philosophical Stahlian†.

We need not dwell on the opposition which was offered to this theory, first by Hoffman, and afterwards by Haller. The former of these had promulgated, as we have seen, the rival theory of a nervous fluid, the latter was the principal assertor of the doctrine of irritability, an important theory on which we may afterwards have to touch. Haller's animosity against the Stahlian hypothesis is a remarkable feature in one who is in general so tolerant in his judgment of opinions. His arguments are taken from the absence of the control of the will over the vital actions, from the want of consciousness accompanying these actions, from the uniformity of them in different conditions of the mind, and from the small sensibility of the heart which is the source of the vital actions. These objections, and the too decided distinction which Haller made between voluntary and involuntary muscles, were very satisfactorily answered by Whytt and Platner. In particular, it was urged that the instinctive actions of brutes are inexplicable by means of mechanism, and may be compared with the necessary vital actions of the human body. Neither kind are accidental, neither kind are voluntary, both are performed without reflection.

Without tracing further the progress of the psychical doctrine, I shall borrow a few reflections upon it from Sprengel‡:—

" When the opponents of the Stahlian system repeat incessantly that the assumption of a psychical cause in corporeal effects is a metaphysical speculation which does not belong to medicine, they talk to no purpose. The states of the soul are objects of our internal experience, and interest the physician too nearly to allow him to

* A.D. 1799. † SPR. v. 360. ‡ Ib. 383.

neglect them. The innumerable unconscious efforts of the soul, the powerful and daily effects of the passions upon the body, too often put to confusion those who would expel into the region of metaphysics the dispositions of the mind. The connexion of our knowledge of the soul, as gathered from experience, with our knowledge of the human body, is far closer than the mechanical and chemical physiologists suspect.

"The strongest objection against the psychical system, and one which has never been sufficiently answered by any of its advocates, is the universality of organic effects in the vegetable kingdom. The comparison of the physiology of plants with the physiology of animals puts the latter in its true light. Without absolutely trifling with the word *soul*, we cannot possibly derive from a soul the organic operations of vegetables. But just as little can we, as some Stahlians have done, draw a sharp line between plants and animals, and ascribe the processes of the former to mere mechanism, while we derive the operations of the latter from an intellectual principle. Not to mention that such a line is not possible, the rise of the sap and the alteration of the fluids of plants cannot be derived entirely from material causes as their highest origin."

Thus, I may add, this psychical theory, however difficult to defend in its detail, does in its generalities express some important truths respecting the vital powers. It not only, like the last theory, gives unity to the living body, but it marks, more clearly than any other theory, the wide interval which separates mechanical and chemical from vital action, and fixes our attention upon the new powers which the consideration of life compels us to assume. It not only reminds us that these powers are elevated above the known laws of the material world, but also that they are closely connected with the world of

thought and feeling, with will and reason; and thus it
carries us in a manner in which none of the preceding
theories have done, to a true conception of a living,
conscious, sentient, active individual.

At the same time we cannot but allow that the life
of plants and of the lower orders of animals shows us very
clearly that, in order to arrive at any sound and consistent
knowledge respecting life, we must form some conception
of it from which all the higher attributes which the term
"soul" involves, are utterly and carefully excluded; and
therefore we cannot but come to the conclusion that the
psychical school are right mainly in this; that in ascribing the functions of life to a soul, they mark strongly and
justly the impossibility of ascribing them to any known
attributes of body.

Chapter III.

ATTEMPTS TO ANALYSE THE IDEA OF LIFE.

1. *Definitions of Life.*—We have seen in the preceding chapter that all attempts to obtain a distinct conception of the nature of Life in general have ended in
failure, and produced nothing beyond a negative result.
And the conjecture may now naturally occur, that the
cause of this failure resides in an erroneous mode of pro-
pounding to ourselves the problem. Instead of contemplating Life as a single Idea, it may perhaps be proper to
separate it into several component notions: instead of
seeking for one cause of all vital operations, it may be
well to look at the separate vital functions, and to seek
their causes. When the view of this possibility opens
upon us, how shall we endeavour to verify it, and to
take advantage of it?

Let us, as one obvious course, take some of the attempts which have been made to *define* Life, and let us see whether they appear to offer to us any analysis of the idea into component parts. Such definitions, when they proceed from men of philosophical minds, are the ultimate result of a long course of thought and observation; and by no means deserve to be slighted as arbitrary selections of conditions, or empty forms of words.

2. Life has been defined by Stahl*, " The condition by which a body resists a natural tendency to chemical changes, such as putrefaction." In like manner, M. von Humboldt† defines living bodies to be " those which, notwithstanding the constant operation of causes tending to change their form, are hindered by a certain inward power from undergoing such change." The first of these definitions amounts only to the assertion, that vital processes are not chemical; a negative result, which we may accept as true, but which is, as we have seen, a barren truth. The second appears to be, in its import, identical with the first. An *inward* principle can only be understood as distinguished from known external powers, such as mechanical and chemical agencies. Or if, by an internal principle, we mean such a principle as that of which we are *conscious* within ourselves, we ascribe a soul to all living things: an hypothesis which we have seen is not more effective than the former in promoting the progress of biological science. Nearly the same criticism applies to such definitions as that of Kant: that " Life is an internal faculty producing change, motion, and action."

Other definitions refer us, not to some property residing in the whole of an organized mass, but to the connexion and relation of its parts. Thus M. von Hum-

* Treviranus. *Biologie*, p. 19. Stahlii, *Theor. Med.*, p. 254.
† *Aphorismen aus d. Chem. Physiol. der Pflanzen*, s. 1.

boldt* has given another definition of a living body: that " it is a whole whose parts, arbitrarily separated, no longer resist chemical changes." But this additional assertion concerning the parts, adds nothing of any value to the definition of the whole. And in some of the lower kinds of plants and animals it is hardly true as a fact.

3. Another definition † places the character of Life in " motions serviceable to the body moved." To this it has been objected ‡, that, on this definition, the earth and the planets are living bodies. Perhaps it would be more philosophical to object to the introduction of so loose a notion as that of a property being *serviceable* to a body. We might also add, that if we speak of all vital functions as *motions*, we make an assumption quite unauthorized, and probably false.

Other definitions refer the idea of Life to the idea of Organization. " Life is the activity of matter according to laws of organization§." We are then naturally led to ask what is Organization. In reply to this is given us the Kantian definition of Organization, which I have already quoted elsewhere‖, "An organized product of nature is that in which all the parts are mutually ends and means¶." That this definition involves exact fundamental ideas, and is capable of being made the basis of sound knowledge, I shall hereafter endeavour to show. But I may observe that such a definition leads us somewhat further. If the parts of organized bodies are known to be means to certain ends, this must be known because they fulfil these ends, and produce certain effects by the operation of a certain cause or causes. The question then

* *Versuche über die gereitzte Muskel und Nervenfäser,* book ii., p. 433.

† ERHARD. RÖSCHLAUB's *Magazin der Heilkunde,* b. i., st. 1, p. 69.

‡ TREVIRANUS, *Biologie,* p. 41.

§ SCHMID, *Physiologie,* b. ii., p. 274.

‖ *Hist. Ind. Sci.,* iii. 470.   ¶ KANT, *Urtheilskraft,* p. 296.

recurs, what is *the cause* which produces such effects as take place in organized or living bodies? and this is identical with the problem of which in the last chapter we traced the history, and related the failure of physiologists in all attempts at its solution.

4. But what has been just said suggests to us that it may be an improvement to put our problem in another shape: not to take for granted that the cause of all vital processes is one, but to suppose that there may be several separate causes at work in a living body. If this be so, life is no longer one kind of activity, but several. We have a number of operations which are somehow bound together, and life is the totality of all these: in short, life is not one function, but a system of functions.

5. We are thus brought very near to the celebrated definition of life given by Bichat*: " Life is the sum of the functions by which death is resisted." But upon the definition thus stated, we may venture to observe ;—first, that the introduction of the notion of *death* in order to define the notion of *life* appears to be unphilosophical. We may more naturally define death with reference to life, as the cessation of life; or at least we may consider life and death as correlative and interdependent notions. Again, the word "sum," used in the way in which it here occurs, appears to be likely to convey an erroneous conception, as if the functions here spoken of were simply added to each other, and connected by co-existence. It is plain that our idea of life involves more than this: the functions are all clearly connected, and mutually depend on each other; nutrition, circulation, locomotion, reproduction, each has its influence upon all the others. These functions not merely co-exist, but exist with many mutual relations and connexions; they are continued so

---

* *Physiological Researches on Life and Death.*

as to form, not merely a *sum*, but a *system*. And thus we are led to modify Bichat's definition, and to say that *Life is the system of vital functions.*

6. But it will be objected that by such a definition we explain nothing: the notion of *vital functions*, it may be said, involves the idea of *life*, and thus brings us round again to our starting point. Or if not, at least it is as necessary to define Vital Functions as to define Life itself, so that we have made little progress in our task.

To this we reply, that if any one seeks, upon such subjects, some ultimate and independent definition from which he can, by mere reasoning, deduce a series of conclusions, he seeks that which cannot be found. In the Inductive Sciences, a definition does not form the basis of reasoning, but points out the course of investigation. The definition must include words; and the meaning of these words must be sought in the progress and results of observations, as I have elsewhere said*. "The meaning of words is to be sought in the progress of thought; the history of science is our dictionary; the steps of scientific induction are our definitions." It will appear, I think, that it is more easy for us to form an idea of a separate Function of the animal frame, as Nutrition or Reproduction, than to comprehend Life in general under any single idea. And when we say that Life is a system of Vital Functions, we are of course directed to study these functions separately, and (as in all other subjects of scientific research) to endeavour to form of them such clear and definite ideas as may enable us to discover their laws.

7. The view to which we are thus led, of the most promising mode of conducting the researches of biology, is one which the greatest and most philosophical physiologists of modern times have adopted. Thus Cuvier considers this as the true office of physiology at present.

* *Hist. Ind. Sci.*, iii. 100.

" It belongs to modern times," he says, " to form a just classification of the vital phenomena; the task of the present time is to analyse the forces which belong to each organic element, and upon the zeal and activity which are given to this task depends, according to my judgment, the fortune of physiology*." This classification of the phenomena of life involves, of course, a distinction and arrangement of the vital functions; and the investigation of the powers by which these functions are carried on, is a natural sequel to such a classification.

8. *Classifications of Functions.*—Attempts to classify the vital functions of man were made at an early period, and have been repeated in great number up to modern times. The task of classification is exposed to the same difficulties, and governed by the same conditions, in this as in other subjects. Here, as in the case of other things, there may be many classifications which are moderately good and natural, but there is only one which is the best and the true natural system. Here, as in other cases, one classification brings into view one set of relations; another, another; and each may be valuable for its special purpose. Here, as in other cases, the classes may be well constituted, though the boundary lines which divide them be somewhat indistinct, and the order doubtful. Here, as in other cases, we may have approached to the natural classification without having attained it; and here, as in other cases, to define our classes is the last and hardest of our problems.

The most ancient classification of the Functions of living things†, is the division of them into *Vital, Natural,* and *Animal.* The *Vital* Functions are those which cannot be interrupted without loss of life, as *Circulation, Respiration,* and *Nervous Communication.* The *Natural*

---

* *Hist. Sc. Nat. dep.* 1789, i. 218.

† *Dict. des Sciences Nat.*, art. Fonctions.

Functions are those which without the intervention of
the will operate on their proper occasions to preserve
the bodies of animals; they are *Digestion, Absorption,
Nutrition;* to which was added *Generation.* The *Animal*
Functions are those which involve perception and will,
by which the animal is distinguished from the vegetable;
they are *Sensibility, Locomotion,* and *Voice.*

The two great grounds of this division, the distinction
of functions which operate continually, and those which
operate occasionally; and again, the distinction of functions which involve sensation and voluntary motion from
those which do not, are really of fundamental importance, and gave a real value to this classification. It
was, however, liable to obvious objections: namely, *First,*
that the names of the classes were ill chosen; for all the
functions are natural, all are vital: *Second,* that the
lines of demarcation between the classes are obscure and
vague; Respiration is a *vital* function, as being continually necessary to life; but it is also a *natural* function,
since it concurs in the formation of the nutritive fluid,
and an *animal* function, since it depends in part on the
will. But these objections were not fatal, for a classification may be really sound and philosophical, though
its boundary lines are vague, and its nomenclature ill
selected. The division of the functions we have mentioned kept its ground long; or was employed with a
subdivision of one class, so as to make them four; the
vital, natural, animal and *sexual* functions.

10. I pass over many intermediate attempts to classify the functions, and proceed to that of Bichat as that
which is, I believe, the one most generally assented to in
modern times. The leading principle in the scheme of
this celebrated physiologist is the distinction between
organic and *animal* life. This separation is nearly identical with the one just noticed between the vital and

animal functions; bnt Bichat, by the contrasts which he pointed out between these classes of functions, gave a decided prominence and permanence to the distinction. The Organic Life, which in animals is analogous to the life of vegetables, and the Animal Life, which implies sensation and voluntary motion, have each its system of organs. The centre of the animal life is the brain, of the organic life, the heart. The former is carried on by a symmetrical, the latter, by an unsymmetrical system of organs: the former produces intermitting, the latter continuous actions: and, in addition to these, other differences are pointed out. This distinction of the two *lives*, being thus established, each is subdivided into two orders of Functions. The Animal Functions are passive, as *Sensation:* or active, as *Locomotion* and *Voice*; again, the Organic Functions are those of composition, which are concerned in taking matter into the system, *Digestion, Absorption, Respiration, Circulation, Assimilation;* and those of decomposition, which reject the materials when they have discharged their office in the system, and these are again, *Absorption, Circulation,* and *Secretion.* To these are added *Calorification,* or the production of animal heat. It appears, from what has been said, that *Absorption* and *Circulation,* (and we may add *Assimilation* and *Secretion,* which are difficult to separate,) belong alike to the processes of composition and decomposition; nor in truth, can we, with any rigour, separate the centripetal and centrifugal movements in that vortex which, as we shall see, is an apt image of organic life.

Several objections have been made to this classification; and in particular, to the terms thus employed. It has been asserted to be a perversion of language to ascribe to animals *two lives*, and to call the higher faculties in man, perception and volition, the *animal* functions. But, as we have already said, when a classification is really good, such objections, which bear only upon the mode in which it is presented, are by no means fatal: and it is generally acknowledged, by all the most philosophical cultivators of biology, that this arrangement of the functions is better suited to the purposes of the science than those which preceded it.

11. But according to the principles which we have already laid down, the solidity of such a classification is to be verified by its serving as a useful guide in biological researches. If the arrangement which we have explained be really founded in natural relations, it will be found that in proportion as physiologists have studied the separate functions above enumerated, their ideas of these functions, and of the powers by which they are carried on, have become more and more clear;—have tended more and more to the character of exact and rigorous science.

To examine how far this has been the case with regard to all the separate functions, would be to attempt to estimate the value of all the principal physiological speculations of modern times;—a task far too vast and too arduous for any one to undertake who has not devoted his life to such studies. But it may properly come within the compass of our present plan to shew how, with regard to the broader lines of the above classification, there has been such a progress as we have above described, from more loose and inaccurate notions of some of the vital functions to more definite and precise ideas. This I shall attempt to point out in one or two instances.

Chapter IV.

ATTEMPTS TO FORM IDEAS OF SEPARATE VITAL FORCES, AND FIRST OF ASSIMILATION AND SECRETION.

Sect. I. *Course of Biological Research.*

1. It is to be observed that at present I do not speak of the progress of our knowledge with regard to the detail of the processes which take place in the human body, but of the approach made to some distinct idea of the specially vital part of each process. In the History of Physiology, it has been seen* that all the great discoveries made respecting the organs and motions of the animal frame have been followed by speculations and hypotheses connected with such discoveries. The discovery of the circulation of the blood led to theories of animal heat; the discovery of the motion of the chyle led to theories of digestion; the close examination of the process of reproduction in plants and animals led to theories of generation. In all these cases, the discovery brought to light some portion of the process which was mechanical or chemical, but it also, in each instance, served to show that the process was something more than mechanical or chemical. The theory attempted to explain the process by the application of known causes; but there always remained some part of it which must unavoidably be referred to an unknown cause. But though unknown, such a cause was not a hopeless object of study. As the vital functions became better and better understood, it was seen more and more clearly at what precise points of the process it was necessary to assume a peculiar vital energy, and what sort of properties this energy must be conceived to possess. It was perceived where, in what

* *Hist. Ind. Sci.*, iii., 406.

manner, in what degree, mechanical and chemical agencies were modified, overruled, or counteracted, by agencies which must be hyper-mechanical and hyper-chemical. And thus the discoveries made in anatomy by a laborious examination of facts, pointed out the necessity of introducing new ideas, in order that the facts might be intelligible. Observation taught much; and among other things, she taught that there was something which could not be observed, but which must, if possible, be conceived. I shall notice a few instances of this.

SECT. II. *Attempts to form a distinct Conception of Assimilation and Secretion.*

2. *The Ancients.*—That plants and animals grow by taking into their substance matter previously extraneous, is obvious to all; but as soon as we attempt to conceive this process distinctly in detail, we find that it involves no inconsiderable mystery. How does the same food become blood and flesh, bone and hair? Perhaps the earliest attempt to explain this mystery, is that recorded by Lucretius* as the opinion of Anaxagoras, that food contains some bony, some fleshy particles, some of blood, and so on. We might, on this supposition, conceive that the mechanism of the body appropriates each kind of particle to its suitable place.

But it is easy to refute this essay at philosophizing (as Lucretius refutes it) by remarking that we do not find milk in grass, or blood in fruit, though such food gives such products in cattle and in men. In opposition to this "Homoiomereia," the opinion that is forced upon us by the facts is, that the process of nutrition is not a selection merely, but an *assimilation;* the organized system does not *find,* but *make,* the additions to its structure.

3. *Buffon.*—This notion of *assimilation* may be variously expressed and illustrated; and all that we can do

* LUCR., i. 855.   Nunc et Anaxagoræ scrutemur ὁμοιομέρειαν.

here, in order to show the progress of thought, is to adduce the speculations of those writers who have been most
successful in seizing and marking its peculiar character.
Buffon may be taken as an example of the philosophy of
his time on this subject. "The body of the animal,"
says he*, "is a kind of *interior mould*, in which the
matter subservient to its increase is modelled and
assimilated to the whole, in such a way that, without
occasioning any change in the order and proportion of
the parts, there results an augmentation in each part
taken separately. This increase, this development, if we
would have a *clear idea* of it, how can we obtain it, except
by considering the body of· the animal, and each of the
parts which is to be developed, as so many interior moulds
which only receive the accessory matter in the order
which results from the position of all their parts? This
development cannot take place, as persons sometimes
persuade themselves, by an addition to the outside; on
the contrary, it goes on by an intimate susception which
penetrates the mass; for, in the part thus developed, the
size increases in all parts proportionally, so that the new
matter must penetrate it in all its dimensions: and it is
quite necessary that this penetration of substance must
take place in a certain order, and according to a certain
measure; for if this were not so, some parts would de-
velope themselves more than others. Now what can
there be which shall prescribe such a rule to the accessory matter except the *interior mould?*"

To speak of a. *mould* simply, would convey a coarse
mechanical notion, which could not be received as any
useful contribution to physiological speculation. But
this *interior* mould is, of course, to be understood figuratively, not as an assemblage of cavities, but as a collection of laws, shaping, directing, and modifying the new

* *Hist. Nat.,*. b. i. c. 3.

matter; giving it not only form, but motion and activity, such as belong to the parts of an organic being.

3. It must be allowed, however, that even with this explanation, the comparison is very loose and insufficient. A *mould* may be permitted to mean a collection of laws, but still it can convey no conception except that of laws regulated by relations of space; and such a conception is very plainly quite inadequate to the purpose. What can we conceive of the interior mould by which chyle is separated from the aliments at the pores of the lacteals, or tears secreted in the lacrymatory gland?

An additional objection to this mode of expression of Buffon is, that it suggests to us only a single marked change in the assimilated matter, not a continuous series of changes. Yet the animal fluids and other substances are, in fact, undergoing a constant series of changes. Food becomes chyme, and chyme becomes chyle; chyle is poured into the blood; from the blood secretions take place, as the bile; the bile is poured into the digestive canal, and a portion of the matter previously introduced is rejected out of the system. Here we must have a series of "interior moulds;" and these must impress matter at its ejection from the organic system as well as at its reception. But, moreover, it is probable that none of the above transformations are quite abrupt. Change is going on between the beginning and the end of each stage of the nutritive circulation. To express the laws of this continuous change, the image of an interior mould is quite unsuited. We must seek a better mode of conception.

4. Vegetable and animal nutrition is, as we have said, a constant circulation. The matter so assumed is not all retained: a perpetual subtraction accompanies a perpetual addition. There is an excretion as well as an intussusception. The matter which is assumed by the

living creature is retained only for a while, and is then parted with. The individual is the same, but its parts are in a perpetual flux: they come and go. For a time the matter which belongs to the organic body is bound to it by certain laws: but before it is thus bound and after it is loose, this matter may circulate about the universe in any other form. Life consists in a permanent influence over a perpetually changing set of particles.

5. *Cuvier.*—This condition also has been happily expressed, by means of a comparison, by another great naturalist. "If," says Cuvier*, "if, in order to obtain a just idea of the essence of life, we consider it in the beings where its effects are most simple, we shall soon perceive that it consists in the faculty which belongs to certain bodily combinations to continue during a determinate time under a determinate form; constantly attracting into their composition a part of the surrounding substances, and giving up in return some part of their own substance.

"Life is thus a *vortex*, more or less rapid, more or less complex, which has a constant direction, and which always carries along its stream particles of the same kinds; but in which the individual particles are constantly entering in and departing out; so that the *form* of the living body is more essential to it than its matter.

"So long as this motion subsists, the body in which it takes place is *alive*; it *lives*. When the motion stops finally, the body *dies*. After death, the elements which compose the body, given up to the ordinary chemical affinities, soon separate, and the body which was alive is dissolved."

This notion of a vortex† which is permanent while

* *Règne Animal*, i. 11.

† The definition of life given by M. de Blainville appears to me not to differ essentially from that of Cuvier. "Un corps vivant est une sorte de foyer chimique où il-y-a à tous momens apport de nou-

the matter which composes it constantly changes,—of peculiar forces which act in this vortex so long as it exists,
and which give place to chemical forces when the circu-
latory motion ceases,—appears to express some of the
leading conditions of the assimilative power of living
things in a simple and general manner, and thus tends
to give distinctness to the notion of this vital function.

6. But we may observe that this notion of a vortex
is still insufficient. Particles are not only taken into the
system and circulated through it for a time, but, as we
have seen, they are altered in character in a manner to
us unintelligible, both at their first admission into the
system and at every period of their progress through it.
In the vortex each particle is constantly transformed
while it whirls.

It may be said, perhaps, that this transformation may
be conceived to be merely a new arrangement of particles,
and that thus all the changes which take place in the
circulating substances are merely so many additional
windings in the course of the whirling current. But to
say this, is to take for granted the atomic hypothesis in
its rudest form. What right have we to assume that
blood and tears, bile and milk, consist of like particles of
matter differently arranged? What can arrangement, a
mere relation of space, do towards explaining such differences? Is not the insufficiency, the absurdity of such
an assumption proved by the whole course of science?
Are not even chemical changes, according to the best
views hitherto obtained, something more than a mere
new arrangement of particles? And are not vital as
much beyond chemical, as chemical are beyond geome-

velles molecules et depart de molecules anciennes; où la composition
n'est jamais fixe (si ce n'est d'un certain nombre de parties verita-
blement mortes ou en depôt), mais toujours pour ainsi dire *in nisu*,
d'on mouvement plus ou moins lent et quelquefois chaleur."—*Principes
d'Anat. Comp.*, 1822, t. i. p. 16.

trical modifications? It is not enough, then, to conceive
life as a vortex. The particles which are taken into the
organic frame do more than circulate there. They are,
at every point of their circulation, acted upon by laws of
an unknown kind, changing the nature of the substance
which they compose. Life is a vortex in which vital
forces act at every point of the stream: it is not only a
current of whirling matter, but a cycle of recurring
powers.

7. *Matter and Form.*—This image of a vortex is
closely connected with the representation of life offered
us by writers of a very different school. In SCHELLING'S
Lectures on Academic Study, he takes a survey of the
various branches of human knowledge, determining according to his own principles the shape which each·science
must necessarily assume. The peculiar character of organization, according to him*, is that the matter is only an
accident of the thing itself, and the organization consists
in form alone. But this form, by its very opposition to
matter, ceases to be independent of it, and is only ideally
separable. In organization, therefore, substance and
accident, matter and form, are completely identical†.
This notion, that in organization the form is essential and
the matter accidental, or, in other words, that the form is
permanent and the matter fluctuating and transitory,
agrees, if taken in the grossest sense of matter and form,
with Cuvier's image of a vortex. In a whirlpool, or in a
waterfall, the form remains, the matter constantly passes
away and is renewed. But we have already seen‡ that in
metaphysical speculations in which *matter* and *form* are
opposed, the word *form* is used in a far more extensive
sense than that which denotes a relation of space. It may

* Lect. xiii., p. 288.

† I have not translated Schelling's words, but given their import
as far as I could.

‡ Book i.

indeed designate any change which matter can undergo; and we may very allowably say that food and blood are the same matter under different forms. Hence if we assert that *life is a constant form of circulating matter*, we express Cuvier's notion in a mode free from the false suggestion which " vortex " conveys.

8. We may, however, still add something to this account of life. The circulating parts of the system not only circulate, but they form the non-circulating parts. Or rather, there are no non-circulating parts: all portions of the frame circulate more or less rapidly. The food which we take circulates rapidly in the fluids, more slowly in the flesh, still more slowly in the bones; but in all these parts it is taken into the system, retained there for some time, and finally replaced by other matter. But while it remains in the body, it exercises upon the other circulating parts the powers by which their motion is produced. Nutriment forms and supports the organs, and the organs carry fresh nutriment to its destination. The peculiar forces of the living body, and its peculiar structure, are thus connected in an indescribable manner. The forces produce the structure; the structure, again, is requisite for the exertion of the forces. The idea of an organic or living being includes this peculiar condition—that its construction and powers are such, that it constantly appropriates to itself new portions of substance which, so appropriated, become indistinguishable parts of the whole, and serve to carry on subsequently the same functions by which they were assimilated. And thus *organic life is a constant form of circulating matter, in which the matter and the form determine each other by peculiar laws* (*that is, by vital forces*).

SECT. III. *Attempts to conceive the Forces of Assimilation and Secretion.*

9. I have already stated that in our attempts to obtain

clear and scientific Ideas of vital forces, we have, in the first place, to seek to understand the course of change and motion in each function, so as to see at what points of the process peculiar causes come into play; and next, to endeavour to obtain some insight into the peculiar character and attributes of these causes. Having spoken of the first part of this mode of investigation in regard to the general nutrition of organic bodies, I must now say a few words on the second part.

The forces here spoken of are *vital* forces. From what has been said, we may see in some measure the distinction between forces of this kind and mechanical or chemical forces; the latter tend constantly to produce a final condition, after which there is no further cause of change : mechanical forces tend to produce equilibrium ; chemical forces tend to produce composition or decomposition ; and this point once reached, the matter in which these forces reside is altogether inert. But an organic body tends to a constant motion, and the highest activity of organic forces shows itself in continuous change. Again, in mechanical and chemical forces, the force of any aggregate is the sum of the forces of all the parts : the sum of the forces corresponds to the sum of the matter. But in organic bodies the amount of effect does not depend on the matter, but on the form : the particles lose their separate energy, in order to share in that of the system ; they are not added, they are assimilated.

10. It is difficult to say whether anything has been gained to science by the various attempts to assign a fixed *name* to the vital force which is thus the immediate cause of assimilation. It has been called *organic attraction* or *vital attraction, organic affinity* or *vital affinity*, being thus compared with mechanical attraction or chemical affinity. But, perhaps, as the process is certainly neither mechanical nor chemical, it is desirable to appro-

priate to it a peculiar name; and the name *Assimilation*, or *Organic Assimilation*, by the usage of good biological writers, is generally employed for this purpose, and may be taken as the standard name of this vital force. To illustrate this, I will quote a passage from the excellent *Elements of Physiology* of Professor Müller. "In the process of nutrition is exemplified the fundamental principle of *organic assimilation*. Each elementary particle of an organ attracts similar particles from the blood, and by the changes it produces in them, causes them to participate in the vital principle of the organ itself. Nerves take up nervous substance, muscles muscular substance: even morbid structures have the assimilating power; warts in the skin grow with their own peculiar structure; in an ulcer, the base and border are nourished in a way conformable to the mode of action and secretion determined by the disease."

11. The force of organic assimilation spoken of in the last paragraph denotes peculiarly the force by which each organ appropriates to itself a part of the nutriment received into the system, and thus is maintained and augmented with the growth of the whole. But the growth of the solid parts is only one portion of the function of nutrition; besides this, we must consider the motion and changes of the fluids, and must ask what kind of forces may be conceived to produce these. What are the powers by which chyle is *absorbed* from the food, by which bile is *secreted* from the blood, by which the circulating *motion* of these and all other fluids of the body are constantly maintained? To the questions,—What are the forces by which *absorption, secretion*, and the *vital motions*, of fluids are produced?—no satisfactory answer has been returned. Yet still some steps have been made, which it may be instructive to point out.

12. In *absorption* it would appear that a part of the

agency is inorganic; for not only dead membranes, but inorganic substances, absorb fluids, and even absorb them with elective forces, according to the ingredients of the fluid. A force which is of this kind, and which has been termed *endosmose,* has been found to produce very curious effects. When a membrane separates two fluids, holding in solution different ingredients, the fluids pass through the membrane in an imperceptible manner, and mix or exchange their elements. The force which produces these effects is capable of balancing a very considerable pressure. It appears, moreover, to depend, at least among other causes, upon attractions operating between the elements of the solids and the fluids, as well as between the different fluids; and this force, though thus apparently of a mechanical and chemical nature, probably has considerable influence in vital phenomena.

13. But still, though endosmose may account in part for absorption in some cases, it is certain that there is some other vital force at work in this process. There must be, as Müller says*, " an organic attraction of a kind hitherto unknown." " If absorption," he adds†, " is to be explained in a manner analogous to the laws of endosmose, it must be supposed that a chemical affinity, resulting from the vital process itself, is exerted between the chyme in the intestines and the chyle in the lacteals, by which the chyle is enabled to attract the chyme without being itself attracted by it. But such affinity or attraction would be of a *vital* nature, since it does not exist after death."

14. If the force of absorption be thus mysterious in its nature, the force of *secretion* is still more so. In this case we have an organ filled with a fine net-work of blood-vessels, and in the cavities of some *gland,* or open part, we have a new fluid formed, of a kind altogether

---

* *Physiology,* p. 299.   † *Ib.,* p. 301.

different from the blood itself. It is easily shown that this cannot be explained by any action of pores or capillary tubes. But what conception can we form of the forces by which such a change is produced? Here, again, I shall borrow the expressions of Müller, as presenting the last result of modern physiology. He says*, "The more probable supposition is, that by virtue of imbibition, or the general organic porosity, the fluid portion of the blood becomes diffused through the tissue of the secreting organ; that the external surface of the glandular canals exerts a chemical attraction on the elements of the fluid, infusing into them at the same time a tendency to unite in new combinations; and then repels them in a manner which is certainly quite inexplicable, towards the inner surface of the secreting membrane, or glandular canals." "Although quite unsupported by facts," he adds, "this theory of attraction and repulsion is not without its analogy in physical phenomena; and it would appear that very similar powers effect the elimination of the fluid in secretion, and cause it to be taken up by the lymphatics in absorption." He elsewhere says†, "Absorption seems to depend on an attraction the nature of which is unknown, but of which the very counterpart, as it were, takes place in secretion; the fluids altered by the secreting action being repelled towards the free side or open surface only of the secreting membranes, and then pressed forwards by the successive portions of the fluids secreted."

15. With regard to the forces which produce the *motion* of absorbed or secreted fluids along their destined course, it may be seen, from the last quoted sentence, that the same vital force which changes the nature, also produces the movement of the substance. The fluids are pressed forwards by the successive portions absorbed or secreted. That this is the sole cause, or at least a very

* *Physiology*, p. 464.　　　　　† *Ib.*, p. 301.

powerful cause, of the motion of the nutritive fluids in organic bodies, is easily shown by experience. It is found[*] that the organs which effect the ascent of the sap in trees during the spring are the terminal parts of the roots; that the whole force by which the sap is impelled upwards is the *vis a tergo*, as it has been called, the force pushing from behind, exerted in the roots. And thus the force which produces this motion is exerted exactly at those points where the organic body selects from the contiguous mass those particles which it absorbs and appropriates. And the same may most probably be taken for the cause of the motion of the lymph and chyle; at least, Müller says[†] that no other motive power has been detected which impels those fluids in their course.

Thus, though we must confess the Vital Force concerned in Assimilation and Secretion to be unknown in its nature, we still obtain a view of some of the attributes which it involves. It has mechanical efficacy, producing motions, often such as would require great mechanical force. But it exerts at the same point both an attraction and a repulsion, attracting matter on one side, and repelling it on the other; and in this circumstance it differs entirely from mechanical forces. Again, it is not only mechanical but chemical, producing a complete change in the nature of the substance on which it acts; to which we must add that the changes produced by the vital forces are such as, for the most part, our artificial chemistry cannot imitate. But, again, by the action of the vital force at any point of an organ, not only are fluids made to pass, and changed as they pass, but the organ itself is maintained and strengthened, so as to continue or to increase its operation: and thus the vital energy supports its activity by its action, and is augmented by being exerted.

* Müller, p. 300.   † *Ib.*, p. 254.

We have thus endeavoured to obtain a view of some of the peculiar characters which belong to the Force of Organic Assimilation;—the Force by which life is kept up, conceived in the most elementary form to which we can reduce it by observation and contemplation. It appears that it is a force which not only produces motion and chemical change, but also *vitalizes* the matter on which it acts, giving to it the power of producing like changes on other matter, and so on indefinitely. It not only circulates the particles of matter, but puts them in a stream of which the flow is development as well as movement.

The force of Organic Assimilation being thus conceived, it becomes instructive to compare it with the force concerned in Generation, which we shall therefore endeavour to do.

Sect. IV. *Attempts to conceive the Process of Generation.*

16. At first sight the function of Nutrition appears very different from the function of Generation. In the former case we have merely the existing organs maintained or enlarged, and their action continued; in the latter, we have a new individual produced and extricated from the parent. The term *reproduction* has, no doubt, been applied, by different writers, to both these functions;—to the processes by which an organ when mutilated, is restored by the forces of the living body, and to the process by which a new generation of individuals is produced which may be considered as taking the place of the old generation, as these are gradually removed by death. But these are obviously different senses of the word. In the latter case, the term reproduction is figuratively used; for the *same* individuals are not reproduced; but the species is kept up by the propagation of new individuals, as in nutrition the organ is kept up by the assimilation of new matter. To escape ambiguity, I shall avoid using the term *reproduction* in the sense of *propagation.*

17. In Nutrition, as we have seen, the matter, which from being at first extraneous, is appropriated by the living system, and directed to the sustentation of the organs, undergoes a series of changes of which the detail eludes our observation and apprehension. The nutriment which we receive contributes to the growth of flesh and bone, viscera and organs of sense. But we cannot trace in its gradual changes a visible preparation for its final office. The portion of matter which is destined to repair the waste of the eye or the skin, is not found assuming a likeness to the parts of the eye or the structure of the skin, as it comes near the place where it is moulded into its ultimate form. The new parts are insinuated among the old ones, in an obscure and imperceptible matter. We can trace their progress only by their effects. The organs *are* nourished, and that is almost all we can learn : we cannot discover *how* this is done. We cannot follow nature through a series of manifest preparations and processes to this result.

18. In Generation the case is quite different. The young being is formed gradually and by a series of distinguishable processes. It is included within the parent before it is extruded, and approaches more or less to the likeness of the parent before it is detached. While it is still an embryo, it shares in the nutriment which circulates through the system of the mother; but its destination is already clear. While the new and the old parts, in every other portion of the mother, are undistinguishably mixed together, this new part, the foetus, is clearly distinct from the rest of the system, and becomes rapidly more and more so, as the time goes on. And thus there is formed, not a new part, but a new whole; it is not an organ which is kept up, but an offspring which is prepared. The progeny is included in the parent, and is gradually fitted to be separated from it The young is at

first only the development of a part of the organization of the mother;—of a germ, an ovule. But it is not developed like other organs, retaining its general form. It does not become merely a larger bud, a larger ovule; it is entirely changed ; it becomes from a bud, a blossom, a flower, a fruit, a seed; from an ovule it becomes an egg, a chick, a bird ; or it may be, a fœtus, a child. The original rudiment is not merely nourished, but unfolded and transformed through the most marked and remote changes, gradually tending to the form of the new individual.

19. But this is not all. The fœtus is, as we have said, a development of a portion of the mother's organization. But the fœtus (supposing it female) is a likeness of the mother. The mother, even before conception, contains within herself the germs of her progeny; the female fœtus, therefore, at a certain stage of development, will contain also the germs of possible progeny; and thus we may have the germs of future generations, pre-existing and included successively within one another. And this state of things, which thus suggests itself to us as possible, is found to be the case in facts which observation supplies. Anatomists have traced ovules in the unborn fœtus, and thus we have three generations included one within another.

20. Supposing we were to stop here, the process of propagation might appear to be altogether different from that of nutrition. The latter, as we have seen, may be in some measure illustrated by the image of a *vortex ;* the former has been represented by the image of a series of germs, *sheathed* one within another successively, and this without any limit. This view of the subject has been termed the doctrine of the *pre-existence* of *germs ;* and has been designated by German writers by a term (einschachtelungs-theorie) descriptive of the successive

sheathing of which I have spoken. Imitating this term, we may call it *the theory of successive inclusion*. It has always had many adherents; and has been, perhaps, up to the present time, the most current opinion on the subject of generation. Cuvier inclines to this opinion *. " Fixed forms perpetuating themselves by generation distinguish the species of living things. These forms do not produce themselves, do not change themselves. Life supposes them to exist already; its flame can be lighted only in organization previously prepared; and the most profound meditations and the most delicate researches terminate alike .in the mystery of the pre-existence of germs."

21. Yet this doctrine is full of difficulty. It is, as Cuvier says, a mysterious view of the subject;—so mysterious that it can hardly be accepted by us, who seek distinct conceptions as the basis of our philosophy. Can it be true, not only that the germ of the offspring is originally included in the parent, but also the germs of *its* progeny, and so on without limit:—So that each fruitful individual contains in itself an infinite collection of future possible individuals;—a reserve of infinite succeeding generations? This is hard to admit. Have we no alternative? What is the opposite doctrine?

22. The opposite doctrine deserves at least some notice. It extends to the production of a new individual, the conception of growth by nutrition. According to this view, we suppose propagation to take place, not as in the view just spoken of, by inclusion and extrusion, but by assimilation and development;—not by the material pre-existence of germs, but by the communication of vital forces to new matter. This opinion appears to be entertained by some of the most eminent physiologists of the present time. Thus, Müller says, " The organic

---

* *Règne Animal*, p. 20.

force is also creative. The organic force which resides in
the whole, and on which the existence of each part
depends, has also the property of generating from organic
matter the parts necessary to the whole." Life, he adds,
is not merely a harmony of the parts.   On the contrary,
the harmonious action of the parts subsists only by the
influence of a force pervading† all parts of the body.
"This force exists before the†harmonizing parts, which
are in fact formed by it during the development of the
embryo." And again; "The creative force exists in the
germ, and creates in it the essential force of the future
animal.   The germ is *potentially* the whole animal : during
the development of the germ the parts which constitute
the actual whole are produced."

23. In this view, we extend to the reproduction of
an individual the same conception of organic assimilation
which we have already arrived at, as the best notion we
can form of the force by which the reproduction and sustentation of parts takes place.   And is not such an
extension really very consistent ?   If a living thing can
appropriate to itself extraneous matter, invest it with its
own functions, and thus put it in the stream of constant
development, may we not conceive the development of a
new *whole* to take place in this way as well as of a *part ?*
If the organized being can infuse into new matter its vital
forces, is there any contradiction in supposing this infu-
sion to take place in the full measure which is requisite
for the production of a new individual?   The force of
organic assimilation is transferred to the very matter on
which it acts; it may be transferred so that the operation
of the forces produces not only an organ, but a system of
organs.

24. This identification of the forces which operate
in Nutrition and Generation may at first seem forced
and obscure, in consequence of the very strong apparent

differences of the two processes which we have already noticed. But this defect in the doctrine is remedied by the consideration of what may be considered as intermediate cases. It is not true that, in the nutrition of special organs, the matter is always conveyed to its ultimate destination without being on its way moulded into the form which it is finally to bear, as the embryo is moulded into the form of the future individual. On the contrary, there are cases in which the waste of the organs is supplied by the growth of new ones, which are prepared and formed before they are used, just as the offspring is prepared and formed before it is separated from the parent. This is the case with the teeth of many animals, and especially with the teeth of animals of the crocodile kind. Young teeth grow near the root of the old ones, like buds on the stem of a plant; and as these become fully developed, they take the place of the parent tooth when that dies and is cast away. And these new teeth in their turn are succeeded by others which germinate from them. Several generations of such teeth, it is said as many as four, have been detected by anatomists, visibly existing at the same time; just as several generations of germs of individuals have been, as we already stated, observed included in one another. But this case of the teeth appears to show very strikingly how insufficient such observations are to establish the doctrine of successive inclusion, or of the pre-existence of germs. Are we to suppose that every crocodile's tooth includes in itself the germs of an infinite number of possible teeth, as in the theory of pre-existing germs, every individual includes an infinite number of individuals? If this be true of teeth, we must suppose that organ to follow laws entirely different from almost every other organ; for no one would apply to the other organs in general such a theory of reproduction. But if such a theory be not maintained respecting the teeth, how can we maintain the theory of the pre-existing germs of individuals, which has no recommendation except that of accounting for exactly the same phenomena?

It would seem, then, that we are, by the closest consideration of the subject, led to conceive the forces by which generation is produced as forces which vitalize certain portions of matter, and thus prepare them for development according to organic forms; and thus the conception of this Generative Force is identified with the conception of the Force of Organic Assimilation, to which we were led by the consideration of the process of nutrition.

I shall not attempt to give further distinctness and fixity to this conception of one of the vital forces; but I shall proceed to exemplify the same analysis of life by some remarks upon another Vital Process, and the Forces of which it exhibits the operation.

CHAPTER V.

ATTEMPTS TO FORM IDEAS OF SEPARATE VITAL FORCES, *Continued.*—VOLUNTARY MOTION.

1. WE formerly noticed the distinctions of *organic* and *animal* functions, organic and animal forces, as one of the most marked distinctions to which physiologists have been led in their analysis of the vital powers. I have now taken one of the former, the organic class of functions, namely, nutrition; and have endeavoured to point out in some measure the peculiar nature of the vital forces by which this function is carried on. It may serve to show the extent and the difficulty of this subject,

if, before quitting it, I offer a few remarks suggested by a function belonging to the other class, the animal functions. This I shall briefly do with respect to *voluntary motion*.

2. In the History of Physiology, I have already related the progress of the researches by which the organs employed in voluntary motion became known to anatomists. It was ascertained to the satisfaction of all physiologists, that the immediate agents in such motion are the muscles; that the muscles are in some way contracted, when the nerves convey to them the agency of the will; and that thus the limbs are moved. It was ascertained, also, that the nerves convey sensations from the organs of sense inwards, so as to make these sensations the object of the animal's consciousness. In man and the higher animals, these impressions upon the nerves are all conveyed to one internal organ, the brain; and from this organ all impressions of the will appear to proceed; and thus the brain is the centre of animal life, towards which sensations converge, and from which volitions diverge.

But this being the process, we are led to inquire how far we can obtain any knowledge, or form any conception of the vital forces by means of which the process is carried on. And here I have further stated in the History*, that the transfer of sensations and volitions along the nerves was often represented as consisting in the motion of a Nervous Fluid. I have related that the hypothesis of such a fluid, conveying its impressions either by motions of translation or of vibration, was countenanced by many great names, as Newton, Haller, and even Cuvier. But I have ventured to express my doubt whether this hypothesis can have much value: "for," I have said, "this principle cannot be mechanical, chemical, or physical, and therefore cannot be better understood by embodying

* *Hist. Ind. Sci.*, iii. 428.

it in a fluid. The difficulty we have in conceiving what the force *is*, is not got rid of by explaining the machinery by which it is *transferred*."

3. I may add, that no succeeding biological researches appear to have diminished the force of these considerations. In modern times, attempts have repeatedly been made to identify the nervous fluid with electricity or galvanism. But these attempts have not been satisfactory or conclusive of the truth of such an identity : and Professor Müller probably speaks the judgment of the most judicious physiologists, when he states it as his opinion, after examining the evidence*, "That the vital actions of the nerves are not attended with the development of any galvanic currents which our instruments can detect; and that the laws of action of the nervous principle are totally different from those of electricity."

That the powers by which the nerves are the instruments of sensation, and the muscles of motion, are vital endowments, incapable of being expressed or explained by any comparison with mechanical, chemical, and electrical forces, is the result which we should expect to find, judging from the whole analogy of science ; and which thus is confirmed by the history of physiology up to the present time. We naturally, then, turn to inquire whether such peculiar vital powers have been brought into view with any distinctness and clearness.

4. The property by which muscles, under proper stimulation, contract and produce motion, has been termed *irritability* or *contractility;* the property by which nerves are susceptible of their appropriate impressions has been termed *sensibility.* A very few words on each of these subjects must suffice.

Irritability.—I have, in the History of Physiology†, noticed that Glisson, a Cambridge professor, distinguished the irritation of muscles as a peculiar property, different

* *Elem. Phys.*, p. 640.   † *Hist. Ind. Sci.*, iii. 427.

from any merely mechanical or physical action. I have mentioned, also, that he divides irritation into natural, vital, and animal; and points out, though briefly, the graduated differences of irritability in different organs. Although these opinions did not at first attract much notice, about seventy years afterwards attention was powerfully called to this vital force, *irritability*, by Haller I shall borrow Sprengel's reflections on this subject.

"Hitherto men had been led to see more and more clearly that the cause of the bodily functions, the fundamental power of the animal frame, is not to be sought in the mechanism, and still less in the mixture of the parts. In this conviction, they had had recourse partly to the quite supersensuous principle of the soul, partly to the half-material principle of the animal spirits, in order to explain the bodily motions. Glisson alone saw the necessity of assuming an original power in the fibres, which, independent of the influence of the animal spirits, should produce contraction in them. And Gorter first held that this original power was not to be confined to the muscles, but to be extended to all parts of the living body.

"But as yet the laws of this power were not known, nor had men come to an understanding whether it were fully distinct from the elasticity of the parts, or by what causes it was put in action. They had neither instituted observations nor experiments which established its relation to other assumed forces of the body. There was still wanting a determination of the peculiar seat of this power, and experiments to trace its gradual differences in different parts of the body. In addition to other causes, the necessity of the assumption of such a power was felt the more, in consequence of the prevalence of Leibnitz's doctrine of the activity of matter; but it was an occult quality, and remained so till Haller, by numerous experiments and solid observations, placed in a clear light the peculiarities of the powers of the animal body."

5. Perhaps, however, Haller did more in the way of determining experimentally the limits and details of the application of this idea of Irritability as a peculiar attribute, than in developing the idea itself. In this way his merits were great. As early as the year 1739, he pub⁺lished his opinion upon irritability as the cause of muscular motion, which he promulgated again in 1743. But from the year 1747 he was more attentive to the peculiarities of irritability, and its difference from the effect of the nerves. In the first edition of his *Physiology*, which appeared in 1747, he distinguished three kinds of force in muscles,—the dead force, the innate force, and the nervous power. The first is identical with the elastic force of dead matter, and remains even after death. The *innate force* continues only a short time after death, and discloses itself especially by alternate oscillations; the motions which arise from this are much more lively than those which arise from mere elasticity: they are not excited by tension, nor by pressure, nor by any mechanical alteration, but only by *irritation*. The *nervous force* of the muscle is imparted to it from without by the nerves; it preserves the *irritability*, which cannot long subsist without the influence of the nervous force, but is not identical with it.

In the year 1752, Haller laid before the Society of Göttingen the result of one hundred and ninety experiments; from which it appears to what parts of the animal system irritability and nervous power belong. These I need not enumerate. He also investigated with care its gradations in those parts which do possess it. Thus the heart possesses it in the highest degree, and other organs follow in their order.

6. Haller's doctrine was, that there resides in the muscles a peculiar vital power by which they contract and that this power is distinct from the attributes of the nerves. And this doctrine has been accepted by the best physiologists of modern times. But this distinction of the *irritability* of the muscles from the *sensibility* of the nerves became somewhat clearer by giving to the former attribute the name of *contractility*. This accordingly was done; it is, for example, the phraseology used by Bichat. By speaking of *animal sensibility* and *animal contractility*, the passive and the active element of the processes of animal life are clearly separated and opposed to each other. The sensations which we feel, and the muscular action which we exert, may be closely and inseparably connected, yet still they are clearly distinguishable. We can easily in our apprehension separate the titillation felt in the nose on taking snuff, from the action of the muscles in sneezing, or the perception of an object falling towards the eye, from the exertion which shuts the eye-lid; although in these cases the passive and active part of the process are almost or quite inseparable in fact. And this clear separation of the active from the passive power is something, it would seem, peculiar to the Animal Vital Powers; it is a character by which they differ, not only from mechanical, chemical, and all other merely physical forces, but even from Organic Vital Powers.

7. But this difference between the Animal and the Organic Vital Powers requires to be further insisted upon, for it appears to have been overlooked or denied by very eminent physiologists. For instance, Bichat classifies the Vital Powers as Animal Sensibility, Animal Contractility, Organic Sensibility, Organic Contractility.

Now the view which suggests itself to us, in agreement with what has been said, is this:—that though Animal Sensibility and Animal Contractility are clearly and certainly distinct, Organic Sensibility and Organic Contractility are neither separable in fact nor in our conception, but together make up a single Vital Power. That they are not separable in fact is, indeed, acknowledged by Bichat himself. "The organic contractility," he says*, " can never be separated from the sensibility of the same kind; the reaction of the excreting tubes is immediately connected with the action which the secreted fluids exercise upon them: the contraction of the heart must necessarily succeed the influx of the blood into it." It is not wonderful, therefore, that it should have happened, as he complains, that " authors have by no means separated these two things, either in their consideration or in language." We cannot avoid asking, Are Organic Sensibility and Organic Contractility really anything more than two different aspects of the same thing, like action and reaction in mechanics, which are only two ways of considering the action which takes place at a point; or like the positive and negative electricities, which, as we have seen, always co-exist and correspond to each other?

8. But we may observe, moreover, that Bichat, by his use of the term Contractility, includes in it powers to which it cannot with any propriety be applied. Why should we suppose that the vital powers of absorption, secretion, assimilation, are of such a nature that the name *contractility* may be employed to describe them? We have seen, in the last chapter, that the most careful study of these powers leads us to conceive them in a manner altogether removed from any notion of contraction. Is it not then an abuse of language which cannot possibly lead to anything but confusion, to write thus†: " The insensible organic contractility is that, by virtue of which the excreting tubes react upon their respective fluids, the secreting organs upon the blood which flows into

* *Life and Death*, p. 94. † *Ib.* p. 95.

them, the parts where nutrition is performed upon the nutritive juices, and the lymphatics upon the substances which excite their open extremities." In the same manner he ascribes* to the peculiar sensibility of each organ the peculiarity of its products and operations. An increased absorption is produced by an increased susceptibility of the "absorbent orifices." And thus, in this view, each organic power may be contemplated either as sensibility or as contractility, and may be supposed to be rendered more intense by magnifying either of these its aspects; although, in fact, neither can be conceived to be increased without an exactly commensurate increase of the other.

9. This opinion, unfounded as it thus appears to be, that all the different organic vital powers are merely different kinds of contractility or excitability, was connected with the doctrines of Brown and his followers, which were so celebrated in the last century, that all diseases arise from increase or from diminution of the vital force. The considerations which have already offered themselves would lead us to assent to the judgment which Cuvier has pronounced upon this system. "The theory of excitation," he says, "so celebrated in these later times by its influence upon pathology and therapeutick, is at bottom only a modification of that, in which, including under a common name sensibility and irritability," and we may add, applying this name to all the vital powers, "the speculator takes refuge in an abstraction so wide, that if, by it, he simplifies medicine, he annihilates all positive physiology†."

10. The separation of the nervous influence and the muscular irritability, although it has led to many highly instructive speculations, is not without its difficulties,

* *Life and Death*, p. 90.
† *Hist. des Sci. Nat. depuis* 1789, i. 219.

when viewed with reference to the Idea of **Vital Power**. If the irritability of each muscle reside in the muscle itself, how does it differ from a mere mechanical force, as elasticity? But, in point of fact, it is certain that the muscular irritability of the animal body is not an attribute of the muscle itself independent of its connexion with the system. No muscle, or other part, removed from the body, *long* preserves its irritability. This power cannot subsist permanently, except in connexion with an organic whole. This condition peculiarly constitutes irritability a living force: and this condition would be satisfied by considering the force as derived from the nervous system; but it appears that though the nervous system has the most important influence upon all vital actions, the muscular irritability must needs be considered as something distinct. And thus the Irritability or Contractility of the muscle is a peculiar endowment of the texture, but it is at the same time an endowment which can only co-exist with life; it is, in short, a peculiar Vital Power.

11. This necessity of the union of the muscle with the whole nervous system, in order that it may possess irritability, was the meaning of the true part of Stahl's psychical doctrine; and the reason why he and his adherents persisted in asserting the power of the soul even over involuntary motions. This doctrine was the source of much controversy in later times.

"But," says Cuvier *, "this opposition of opinion may be reconciled by the intimate union of the nervous substance with the fibre and the other contractile organic elements, and by their reciprocal action;—doctrines which had been presented with so much probability by physiologists of the Scotch school, but which were elevated above the rank of hypotheses only by the observations of more recent times.

* *Hist. des Sci. Nat. depuis* 1789, i. 213.

" The fibre does not contract by itself, but by the influence of the nervous filaments, which are always united with it. The change which produces the contraction cannot take place without the concurrence of both these substances; and it is further necessary that it should be occasioned each time by an exterior cause, by a stimulant.

" The will is one of these stimulants; but it only excites the irritability, it does not constitute it; for in the case of persons paralytic from apoplexy, the irritability remains, though the power of the will over it is gone. Thus *irritability* depends in part on the nerve, but not on the *sensibility:* this last is another property, still more admirable and occult than the irritability; but it is only one among several functions of the nervous system. It would be an abuse of words to extend this denomination to functions unaccompanied by perception."

12. Supposing, then, that Contractility is established as a peculiar Vital Power residing in the muscles, we may ask whether we can trace with any further exactness the seat and nature of this power. It would be unsuitable to the nature of the present work to dwell upon the anatomical discussions bearing upon this point. I will only remark that some anatomists maintain* that muscles are contracted by those fibres assuming a zigzag form, which at first were straight. Others (Professor Owen and Dr. A. Thompson,) doubt the accuracy of this observation; and conceive that the muscular fibre becomes shorter and thicker, but does not deviate from a right line. We may remark that the latter kind of action appears to be more elementary in its nature. We can conceive a straight line thrown into a zigzag shape by muscular contractions taking place between remote parts of it; but it is difficult to conceive by what *elementary* mode of

* MULLER, *Elem. Phys.,* p. 887.

action a straight fibre could bend itself at certain points, and at certain points only; since the elementary force must act at every point of the fibre, and not at certain selected points.

13. A circumstance which remarkably marks the difference between the vital force of contractility, inherent in muscles, and any merely dead or mechanical force, is this; that in assuming their contractile state, muscles exert a tension which they could not themselves support or convey if not strengthened by their vital irritability. They are capable of raising weights by their exertion, which will tear them asunder when the power of contraction is lost by death. This has induced Cuvier and other physiologists* to believe "that in the moment of action, the particles that compose a fibre, not only approach towards each other longitudinally, but that their cohesive attraction becomes instantaneously much greater than it was before: for without such an increase of cohesive force, the tendency to shorten could not, as it would appear, prevent the fibre from being torn." We see here the difficulty, or rather the impossibility, of conceiving muscular contractility as a mere mechanical force; and perhaps there is little hope of any advantage by calling in the aid of chemical hypothesis to solve the mechanical difficulty. Cuvier conjectures that a sudden change in the chemical composition may thus so quickly and powerfully augment the cohesion. But we may ask, are not a chemical synthesis and analysis, suddenly performed by a mere act of the will, as difficult to conceive as a sudden increase and decrease of mechanical power directly produced by the same cause?

15. *Sensibility.* The nerves are the organs and channels of sensibility. By means of them we receive our sensations, whether of mere pleasure and pain, or of qua-

* PRICHARD, *Vital Prin.*, p. 126.

lities which we ascribe to external objects, as a bitter taste, a sweet odour, a shrill sound, a red colour, a hard or a hot object of touch. Some of these sensations are but obscurely the objects of our consciousness; as for example the feeling which our feet have of the ground, or the sight which our eyes have of neighbouring objects, when we walk in a reverie. In these cases the sensations, though obscure, exist; for they serve to balance and guide us as we walk. In other cases, our sensations are distinctly and directly the objects of our attention.

But our sensations, as we have already said, we ascribe as qualities to external objects. By our senses we perceive objects, and thus our *sensations* become *perceptions*. We have not only the sensation of round, purple, and green, repeated and varied, but the perception of a bunch of grapes partly ripe and partly unripe. We have not only sensations of noise and of variously-coloured specks rapidly changing their places, but we have perceptions, by sound and sight, of a stone rolling down the hill and crushing the shrubs in its path. We scarcely ever dwell upon our sensations; our thoughts are employed upon objects. We regard the impressions upon our nerves, not for what they *are*, but for what they *tell* us.

But in what language is it that the impressions upon the nerves thus speak to us of an external world, of the forms and qualities and actions of objects? How is it that by the aid of our nervous system we become acquainted not only with impressions but with things; that we learn not only the relation of objects to us, but to one another?

16. It has been shown at some length in the previous Books, that the mode in which sensations are connected in our minds so as to convey to us the knowledge of objects and their relations, is by being contemplated with reference to *Ideas*. Our sensations, connected by the

Idea of Space, become figures; connected by the Idea of
Time, they become causes and effects; connected by the
Idea of Resemblance, they become individuals and kinds;
connected by the Idea of Organization, they become
living things. It has been shown that without these
Ideas there can be no connexion among our sensations,
and therefore no perception of Figure, Action, Kind, or
in short of bodies under any aspect whatever. Sensations
are the rude *Matter* of our perceptions; and are nothing,
except so far as they have *Form* given them by Ideas.
But thus moulded by our Ideas, sensation becomes the
source of an endless store of important knowledge of
every possible kind.

17. But one of the most obvious uses of our perceptions and our knowledge is to direct our actions. It is
suitable to the condition of our being that when we perceive a bunch of grapes, we should be able to pluck and
eat the ripe ones; that when we perceive a stone rushing
down the side of a hill, we should be able to move so as to
avoid it. And this must be done by moving our limbs;
in short, by the use of our muscles. And thus sensation
leads, not directly, but through the medium of Ideas, to
muscular contraction. I say that sensation and muscular
action are in such cases connected through the medium
of Ideas. For when we proceed to pluck the grape
which we see, the *sensation* does not determine the
motion of the hand by any necessary geometrical or
mechanical conditions, as an impression made upon a
machine determines its motions; but the *perception* leads
us to stretch forth the hand to that part of space, where-
ever it is, where we *know* that the grape is, and this, not in
any determinate path, but, it may be, avoiding or removing
intervening obstacles, which we also *perceive*. There is in
every such case a connexion between the sensation and
the resulting action, not of a material but of a mental

kind. The cause and the effect are bound together, not by physical but by intellectual ties.

18. And thus in such cases, between the two vital operations, sensation and muscular action, there intervenes, as an intermediate step, perception or knowledge, which is not merely vital but ideal. But this is not all; there is still another mental part of the process which may be readily distinguished from that which we have described. An act of the *will*, a volition, is that in the mind which immediately determines the action of the muscles of the body. And thus Will intervenes between Knowledge and Action; and the cycle of operations which take place when animals act with reference to external objects is this:—Sensation, Perception, Volition, Muscular Contraction.

19. To attempt further to analyse the mental part of this cycle does not belong to the present part of our work. But we may remark here, as we have already remarked in the History*, how irresistibly we are led by physiological researches into the domain of thought and mind. We pass from the body to the soul, from physics to metaphysics; from biology to psychology; from things to persons; from nouns to pronouns. I have there noticed the manner in which Cuvier expresses this transition by the introduction of the pronoun: "The impression of external objects upon the ME, the production of a sensation, of an image, is a mystery impenetrable to our thoughts."

20. But to return to the merely biological part of our speculations. We have arrived, it will be perceived, at this result: that in animal actions there intervenes between the two terms of Sensation and Muscular Contraction, an intermediate process; which may be described as a communication to and from a centre. This centre is

* *Hist. Ind. Sci.*, iii., 430.

the seat of the sentient and volent faculties, and is of a hyperphysical nature. But the existence of such a centre as a necessary element in the functions of the animal life is a truth which is important in biology. This indeed may be taken as the peculiar character of animal, as distinguished from merely organic powers. Accordingly it is so stated by Bichat. For although he superfluously, as it would seem, introduces into his list of vital powers an organic sensibility, he still draws the distinction of which I have spoken; "in the animal life, sensibility is the faculty of receiving an impression *plus* that of referring it to a common centre*."

21. But since Sensibility and Contractility are thus connected by reference to a common Centre, we may ask, before quitting the subject, what are the different forms which this reference assumes. Is the connexion always attended by the distinct steps of knowledge and will,—by a clear act of consciousness, as in the case which we have taken, of plucking a grape; or may these steps become obscure, or vanish altogether?

We need not further illustrate the former connexion. Such actions as we have described are called *voluntary* actions. In extreme cases, the mental part of the process is obvious enough. But we may gradually pass from these to cases in which the mental operation is more and more obscure.

In walking, in speaking, in eating, in breathing, our muscular exertions are directed by our sensations and perceptions: yet in such processes, how dimly are we conscious of perceptive and directive power! How the mind should be able to exercise such a power, and yet should be scarcely or not at all conscious of its exercise, is a very curious problem. But in all or in most of the above instances, the solution of this problem appears to

* *Life and Death*, p. 84.

depend upon psychological rather than biological principles, and therefore does not belong to this place.

22. But in cases at the other extreme, the mental
part of the operation vanishes altogether. In many
animals, even after decapitation, the limbs shrink when
irritated. The motions of the iris are determined by the
influence of light on our eyes, without our being aware
of the motions. Here sensations produce motions, but
with no trace of intervening perception or will. The
sensation appears to be *reflected* back from the central
element of animal life, in the form of a muscular contraction; but in this case the sensation is not modified or
regulated by any *idea*. These reflected motions have no
reference to relations of space or force among surrounding objects. They are blind and involuntary, like the
movements of convulsion, depending for direction and
amount only on the position and circumstances of the
limb itself with its muscles. Here the Centre from which
the reflection takes place is merely animal, not intellectual.

In this case some physiologists have doubted whether
the reflection of the sensation in the form of a muscular
contraction does really take place from the Centre; and
have conceived that sensorial impressions might affect
motor nerves without any communication with the nervous Centre. But on this subject we may, I conceive,
with safety adopt the decision of Professor Müller, deli-
berately given after a careful examination of the subject.
" When impressions made by the action of external stimuli
on sensitive nerves give rise to motions in other parts,
these motions are never the result of the direct reaction
of the sensitive and motor fibres of the nerves on each
other; the irritation is conveyed by the sensitive fibres
to the brain and spinal cord, and is by these communi-
cated to the motor fibres."

23. Thus we have two extreme cases of the connexion of sensation with muscular action; in one of which the connexion clearly *is*, and in the other it as clearly *is not*, determined by relations of Ideas, in its transit through the nervous Centre.   There is another highly curious case, standing intermediate between these two, and extremely difficult to refer to either.   I speak of the case of *Instinct*.

Instinct leads to actions which are *such as if they were determined by Ideas*.   The lamb follows its mother by instinct; but the motions by which it does this, the special muscular exertions, depend entirely upon the geometrical and mechanical relations of external bodies, as the form of the ground, and the force of the wind. The contractions of the muscles which are requisite in order that the creature may obey its instinct, vary with every variation of these external conditions;—are not determined by any rule or necessity, but by properties of space and force.   Thus the action is not governed by sensations directly, but by sensations moulded by ideas. And the same is the case with other cases of instinct. The dog hunts by instinct; but he hunts certain kinds of animals merely, thus showing that his instinct acts according to resemblances and differences; he crosses the field repeatedly to find the track of his prey by scent; thus recognising the relations of space with reference to the track; he leaps, adjusting his force to the distance and height of the leap with mechanical precision; and thus he practically recognises the Ideas of Resemblance, Space, and Force.

But have animals such Ideas?   In any proper sense in which we can speak of possessing Ideas, it appears plain that they have not.   Animals cannot, at any time, be said properly to possess ideas, for ideas imply the possibility of *speculative* knowledge.

24. But if we allow to animals the *practical* possession of ideas, we have still a great difficulty remaining.
In the case of man, his ideas are unfolded gradually by
his intercourse with the external world. The child learns
to distinguish forms and positions by a repeated and
incessant use of his hands and eyes; he learns to walk,
to run, to leap, by slow and laborious degrees; he dis-
tinguishes one man from another, and one animal from
another, only after repeated mistakes. Nor can we conceive this to be otherwise. How should the child know
at once what muscles he is to exert in order to touch
with his hand a certain visible object? How should he
know what muscles to exert that he may stand and not
fall, till he has tried often? How should he learn to
direct his attention to the differences of different faces
and persons, till he is roused by some memory, or hope
which implies memory? It seems to us as if the sensations could not, without considerable practice, be rightly
referred to Ideas of Space, Force, Resemblance, and the
like.

Yet that which thus appears impossible, is in fact
done by animals. The lamb almost immediately after its
birth follows its mother, accommodating the actions of its
muscles to the form of the ground. The chick, just
escaped from the shell, picks up a minute insect, directing
its beak with the greatest accuracy. Even the human
infant seeks the breast and exerts its muscles in sucking,
almost as soon as it is born. Hence, then, we see that
Instinct produces at once actions regulated by Ideas, or,
at least, which take place *as if* they were regulated by
Ideas; although the Ideas cannot have been developed
by exercise, and only appear to exist so far as such
actions are concerned.

25. The term *Instinct* may properly be opposed to *Insight*. The former implies an inward principle of action,

implanted within a creature and practically impelling it, but not capable of being developed into a subject of contemplation. While the instinctive actions of animals are directed by such a principle, the deliberate actions of man are governed by insight: he can contemplate the ideal relations on which the result of his action depends. He can in his mind map the path he will follow, and estimate the force he will exert, and class the objects he has to deal with, and determine his actions by the relations which he thus has present to his mind. He thus possesses Ideas not only practically, but speculatively. And knowing that the Ideas by which he commonly directs his actions, Space, Cause, Resemblance, and the like, have been developed to that degree of clearness in which he possesses them by the assiduous exercise of the senses and the mind from the earliest stage of infancy, and that these Ideas are capable of being still further unfolded into long trains of speculative truth, he is unable to conceive the manner in which animals possess such Ideas as their instinctive actions disclose:—Ideas which neither require to be unfolded nor admit of unfolding; which are adequate for practical purposes without any previous exercise, and inadequate for speculative purposes with whatever labour cultivated.

I have ventured to make these few remarks on Instinct since it may, perhaps, justly be considered as the last province of Biology, where we reach the boundary line of Psychology. I have now, before quitting this subject, only one other principle to speak of.

Chapter VI.

OF THE IDEA OF FINAL CAUSES.

1. BY an examination of those notions which enter into all our reasonings and judgments on living things, it appeared that we conceive animal life as a vortex or cycle of moving matter in which the form of the vortex determines the motions, and these motions again support the form of the vortex: the stationary parts circulate the fluids, and the fluids nourish the permanent parts. Each portion ministers to the others, each depends upon the other. The parts make up the whole, but the existence of the whole is essential to the preservation of the parts. But parts existing under such conditions are *organs*, and the whole is *organized*. This is the fundamental conception of organization. "Organized beings," says the physiologist*, "are composed of a number of essential and mutually dependent parts." "An organized product of nature," says the great metaphysician†, "is that in which all the parts are mutually ends and means."

2. It will be observed that we do not content ourselves with saying that in such a whole, all the parts are *mutually dependent*. This might be true even of a mechanical structure; it would be easy to imagine a framework in which each part should be necessary to the support of each of the others; for example, an arch of several stones. But in such a structure the parts have no properties which they derive from the whole. They are beams or stones when separate; they are no more when joined. But the same is not the case in an organized whole. The limb of an animal separated from

---

* Müller, *Elem.*, p. 18.

† Kant, *Urtheilskraft*, p. 296.

the body, loses the properties of a limb and soon ceases
to retain even its form.

3. Nor do we content ourselves with saying that the
parts are *mutually causes and effects.* This is the case
in machinery. In a clock, the pendulum by means of
the escapement causes the descent of the weight, the
weight by the same escapement keeps up the motion of
the pendulum. But things of this kind may happen by
accident. Stones slide from a rock down the side of a
hill and cause it to be smooth; the smoothness of the
slope causes stones still to slide. Yet no one would call
such a slide an organized system. The system is organized, when the effects which take place among the parts
are *essential to our conception of the whole;* when the
whole would not *be* a whole, nor the parts, parts, except
these effects were produced; when the effects not only
happen in fact, but are included in the idea of the object;
whenithey are not only seen, but foreseen; not only expected, but intended: in short when, instead of being
causes and effects, they are *ends* and *means,* as they are
termed in the above definition.

Thus we necessarily include, in our Idea of Organization, the notion of an end, a purpose, a design; or, to
use another phrase which has been peculiarly appropriated in this case, a *Final Cause.* This idea of a Final
Cause is an essential condition in order to the pursuing
our researches respecting organized bodies.

4. This Idea of Final Cause is not deduced from the
phenomena by reasoning, but is assumed as the only condition under which we can reason on such subjects at all.
We do not deduce the Idea of Space, or Time, or efficient Cause from the phenomena about us, but necessarily
look at phenomena as subordinate to these Ideas from
the beginning of our reasoning. It is true, our ideas of
relations of Space, and Time, and Force may become

much more clear by our familiarizing ourselves with particular phenomena: but still, the Fundamental Ideas are not generated but unfolded; not extracted from the external world, but evolved from the world within. In like manner, in the contemplation of organic structures, we consider each part as subservient to some use, and we cannot study the structure as organic without such a conception. This notion of adaptation,—this Idea of an End,—may become much more clear and impressive by seeing it exemplified in particular cases. But still, though suggested and evoked by special cases, it is not furnished by them. If it be not supplied by the mind itself, it can never be logically deduced from the phenomena. It is not a portion of the facts which we study, but it is a principle which connects, includes, and renders them intelligible; as our other Fundamental Ideas do the classes of facts to which they respectively apply.

5. This has already been confirmed by reference to fact; in the History of Physiology, I have shown that those who studied the structure of animals were irresistibly led to the conviction that the parts of this structure have each its end or purpose;—that each member and organ not merely produces a certain effect or answers a certain use, but is so framed as to impress us with the persuasion that it was constructed *for* that use:—that it was *intended* to produce the effect. It was there seen that this persuasion was repeatedly expressed in the most emphatic manner by Galen;—that it directed the researches and led to the discoveries of Harvey;—that it has always been dwelt upon as a favourite contemplation, and followed as a certain guide, by the best anatomists;—and that it is inculcated by the physiologists of the profoundest views and most extensive knowledge of our own time. All these persons have deemed it a most certain and important principle of physiology, that in every organized

structure, plant or animal, each intelligible part has its allotted office:—each organ is designed for its appropriate function:—that nature, in these cases, produces nothing in vain: that, in short, each portion of the whole arrangement has its *final cause;* an end to which it is adapted, and in this end, the reason that it is where and what it is.

6. This Notion of Design in organized bodies must, I say, be supplied by the student of organization out of his own mind: a truth which will become clearer if we attend to the most conspicuous and acknowledged instances of *design.* The structure of the eye, in which the parts are curiously adjusted so as to produce a distinct image on the retina, as in an optical instrument;—the trochlear muscle of the eye, in which the tendon passes round a support and turns back, like a rope round a pulley;—the prospective contrivances for the preservation of animals, provided long before they are wanted, as the milk of the mother, the teeth of the child, the eyes and lungs of the fœtus:—these arrangements, and innumerable others, call up in us a persuasion that Design has entered into the plan of animal form and progress. And if we bring in our minds this conception of Design, nothing can more fully square with and fit it, than such instances as these. But if we did not already possess the Idea of Design;—if we had not had our notion of mechanical contrivance awakened by inspection of optical instruments, or pulleys, or in some other way;—if we had never been conscious ourselves of providing for the future;—if this were the case, we could not recognise contrivance and prospectiveness in such instances as we have referred to. The facts are, indeed, admirably in accordance with these conceptions, when the two are brought together: but the facts and the conceptions come together from different quarters—from without and from within.

7. We may further illustrate this point by referring to the relations of travellers who tell us that when consummate examples of human mechanical contrivance have been set before savages, they have appeared incapable of apprehending them as proofs of design. This shows that in such cases the Idea of Design had not been developed in the minds of the people who were thus unintelligent: but it no more proves that such an idea does not naturally and necessarily arise, in the progress of men's minds, than the confused manner in which the same savages apprehend the relations of space, or number, or cause, proves that these ideas do not naturally belong to their intellects. All men have these ideas; and it is because they cannot help referring their sensations to such ideas, that they apprehend the world as existing in time and space, and as a series of causes and effects. It would be very erroneous to say that the belief of such truths is obtained by logical reasoning from facts. And in like manner we cannot logically deduce design from the contemplation of organic structures; although it is impossible for us, when the facts are clearly before us, not to find a reference to design operating in our minds.

8. Again; the evidence of the doctrine of Final Causes as a fundamental principle of Biology may be obscured and weakened in some minds by the constant habit of viewing this doctrine with suspicion as unphilosophical and at variance with morphology. By cherishing such views it is probable that many persons, physiologists and others, have gradually brought themselves to suppose that many or most of the arrangements which are familiarly adduced as instances of design may be accounted for, or explained away;—that there is a certain degree of prejudice and narrowness of comprehension in that lively admiration of the adaptation of

means to ends which common minds derive from the
spectacle of organic arrangements. And yet, even in
persons accustomed to these views, the strong and natural influence of the Idea of a Final Cause, the spontaneous recognition of the relation of means to an end
as the assumption which makes organic arrangements
intelligible, breaks forth when we bring before them a
new case, with regard to which their genuine convictions
have not yet been modified by their intellectual habits.
I will offer, as an example which may serve to illustrate
this, the discoveries recently made with regard to the
process of suckling of the kangaroo. In the case of this,
as of other pouched animals, the young animal is removed, while very small and imperfectly formed, from
the womb to the pouch, in which the teats are, and is
there placed with its lips against one of the nipples. But
the young animal taken altogether is not so large as
the nipple, and is therefore incapable of sucking after
the manner of common mammals. Here is a difficulty:
how is it overcome?—By an appropriate *contrivance:* the
nipple, which in common mammals is not furnished with
any muscle, is in the kangaroo provided with a powerful
extrusory muscle by which the mother can inject the
milk into the mouth of her offspring. And again; in
order to give attachment to this muscle there is a bone
which is not found in animals of other kinds. But this
mode of solving the problem of suckling so small a
creature introduces another difficulty. If the milk is
injected into the mouth of the young one, without any
action of its own muscles, what is to prevent the fluid
entering the windpipe and producing suffocation? How
is this danger avoided?—By another appropriate *contrivance:* there is a funnel in the back of the throat by
which the air passage is completely separated from the
passage for nutriment, and the injected milk passes in a

divided stream on each side of the larynx to the œso-
phagus*. And as if to show that this apparatus is really
formed with a view to the wants of the young one, the
structure alters in the course of the animal's growth; and
the funnel, no longer needed, is modified and disappears.

With regard to this and similar examples, the remark
which I would urge is this:—that no one, however prejudiced or unphilosophical he may in general deem the
reference to Final Causes, can, at the first impression,
help regarding this curious system of arrangement as
the means to an end. So contemplated, it becomes
significant, intelligible, admirable: without such a principle, it is an unmeaning complexity, a collection of con-
tradictions, producing an almost impossible result by a
portentous conflict of chances. The parts of this apparatus cannot have produced one another; one part is in
the mother; another part in the young one: without
their harmony they could not be effective; but nothing
except design can operate to make them harmonious.
They are *intended* to work together; and we cannot resist
the conviction of this intention when the facts first come
before us. Perhaps there may hereafter be physiologists who, tracing the gradual development of the parts
of which we have spoken, and the analogies which connect them with the structures of other animals, may think
that this development, these analogies, account for the
conformation we have described; and may hence think
lightly of the explanation derived from the reference
to Final Causes. Yet surely it is clear, on a calm consideration of the subject, that the latter explanation is
not disturbed by the former; and that the observer's
first impression, that this is "an irrefragable evidence
of creative foresight†," can never be obliterated; how-

* Mr. Owen, in *Phil. Trans.*, 1834, p. 348.   † *Ib.* p. 349.

ever much it may be obscured in the minds of those
who confuse this view by mixing it with others which
are utterly heterogeneous to it, and therefore cannot be
contradictory.

9. I have elsewhere* remarked how physiologists,
who thus look with suspicion and dislike upon the
introduction of Final Causes into physiology, have still
been unable to exclude from their speculations causes
of this kind. Thus Cabanis says†, "I regard with
the great Bacon, the philosophy of Final Causes as
sterile; but I have elsewhere acknowledged that it was
very difficult for the most cautious man never to have
recourse to them in his explanations." Accordingly, he
says, "The partisans of Final Causes nowhere find arguments so strong in favour of their way of looking at
nature as in the laws which preside and the circumstances
of all kinds which concur in the reproduction of living
races. In no case do the means employed appear so
clearly relative to the end." And it would be easy to find
similar acknowledgments, express or virtual, in other writers of the same kind. Thus Bichat, after noting the difference between the organic sensibility by which the organs
are made to perform their offices, and the animal sensibility of which the nervous centre is the seat, says‡, "No
doubt it will be asked, *why*"—that is, as we shall see, for
what *end*—"the organs of internal life have received from
nature an inferior degree of sensibility only, and why
they do not transmit to the brain the impressions which
they receivé, while all the acts of the animal life imply
this transmission? The reason is simply this, that all the
phenomena which establish our connexions with surrounding objects *ought to be*, and are in fact, under the influence

---

* *Bridgewater Treatise*, p. 352.

† *Rapports de Physique et du Moral*, i. 299.

‡ *Life and Death*, (trans.) p. 32.

of the will; while all those which serve for the purpose of assimilation only, escape, and *ought* indeed to escape, such influence." The *reason* here assigned is the Final Cause; which, as Bichat justly says, we cannot help asking for.

10. Again; I may quote from the writer last mentioned another remark, which shows that in the organical sciences, and in them alone, the Idea of forces as Means acting to an End, is inevitably assumed and acknowledged as of supreme authority. In Biology alone, observes Bichat*, have we to contemplate the state of *disease.* " Physiology is to the movements of living bodies, what astronomy, dynamics, hydraulics, &c., are to those of inert matter: but these latter sciences have no branches which correspond to them as pathology corresponds to physiology. For the same reason all notion of a medicament is repugnant to the physical sciences. A medicament has for its object to bring the properties of the system back to their natural type; but the physical properties never depart from this type, and have no need to be brought back to it: and thus there is nothing in the physical sciences which holds the place of therapeutick in physiology." Or, as we might express it otherwise, of inert forces we have no conception of what they *ought to do*, except what they *do*. The forces of gravity, elasticity, affinity, never act in a *diseased* manner; we never conceive them as failing in their purpose; for we do not conceive them as having any purpose which is answered by one mode of their action rather than another. But with organical forces the case is different; they are necessarily conceived as acting for the preservation and development of the system in which they reside. If they do not do this, they fail, they are deranged, diseased. They have for their object to conform the living being to a

* *Anatomie Generale*, i. liij.

certain type; and if they cause or allow it to deviate from this type, their action is distorted, morbid, contrary to the ends of nature. And thus this conception of organized beings as susceptible of disease, implies the recognition of a state of health, and of the organs and the vital forces as means for preserving this normal condition. The state of health and of perpetual development is necessarily contemplated as the Final Cause of the processes and powers with which the different parts of plants and animals are endowed.

11. This Idea of a Final Cause is applicable as a fundamental and regulative idea to our speculations concerning organized creatures only. That there is a purpose in many other parts of the creation, we find abundant reason to believe from the arrangements and laws which prevail around us. But this persuasion is not to be allowed to regulate and direct our reasonings with regard to inorganic matter, of which conception the relation of means and end forms no essential part. In mere Physics, Final Causes, as Bacon has observed, are not to be admitted as a principle of reasoning. But in the organical sciences, the assumption of design and purpose in every part of every whole, that is, the pervading idea of Final Cause, is the basis of sound reasoning and the source of true doctrine.

12. The Idea of Final Cause, of end, purpose, design, intention, is altogether different from the Idea of Cause, as efficient cause, which we formerly had to consider; and on this account the use of the word *Cause* in this phrase has been objected to. If the idea be clearly entertained and steadily applied, the word is a question of subordinate importance. The term Final Cause has been long familiarly used, and appears not likely to lead to confusion.

13. The consideration of Final Causes, both in physiology and in other subjects, has at all times attracted

much attention, in consequence of its bearing upon the belief of an Intelligent Author of the Universe. I do not intend, in this place, to pursue the subject far in this view: but there is one antithesis of opinion, already noticed in the History of Physiology, on which I will again make a few remarks*.

It has appeared to some persons that the mere aspect of order and symmetry in the works of nature—the contemplation of comprehensive and consistent law—is sufficient to lead us to the conception of a design and intelligence producing the order and carrying into effect the law. Without here attempting to decide whether this is true, we may discern, after what has been said, that the conception of design, arrived at in this manner, is altogether different from that idea of design which is suggested to us by organized bodies, and which we describe as the doctrine of Final Causes. The regular form of a crystal, whatever beautiful sym‑ metry it may exhibit, whatever general laws it may exemplify, does not prove design in the same manner in which design is proved by the provisions for the preser‑ vation and growth of the seeds of plants, and of the young of animals. The law of universal gravitation, however wide and simple, does not impress us with the belief of a purpose, as does that propensity by which the two sexes of each animal are brought together. If it could be shown that the symmetrical structure of a flower results from laws of the same kind as those which determine the regu‑ lar forms of crystals, or the motions of the planets, the discovery might be very striking and important, but it would not at all come under our idea of Final Cause.

14. Accordingly, there have been, in modern times, two different schools of physiologists, the one proceeding

* *Hist. Ind. Sci.*, b. xvii., chap. 8. On the Doctrine of Final Causes in Physiology.

upon the idea of Final Causes, the other school seeking
in the realm of organized bodies wide laws and analogies
from which that idea is excluded.   All the great biolo-
gists of preceding times, and some of the greatest of
modern times, have belonged to the former school; and
especially Cuvier, who may be considered as the head of
it.   It was solely by the assiduous application of this
principle of Final Cause, as he himself constantly declared,
that he was enabled to make the discoveries which have
rendered his name so illustrious, and which contain a far
larger portion of important anatomical and biological
truth than it ever before fell to the lot of one man to
contribute to the science.

15. The opinions which have been put in opposition
to the principle of Final Causes have, for the most part,
been stated vaguely and ambiguously.   Among the most
definite of such principles, is that which, in the History
of the subject, I have termed the Principle of metamorphosed and developed Symmetry, upon which has been
founded the science of Morphology.

The reality and importance of this principle are not
to be denied by us: we have shown how they are proved
by its application in various sciences, and especially in
botany.   But those advocates of this principle who have
placed it in antithesis to the doctrine of Final Causes,
have by this means done far more injustice to their own
favourite doctrine than damage to the one which they
opposed.   The adaptation of the bones of the skeleton to
the muscles, the provision of fulcrums, projecting processes, channels, so that the motions and forces shall
be such as the needs of life require, cannot possibly
become less striking and convincing, from any discovery
of general analogies of one animal frame with another,
or of laws connecting the development of different parts.
Whenever such laws are discovered, we can only consider

them as the means of producing that adaptation which we so much admire. Our conviction that the Artist works intelligently, is not destroyed, though it may be modified and transferred, when we obtain a sight of his tools. Our discovery of laws cannot contradict our persuasion of ends; our Morphology cannot prejudice our Teleology.

16. The irresistible and constant apprehension of a purpose in the forms and functions of animals has introduced into the writings of speculators on these subjects various forms of expression, more or less precise, more or less figurative; as, that animals are framed with a view to the part which they have to play;—that nature does nothing in vain; that she employs the best means for her ends; and the like. However metaphorical or inexact any of these phrases may be in particular, yet taken altogether, they convey, clearly and definitely enough to preclude any serious error, a principle of the most profound reality and of the highest importance in the organical sciences. But some adherents of the morphological school of which I have spoken reject, and even ridicule, all such modes of expression. " I know nothing," says M. Geoffroy Saint Hilaire, " of animals which have to play a part in nature. I cannot make of nature an intelligent being who does nothing in vain; who acts by the shortest mode; who does all for the best." The philosophers of this school, therefore, do not, it would seem, feel any of the admiration which is irresistibly excited in all the rest of mankind at the contemplation of the various and wonderful adaptations for the preservation, the enjoyment, the continuation of the creatures which people the globe;—at the survey of the mechanical contrivances, the chemical agencies, the prospective arrangements, the compensations, the minute adaptations, the comprehensive interdependencies, which zoology and physiology have brought

into view, more and more, the further their researches
have been carried. Yet the clear and deep-seated conviction of the reality of these provisions, which the study
of anatomy produces in its most profound and accurate
cultivators, cannot be shaken by any objections to the
metaphors or terms in which this conviction is clothed.
In regard to the Idea of a Purpose in organization, as in
regard to any other idea, we cannot fully express our
meaning by phrases borrowed from any extraneous source;
but that impossibility arises precisely from the circumstance of its being a Fundamental Idea which is inevitably assumed in our representation of each special fact.
The same objection has been made to the idea of mechanical *force*, on account of its being often expressed in
metaphorical language; for writers have spoken of an
energy, effort, or *solicitation* to motion; and bodies have
been said to be *animated* by a force. Such language, it
has been urged, implies volition, and the act of animated
beings. But the idea of force as distinct from mere
motion,—as the cause of motion, or of tendency to
motion,—is not on that account less real. We endeavour
in vain to conduct our mechanical reasonings without the
aid of this idea, and must express it as we can. Just as
little can we reason concerning organized beings without
assuming that each part has its function, each function
its purpose; and so far as our phrases imply this, they
will not mislead us, however inexact, or however figurative they be.

17. The doctrine of a purpose in organization has
been sometimes called the doctrine of *the Conditions of
Existence;* and has been stated as teaching that each
animal must be so framed as to contain in its structure
the conditions which its existence requires. When expressed in this manner, it has given rise to the objection,
that it merely offers an identical proposition; since no

animal *can* exist without such conditions. But in reality, such expressions as those just quoted give an inadequate statement of the Principle of a Final Cause. For we discover in innumerable cases, arrangements in an animal, of which we see, indeed, that they are subservient to its well being; but the nature of which we never should have been able at all to conjecture, from considering what was necessary to its existence, and which strike us, no less by their unexpectedness than by their adaptation: so far are they from from being presented by any perceptible necessity. Who would venture to say that the trochlear muscle, or the power of articulate speech, must occur in man, because they are the necessary conditions of his existence? When, indeed, the general scheme and mode of being of an animal are known, the expert and profound anatomist can reason concerning the proportions and form of its various parts and organs, and prove in some measure what their relations must be. We can assert, with Cuvier, that certain forms of the viscera require certain forms of the teeth, certain forms of the limbs, certain powers of the senses. But in all this, the functions of self-nutrition and digestion are supposed already existing as ends: and it being taken for granted, as the only conceivable basis of reasoning, that the organs are means to these ends, we may discover what modifications of these organs are necessarily related to and connected with each other. Instead of terming this rule of speculation merely " the principle of the conditions of existence," we might term it " the principle of the conditions of organs as *means* adapted to animal existence as their *end.*" And how far this principle is from being a mere barren truism, the extraordinary discoveries made by the great assertor of the principle, and universally assented to by naturalists, abundantly prove. The vast extinct creation which is recalled to life in

Cuvier's great work, the *Ossemens Fossiles*, cannot be the consequence of a mere identical proposition.

18. It has been objected, also, that the doctrine of Final Causes supposes us to be acquainted with the intentions of the Creator; which, it is insinuated, is a most presumptuous and irrational basis for our reasonings. But there can be nothing presumptuous or irrational in reasoning on that basis, which if we reject, we cannot reason at all. If men really can discern, and cannot help discerning, a design in certain portions of the works of creation, this perception is the soundest and most satisfactory ground for the convictions to which it leads. The Ideas which we necessarily employ in the contemplation of the world around us, afford us the only natural means of forming any conception of the Creator and Governor of the Universe; and if we are by such means enabled to elevate our thoughts, however inadequately, towards Him, where is the presumption of doing so? or rather, where is the wisdom of refusing to open our minds to contemplations so animating and elevating, and yet so entirely convincing? We possess the ideas of time and space, under which all the objects of the universe present themselves to us; and in virtue of these ideas thus possessed, we believe the Creator to be eternal and omnipotent. When we find that we, in like manner, possess the idea of a Design in Creation, and that with regard to ourselves, and creatures more or less resembling ourselves, we cannot but contemplate their constitution under this idea, we cannot abstain from ascribing to the Creator the infinite profundity and extent of design to which all these special instances belong as parts of a whole.

19. I have here considered Design as manifest in organization only: for in that field of speculation it is forced upon us as contained in all the phenomena, and

as the only mode of our understanding them. The existence of Final Causes has often been pointed out in other portions of the creation;—in the apparent adaptations of the various parts of the earth and of the solar system to each other and to organized beings. In these provinces of speculation, however, the principle of Final Causes is no longer the basis and guide, but the sequel and result of our physical reasonings. If in looking at the universe, we follow the widest analogies of which we obtain a view, we see, however dimly, reason to believe that all its laws are adapted to each other, and intended to work together for the benefit of its organic population, and for the general welfare of its rational tenants. On this subject, however, not immediately included in the principle of Final Causes as here stated, I shall not dwell. I will only make this remark; that the assertion appears to be quite unfounded, that as science advances from point to point, final causes recede before it, and disappear one after the other. The principle of design changes its mode of application indeed, but it loses none of its force. We no longer consider particular facts as produced by special interpositions, but we consider design as exhibited in the establishment and adjustment of the laws by which particular facts are produced. We do not look upon each particular cloud as brought near us that it may drop fatness on our fields, but the general adaptation of the laws of heat, and air, and moisture, to the promotion of vegetation, does not become doubtful. We do not consider the sun as less intended to warm and vivify the tribes of plants and animals, because we find that, instead of revolving round the earth as an attendant, the earth along with other planets revolves round him. We are rather, by the discovery of the general laws of nature, led into a scene of wider design, of deeper contrivance, of more comprehensive adjustments. Final causes, if they appear driven further from us by such an extension of our views, embrace us only with a vaster and more majestic circuit: instead of a few threads connecting some detached objects, they become a stupendous net-work, which is wound round and round the universal frame of things.

I now quit the subject of Biology, and with it the circle of sciences depending upon separate original Ideas and permanent relations. If from the general relations which permanently prevail and constantly recur among the objects around us, we turn to the inquiry of what has actually happened;—if from Science we turn to History;—we find ourselves in a new field. In this region of speculation we can rarely obtain a complete and scientific view of the connexion between objects and events. The past History of Man, of the Arts, of Languages, of the Earth, of the Solar System, offers a vast series of problems, of which perhaps not one has been rigorously solved. Still man, as his speculative powers unfold themselves, cannot but feel prompted and invited to employ his thoughts even on these problems. He cannot but wish and endeavour to understand the connexion between the successive links of such chains of events. He attempts to form a Science which shall be applicable to each of these Histories; and thus he begins to construct the class of sciences to which I now, in the last place, proceed.

BOOK X.

THE PHILOSOPHY OF PALÆTIOLOGY.

CHAPTER I.

OF PALÆTIOLOGICAL SCIENCES IN GENERAL.

1. I HAVE already stated in the *History of the Sciences*[*], that the class of Sciences which I designate as *Palætiological* are those in which the object is to ascend from the present state of things to a more ancient condition, from which the present is derived by intelligible causes. As conspicuous examples of this class we may take Geology, Glossology or Comparative Philology, and Comparative Archæology. These provinces of knowledge might perhaps be intelligibly described as Histories; the History of the Earth,—the History of Languages,—the History of Arts. But these phrases would not fully describe the sciences we have in view; for the object to which we now suppose their investigations to be directed is not merely to ascertain what the series of events has been, as in the common forms of History, but also how it has been brought about. These sciences are to treat of causes as well as of effects. Such researches might be termed *philosophical history;* or, in order to mark more distinctly that the *causes* of events are the leading object of attention, *ætiological history.* But since it will be more convenient to describe this class of sciences by a single appellation, I have taken the

* iii., 481.

liberty of proposing to call them* the *Palætiological* Sciences.

While *Palæontology* describes the beings which have lived in former ages without investigating their causes, and *Ætiology* treats of causes without distinguishing historical from mechanical causation ; *Palætiology* is a combination of the two sciences ; exploring by means of the second the phenomena presented by the first. The portions of knowledge which I include in this term are palæontological ætiological sciences.

2. All these sciences are connected by this bond ;— that they all endeavour to ascend to a past state, by considering what is the present state of things, and what are the causes of change. Geology examines the existing appearances of the materials which form the earth, infers from them previous conditions, and speculates concerning the forces by which one condition has been made to succeed another. Another science, cultivated with great zeal and success in modern times, compares the languages of different countries and nations, and by an examination of their materials and structure, endeavours to determine their descent from one another : this science has been termed Comparative Philology or Ethnography ; and by the French, *Linguistique,* a word which we might imitate in order to have a single name for the science, but the Greek derivative *Glossology* appears to be more convenient in its form. The progress of the Arts (Architecture and the like) ; how one stage of their culture produced another ; and how far we can trace their ma-

---

* A philological writer, in a very interesting work, (Mr. Donaldson, in his *New Cratylus,* p. 12,) expresses his dislike of this word, and suggests that I must mean *palæ-ætiological.* I think the word is more likely to obtain currency in the more compact and euphonious form in which I have used it. It has been adopted by Mr. Winning, in his *Manual of Comparative Philology.*

turest and most complete condition to their earliest form
in various nations; — are problems of great interest
belonging to another subject, which we may for the
present term Comparative Archæology. I have already
noticed, in the *History**, how the researches into the
origin of natural objects, and those relating to works of
art, pass by slight gradations into each other; how the
examination of the changes which have affected an ancient
temple or fortress, harbour or river, may concern alike
the geologist and the antiquary. Cuvier's assertion that
the geologist is an antiquary of a new order, is perfectly
correct, for both are palætiologists.

3. We are very far from having exhausted, by this
enumeration, the class of sciences which are thus connected. We may easily point out many other subjects
of speculation of the same kind. As we may look back
towards the first condition of our planet, we may in like
manner turn our thoughts towards the first condition of
the solar system, and try whether we can discern any
traces of an order of things antecedent to that which is
now established; and if we find, as some great mathematicians have conceived, indications of an earlier state
in which the planets were not yet gathered into their
present forms, we have, in the pursuit of this train of
research, a palætiological portion of Astronomy. Again,
as we may inquire how languages, and how man, have
been diffused over the earth's surface from place to
place, we may make the like inquiry with regard to the
races of plants and animals, founding our inferences upon
the existing geographical distribution of the animal and
vegetable kingdoms: and thus the Geography of Plants
and of Animals also becomes a portion of Palætiology.
Again, as we can in some measure trace the progress of
Arts from nation to nation and from age to age, we can

* iii., 432.

also pursue a similar investigation with respect to the progress of Mythology, of Poetry, of Government, of Law. Thus the philosophical history of the human race, viewed with reference to these subjects, if it can give rise to knowledge so exact as to be properly called Science, will supply sciences belonging to the class I am now to consider.

4. It is not an arbitrary and useless proceeding, to construct such a class of sciences. For wide and various as their subjects are, it will be found that they have all certain principles, maxims, and rules of procedure in common; and thus may reflect light upon each other by being treated of together. Indeed it will, I trust, appear, that we may by such a juxtaposition of different speculations, obtain most salutary lessons. And questions, which, when viewed as they first present themselves under the aspect of a special science, disturb and alarm men's minds, may perhaps be contemplated more calmly, as well as more clearly, when they are considered as general problems of palætiology.

5. It will at once occur to the reader that, if we include in the circuit of our classification such subjects as have been mentioned,—politics and law, mythology and poetry,—we are travelling very far beyond the material sciences within whose limits we at the outset proposed to confine our discussion of principles. But we shall remain faithful to our original plan; and for that purpose shall confine ourselves in this work to those palætiological sciences which deal with material things. It is true, that the general principles and maxims which regulate these sciences apply also to investigations of a parallel kind respecting the products which result from man's imaginative and social endowments. But although there may be a similarity in the general form of such portions of knowledge, their materials are so different from those

with which we have been hitherto dealing, that we cannot hope to take them into our present account with any profit. Language, Government, Law, Poetry, Art, embrace a number of peculiar Fundamental Ideas, hitherto not touched upon in the disquisitions in which we have been engaged; and most of them involved in far greater perplexity and ambiguity, the subject of controversies far more vehement, than the Ideas we have hitherto been examining. We must therefore avoid resting any part of our philosophy upon sciences, or supposed sciences, which treat of such subjects. To attend to this caution, is the only way in which we can secure the 'advantage we proposed to ourselves at the outset, of taking, as the basis of our speculations, none but systems of undisputed truths, clearly understood and expressed*. We have already said that we must, knowingly and voluntarily, resign that livelier and warmer interest which doctrines on subjects of Polity or Art possess, and content ourselves with the cold truths of the material sciences, in order that we may avoid having the very foundations of our philosophy involved in controversy, doubt, and obscurity.

6. We may remark, however, that the necessity of rejecting from our survey a large portion of the researches which the general notion of Palætiology includes, suggests one consideration which adds to the interest of our task. We began our inquiry with the trust that any sound views which we should be able to obtain respecting the nature of truth in the physical sciences, and the mode of discovering it, must also tend to throw light upon the nature and prospects of knowledge of all other kinds;—must be useful to us in moral, political, and philological researches. We stated this as a

* See vol. i. p. 8.

confident anticipation; and the evidence of the justice
of our belief already begins to appear. We have seen,
in the last Book, that biology leads us to psychology,
if we choose to follow the path; and thus the passage
from the material to the immaterial has already unfolded itself at one point; and we now perceive that
there are several large provinces of speculation which
concern subjects belonging to man's immaterial nature,
and which are governed by the same laws as sciences
altogether physical. It is not our business here to dwell
on the prospects which our philosophy thus opens to our
contemplation; but we may allow ourselves, in this last
stage of our pilgrimage among the foundations of the
physical sciences, to be cheered and animated by the ray
that thus beams upon us, however dimly, from a higher
and brighter region.

But in our reasonings and examples we shall mainly
confine ourselves to the physical sciences; and for the
most part to Geology, which in the *History* I have put
forwards as the best representative of the Palætiological
Sciences.

CHAPTER II.

OF THE THREE MEMBERS OF A PALÆTIO-
LOGICAL SCIENCE.

1. *Divisions of such Sciences.*—In each of the Sciences
of this class we consider some particular order of phenomena now existing: from our knowledge of the causes
of change among such phenomena, we endeavour to infer
the causes which have made this order of things what it
is: we ascend in this manner to some previous stage of
such phenomena; and from that, by a similar course of
inference, to a still earlier stage, and to *its* causes.

Hence it will be seen that each such science will consist of two parts,—the knowledge of the phenomena, and the knowledge of their causes. And such a division is, in fact, generally recognised in such sciences: thus we have History, and the Philosophy of History; we have Comparison of Languages, and the Theories of the Origin and Progress of Language; we have Descriptive Geology, and Theoretical or Physical Geology. In all these cases, the relation between the two parts in these several provinces of knowledge is nearly the same; and it may, on some occasions at least, be useful to express the distinction in a uniform or general manner. The investigation of causes has been termed *Ætiology* by philosophical writers, and this term we may use in contradistinction to the mere *Phenomenology* of each such department of knowledge. And thus we should have Phenomenal Geology and Ætiological Geology, for the two divisions of the science which we have above termed Descriptive and Theoretical Geology.

2. *Study of Causes.*—But our knowledge respecting the causes which actually *have* produced any order of phenomena must be arrived at by ascertaining what the causes of change in such matters *can* do. In order to learn, for example, what share earthquakes, and volcanoes, and the beating of the ocean against its shores, ought to have in our Theory of Geology, we must make out what effects these agents of change are able to produce. And this must be done, not hastily, or unsystematically, but in a careful and connected manner; in short, this study of the causes of change in each order of phenomena must become a distinct body of Science, which must include a large amount of knowledge, both comprehensive and precise, before it can be applied to the construction of a theory. We must have an Ætiology corresponding to each order of phenomena.

3. *Ætiology.*—In the History of Geology, I have spoken of the necessity for such an Ætiology with regard to geological phenomena: this necessity I have compared with that which, at the time of Kepler, required the formation of a separate science of Dynamics, (the doctrine of the causes of motion,) before Physical Astronomy could grow out of Phenomenal Astronomy. In pursuance of this analogy, I have there given the name of *Geological Dynamics* to the science which treats of the causes of geological change in general. But, as I have there intimated, in a large portion of the subject the changes are so utterly different in their nature from any modification of motion, that the term Dynamics, so applied, sounds harsh and strange. For in this science we have to treat, not only of the subterraneous forces by which parts of the earth's crust are shaken, elevated, or ruptured, but also of the causes which may change the climate of a portion of the earth's surface, making a country hotter or colder than in former ages; again, we have to treat of the causes which modify the forms and habits of animals and vegetables, and of the extent to which the effects of such causes can proceed; whether, for instance, they can extinguish old species and produce new. These and other similar investigations would not be naturally included in the notion of *Dynamics;* and therefore it will be better to use the term *Ætiology* when we wish to group together all those researches which have it for their object to determine the laws of such changes. In the same manner the Comparison and History of Languages, if it is to lead to any stable and exact knowledge, must have appended to it an Ætiology, which aims at determining the nature and the amount of the causes which really do produce changes in language; as colonization, conquest, the mixture of races, civilization, literature, and the like. And the same rule applies to all sciences of this class. We

shall now make a few remarks on the characteristics of such branches of science as those to which we are led by the above considerations.

4. *Phenomenology requires Classification. Phenomenal Geology.*—The Phenomenal portions of each science imply Classification, for no description of a large and varied mass of phenomena can be useful or intelligible without classification. A representation of phenomena, in order to answer the purposes of science, must be systematic. Accordingly, in giving the History of Descriptive or Phenomenal Geology, I have called it *Systematic Geology*, just as Classificatory Botany is termed *Systematic Botany*. Moreover, as we have already seen, Classification can never be an arbitrary process, but always implies some natural connexion among the objects of the same class; for if this did not exist, the classes could not be made the subjects of any true assertion. Yet though the classes of phenomena which our system acknowledges must be such as already exist in nature, the discovery of these is, for the most part, very far from obvious or easy. To detect the true principles of natural classes, and to select marks by which these may be recognised, are steps which require genius and good fortune, and which fall to the lot only of the most eminent persons in each science. In the History, I have pointed out Werner, William Smith, and Cuvier, as the three great authors of Systematic Geology of Europe. The mode of classifying the materials of the earth's surface which was found, by these philosophers, fitted to enunciate such general facts as came·under their notice, was to consider the rocks and other materials as divided into successive layers or·strata, superimposed one on another, and variously inclined and broken. The German geologist distinguished his strata for the most part by their mineralogical character; the

other two by the remains of animals and plants which the rocks contained. After a beginning had thus been made in giving a genuine scientific form to phenomenal geology, other steps followed in rapid succession, as has already been related in the History*. The Classification of the Strata was fixed by a suitable nomenclature. Attempts were made to apply to other countries the order of strata which had been found to prevail in that first studied: and in this manner it was ascertained what rocks in distant regions are the synonyms, or *equivalents*†, of each other. The knowledge thus collected and systematized was exhibited in the form of geological maps.

Moreover, among the phenomena of geology we have Laws of nature as well as Classes. The general form of mountain chains; the relations of the direction and inclination of different chains to each other; the general features of mineral veins, faults, and fissures; the prevalent characters of slaty cleavage;—were the subjects of laws established, or supposed to be established, by extensive observation of facts. In like manner the organic fossils discovered in the strata were found to follow certain laws with reference to the climate which they appeared to have lived in; and the evidence which they gave of a regular zoological development. And thus by the assiduous labours of many accomplished and active philosophers, Descriptive or Phenomenal Geology was carried towards a state of completeness.

5. *Phenomenal Uranography.*—In like manner in other palætiological researches, as soon as they approach to an exact and scientific form, we find the necessity of constructing in the first place a science of classification and exact description, by means of which the phenomena may be correctly represented and compared; and of

* *Hist. Ind. Sci.*, iii., 527.　　　　† *Ib.*, iii., 532.

obtaining by this step a solid basis for an inquiry into the causes which have produced them. Thus the Palætiology of the solar system has, in recent times, drawn the attention of speculators; and a hypothesis has been started, that our sun and his attendant planets have been produced by the condensation of a mass of diffused matter, such as that which constitutes the nebulous patches which we observe in the starry heavens. But the sagest and most enlightened astronomers have not failed to acknowledge, that to verify or to disprove this conjecture, must be the work of many ages of observation and thought. They have perceived also that the first step of the labour requisite for the advancement of this portion of science must be to obtain and to record the most exact knowledge at present within our reach; respecting the phenomena of these nebulæ, with which we thus compare our own system; and, as a necessary element of such knowledge, they have seen the importance of a classification of these objects, and of others, such as Double Stars, of the same kind. Sir William Herschel, who first perceived the bearing of the phenomena of nebulæ upon the history of the solar system, made the observation of such objects his business, with truly admirable zeal and skill; and in the account of the results of his labours, gave a classification of Nebulæ; separating them into, first, *Clusters of Stars;* second, *Resolvable Nebulæ;* third, *Proper Nebulæ;* fourth, *Planetary Nebulæ;* fifth, *Stellar Nebulæ;* sixth, *Nebulous Stars*. And since, in order to obtain from these remote appearances, any probable knowledge respecting our own system, we must discover whether they undergo any changes in the course of ages, he devoted himself to the task of forming a record of their number and appearance in his own time, that thus the astronomers of succeeding

* *Phil. Trans.,* 1786 and 1789, and Sir J. HERSCHEL's *Astronomy,* Art. 616.

generations might have a definite and exact standard with which to compare their observations. Still, this task would have been executed only for that part of the heavens which is visible in this country, if this Hipparchus of the Nebulæ and Double Stars had not left behind him a son who inherited all his father's zeal and more than his father's knowledge. Sir John Herschel in 1833 went to the Cape of Good Hope to complete what Sir William Herschel left wanting; and in the course of five years observed with care all the nebulæ and double stars of the Southern hemisphere. This great *Herschelian Survey of the Heavens,* the completion of which is the noblest monument ever erected by a son to a father, must necessarily be, to all ages, the basis of all speculations concerning the history and origin of the solar system; and has completed, so far as at present it can be completed, the phenomenal portion of Astronomical Palætiology.

6. *Phenomenal Geography of Plants and Animals.*— Again, there is another Palætiological Science, closely connected with the speculations forced upon the geologist by the organic fossils which he discovers imbedded in the strata of the earth;—namely, the Science which has for its object the Causes of the Diffusion and Distribution of the various kinds of Plants and Animals. And this science also has for its first portion and indispensable foundation a description and classification of the existing phenomena. Such portions of science have recently been cultivated with great zeal and success, under the titles of the *Geography of Plants,* and the *Geography of Animals.* And the results of the inquiries thus undertaken have assumed a definite and scientific form by leading to a division of the earth's surface into a certain number of botanical and zoological *Provinces,* each province occupied by its own peculiar vegetable and animal population.

We find, too, in the course of these investigations, various general laws of the phenomena offered to our notice; such, for instance, as this:—that the difference of the animals originally occupying each province, which is clear and entire for the higher orders of animals and plants, becomes more doubtful and indistinct when we descend to the lower kinds of organizations; as Infusoria and Zoophytes* in the animal kingdom, Grasses and Mosses among vegetables. Again, other laws discovered by those who have studied the geography of plants are these:—that countries separated from each other by wide tracts of sea, as the opposite shores of the Mediterranean, the islands of the Indian and Pacific Oceans, have usually much that is common in their vegetation:—and again, that in parallel climates, analogous tribes replace each other. It would be easy to adduce other laws, but those already stated may serve to show the great extent of the portions of knowledge which have just been mentioned, even considered as merely Sciences of Phenomena.

7. *Phenomenal Glossology.*—It is not my purpose in the present work to borrow my leading illustrations from any portions of knowledge but those which are concerned with the study of material nature; and I shall, therefore, not dwell upon a branch of research, singularly interesting, and closely connected with the one just mentioned, but dealing with relations of thought rather than of things;—I mean the Palætiology of Language;—the theory, so far as the facts enable us to form a theory, of the causes which have led to the resemblances and differences of human speech in various regions and various ages. This, indeed, would be only a portion of the study of the history and origin of the diffusion of animals, if we were to include *man* among the animals whose dispersion

* Prichard, *Researches into the Physical History of Mankind,* i. 55, 28.

we thus investigate; for language is one of the most clear and imperishable records of the early events in the career of the human race. But the peculiar nature of the faculty of speech, and the ideas which the use of it involves, make it proper to treat *Glossology* as a distinct science. And of this science, the first part must necessarily be, as in the other sciences of this order, a classification and comparison of languages, governed in many respects by the same rules, and presenting the same difficulties, as other sciences of classification. Such, accordingly, has been the procedure of the most philosophical glossologists. They have been led to throw the languages of the earth into certain large classes or *families*, according to various kinds of resemblance; as the *Semitic* family, to which belong Hebrew, Arabic, Chaldean, Syrian, Phœnician, Ethiopian, and the like; the *Indo-European*, which includes Sanskrit, Persian, Greek, Latin, and German; the *Monosyllabic* languages, Chinese, Tibetan, Birman, Siamese; the *Polysynthetic* languages, a class including most of the North-American Indian dialects; and others. And this work of classification has been the result of the labour and study of many very profound linguists, and has advanced gradually from step to step. Thus the Indo-European family was first formed on an observation of the coincidences between Sanskrit, Greek, and Latin; but it was soon found to include the Teutonic languages, and more recently Dr. Prichard* has shown beyond doubt that the Celtic must be included in the same family   Other general resemblances and differences of languages have been marked by appropriate terms: thus August von Schlegel has denominated them *synthetical* and *analytical,* according as they form their conjugations and declensions by auxiliary verbs and

* DR. PRICHARD, *On the Eastern Origin of the Celtic Nations.* 1831.

prepositions, or by changes in the word itself: and the *polysynthetic* languages are so named by M. Duponceau, in consequence of their still more complex mode of inflexion. Nor are there wanting, in this science also, general laws of phenomena; such, for instance, is the curious rule of the interchange of consonants in the cognate words of Greek, Gothic, and German, which has been discovered by James Grimm. All these remarkable portions of knowledge, and the great works which have appeared on Glossology, such, for example, as the *Mithridates* of Adelung and Vater, contain, for their largest, and hitherto probably their most valuable part, the phenomenal portion of the science, the comparison of languages as they now are. And beyond all doubt, until we have brought this comparative philology to a considerable degree of completeness, all our speculations respecting the causes which have operated to produce the languages of the earth must be idle and unsubstantial dreams.

Thus in all Palætiological Sciences, in all attempts to trace back the history and discover the origin of the present state of things, the portion of the science which must first be formed is that which classifies the phenomena, and discovers general laws prevailing among them. When this work is performed, and not till then, we may begin to speculate successfully concerning causes, and to make some progress in our attempts to go back to an origin. We must have a *Phenomenal* science preparatory to each *Ætiological* one.

8. *The Study of Phenomena leads to Theory.*—As we have just said, we cannot, in any subject, speculate successfully concerning the causes of the present state of things, till we have obtained a tolerably complete and systematic view of the phenomena. Yet in reality men have not in any instance waited for this completeness

and system in their knowledge of facts before they have begun to form theories. Nor was it natural, considering the speculative propensities of the human mind, and how incessantly it is endeavouring to apply the Idea of Cause, that it should thus restrain itself. I have already noticed this in the History of Geology. "While we have been giving an account," it is there said, "of the objects with which Descriptive Geology is occupied, it must have been felt how difficult it is, in contemplating such facts, to confine ourselves to description and classification. Conjectures and reasonings respecting the causes of the phenomena force themselves upon us at every step; and even influence our classification and nomenclature. Our Descriptive Geology impels us to construct a Physical Geology." And the same is the case with regard to the other subjects which I have mentioned. The mere consideration of the different degrees of condensation of different nebulæ led Herschel and Laplace to the hypothesis that our solar system is a condensed nebula. Immediately upon the division of the earth's surface into botanical and zoological provinces, and even at an earlier period, the opposite hypotheses of the origin of all the animals of each kind from a single pair, and of their original diffusion all over the earth, were under discussion. And the consideration of the families of languages irresistibly led to speculations concerning the families of the earliest human inhabitants of the earth. In all cases the contemplation of a very few phenomena, the discovery of a very few steps in the history, made men wish for and attempt to form a theory of the history from the very beginning of things.

9. *No sound Theory without Ætiology.*—But though man is thus impelled by the natural propensities of his intellect to trace each order of things to its causes, he does not at first discern the only sure way of

obtaining such knowledge: he does not suspect how much labour and how much method are requisite for success in this undertaking: he is not aware that for each order of phenomena he must construct, by the accumulated results of multiplied observation and distinct thought, a separate Ætiology. Thus, as I have elsewhere remarked*, when men had for the first time become acquainted with some of the leading phenomena of Geology, and had proceeded to speculate concerning the past changes and revolutions by which such results had been produced, they forthwith supposed themselves able to judge what would be the effects of any of the obvious agents of change, as water or volcanic fire. It did not at first occur to them to suspect that their common and extemporaneous judgment on such points was by no means sufficient for sound knowledge. They did not foresee that, before they could determine what share these or any other causes had had in producing the present condition of the earth, they must create a special science whose object should be to estimate the general laws and effects of such assumed causes:—that before they could obtain any sound Geological Theory, they must carefully cultivate Geological Ætiology.

The same disposition to proceed immediately from the facts to the theory, without constructing, as an intermediate step, a science of Causes, might be pointed out in the other sciences of this order. But in all of them this error has been corrected by the failures to which it led. It soon appeared, for instance, that a more careful inquiry into the effects which climate, food, habit and circumstances can produce in animals was requisite in order to determine how the diversities of animals in different countries have originated. The Ætiology of Animal Life (if we may be allowed to give this name to

* *Hist. Ind. Sci.*, iii. 546.

that study of such causes of change which is at present
so zealously cultivated, and which yet has no distinctive
designation,) is now perceived to be a necessary portion of
all attempts to construct a history of the earth and its
inhabitants.

10. *Cause in Palætiology.*—We are thus led to contemplate a class of sciences which are commenced with
the study of Causes.   We have already considered sciences which depended mainly upon the Idea of Cause,
namely, the Mechanical Sciences.   But it is obvious that
the Idea of Cause in the researches now under our consideration must be employed in a very different way from
that in which we applied it formerly.   Force is the cause
of motion, because force at all times and under all circumstances, if not counteracted, produces motion ; but
the cause of the present condition and elevation of the
Alps, whatever it was, was manifested in a series of
events of which each happened but once, and occupied
its proper place in the series of time.   The former is
mechanical, the latter *historical, cause.*   In our present
investigations, we consider the events which we contemplate, of whatever order they be, as forming a chain which
is extended from the beginning of things down to the
present time ; and the causes of which we now speak are
those which connect the successive links of this chain.
Every occurrence which has taken place in the history of
the solar system, or the earth, or its vegetable and animal
creation, or man, has been at the same time effect and
cause ;—the effect of what preceded, the cause of what
succeeded.   By being effect and cause, it has occupied
some certain portion of time ; and the times which have
thus been occupied by effects and causes, summed up
and taken all together, make up the total of Past Time.
The Past has been a series of events connected by this
historical causation, and the Present is the last term of

this series. The problem in the Palætiological Sciences, with which we are here concerned, is, to determine the manner in which each term is derived from the preceding, and thus, if possible, to calculate backwards to the origin of the series.

11. *Various kinds of Cause.*—Those modes by which one term in the natural series of events is derived from another, the forms of historical causation, the kinds of connexion between the links of the infinite chain of time, are very various; nor need we attempt to enumerate them. But these kinds of causation being distinguished from each other, and separately studied, each becomes the subject of a separate Ætiology. Thus the causes of change in the earth's surface, residing in the elements, fire and water, form the main subject of Geological Ætiology. The Ætiology of the vegetable and animal kingdoms investigates the causes by which the forms and distribution of species of plants and animals are affected. The study of causes in Glossology leads to an Ætiology of Language, which shall distinguish, analyse, and estimate the causes by which certain changes are produced in the languages of nations; in like manner we may expect to have an Ætiology of Art, which shall scrutinize the influences by which the various forms of art have each given birth to its successor: by which, for example, there have been brought into being those various forms of architecture which we term Egyptian, Doric, Ionic, Roman, Byzantine, Romanesque, Gothic, Italian, Elizabethan. It is easily seen by this slight survey how manifold and diverse are the kinds of cause which the Palætiological Sciences bring under our consideration. But in each of those sciences we shall obtain solid and complete systems of knowledge, only so far as we study, with steady thought and careful observation, that peculiar kind of cause which is appropriate to the phenomena under our consideration.

12. *Hypothetical Order of Palætiological Causes.—* The various kinds of historical cause are not only connected with each other by their common bearing upon the historical sciences, but they form a kind of progression which we may represent to ourselves as having acted *in succession* in the hypothetical history of the earth and its inhabitants. Thus assuming, merely as a momentary hypothesis, the origin of the solar system by the condensation of a nebula, we have to contemplate, first, the causes by which the luminous incandescent diffused mass of which a nebula is supposed to be constituted, is gradually condensed, cooled, collected into definite masses, solidified, and each portion made to revolve about its axis, and the whole to travel about another body. We have no difficulty in ascribing the globular form of each mass to the mutual attraction of its particles: but when this form was once assumed and covered with a solid crust, are there, we may ask, in the constitution of such a body, any causes at work by which the crust might be again broken up and portions of it displaced, and covered with other matter? Again, if we can thus explain the origin of the earth, can we with like success account for the presence of the atmosphere and the waters of earth and ocean? Supposing this done, we have then to consider by what causes such a body could become stocked with vegetable and animal life; for there have not been wanting persons, extravagant speculators, no doubt, who have conceived that even this event in the history of the world might be the work of natural causes. Supposing an origin given to life upon our earth, we have then, brought before us by geological observations, a series of different forms of vegetable and animal existence; occurring in different strata, and, as the phenomena appear irresistibly to prove, existing at successive periods: and we are com-

pelled to inquire what can have been the causes by which the forms of each period have passed into those of the next. We find, too, that strata, which must have been at first horizontal and continuous, have undergone enormous dislocations and ruptures, and we have to consider the possible effect of aqueous and volcanic causes to produce such changes in the earth's crust. We are thus led to the causes which have produced the present state of things on the earth; and these are causes to which we may hypothetically ascribe, not only the form and position of the inert materials of the earth, but also the nature and distribution of its animal and vegetable population. Man too, no less than other animals, is affected by the operation of such causes as we have referred to, and must, therefore, be included in such speculations. But man's history only begins, where that of other animals ends, with his mere existence. They are stationary, he is progressive. Other species of animals once brought into being, continue the same through all ages; man is changing, from age to age, his language, his thoughts, his works. Yet even these changes are bound together by laws of causation; and these causes too may become objects of scientific study. And such causes, though not to be dwelt upon now, since we permit ourselves to found our philosophy upon the material sciences only, must still, when treated scientifically, fall within the principles of our philosophy, and must be governed by the same general rules to which all science is subject. And thus we are led by a close and natural connexion, through a series of causes, from those which regulate the imperceptible changes of the remotest nebulæ in the heavens, to those which determine the diversities of language, the mutations of art, and even the progress of civilization, polity, and literature.

While I have been speaking of this supposed series

of events, including in its course the formation of the earth, the introduction of animal and vegetable life, and the revolutions by which one collection of species has succeeded another, it must not be forgotten, that though I have thus hypothetically spoken of these events as occurring by force of natural causes, this has been done only that the true efficacy of such causes might be brought under our consideration and made the subject of scientific examination. It may be found, that such occurrences as these are quite inexplicable by the aid of any natural causes with which we are acquainted; and thus the result of our investigations, conducted with strict regard to scientific principles, may be, that we must either contemplate supernatural influences as part of the past series of events, or declare ourselves altogether unable to form this series into a connected chain.

13. *Mode of Cultivating Ætiology :—In Geology.—* In what manner, it may be asked, is Ætiology, with regard to each subject such as we have enumerated, to be cultivated? In order to answer this question, we must, according to our method of proceeding, take the most successful and complete examples which we possess of such portions of science. But in truth, we can as yet refer to few examples of this kind. In Geology, it is only very recently, and principally through the example and influence of Mr. Lyell, that the Ætiology has been detached from the descriptive portion of the science; and cultivated with direct attention: in other sciences the separation has hardly yet been made. But if we examine what has already been done in Geological Ætiology, or as in the History it is termed, Geological Dynamics, we shall find a number of different kinds of investigation which, by the aid of our general principles respecting the formation of sciences, may suffice to supply very useful suggestions for Ætiology in general.

In Geological Ætiology causes have been studied, in many instances, by attending to their action in the phenomena of the present state of things, and by inferring from this the nature and extent of the action which they may have exercised in former times.   This has been done, for example, by Von Hoff, Mr. Lyell and others, with regard to the operations of rivers, seas, springs, glaciers, and other aqueous causes of change.   Again, the same course has been followed by the same philosophers with respect to volcanos, earthquakes, and other violent agents.   Mr. Lyell has attempted to show, too, that there take place, in our own time, not only violent agitations, but slow motions of parts of the earth's crust, of the same kind and order with those which have assisted in producing all anterior changes.

But while we thus seek instruction in the phenomena of the present state of things, we are led to the question, What are the limits of this present period?   For instance, among the currents of lava which we trace as part of the shores of Italy and Sicily, which shall we select as belonging to the existing order of things?   In going backwards in time, where shall we draw the line? and why at such particular point?   These questions are important, for our estimate of the efficacy of known causes will vary with the extent of the effects which we ascribe to them.   Hence the mode in which we group together rocks is not only a step in geological classification, but is also important to Ætiology.   Thus when the vast masses of trap rocks in the Western Isles of Scotland and in other countries, which had been maintained by the Wernerians to be of aqueous origin, were, principally by the sagacity and industry of Macculloch, identified as to their nature with the products of recent volcanos, the amount of effect which might justifiably be ascribed to volcanic agency was materially extended.

In other cases, instead of observing the current effects of our geological causes, we have to estimate the results from what we know of the causes themselves; as when, with Herschel, we calculate the alterations in the temperature of the earth which astronomical changes may possibly produce; or when, with Fourier, we try to calculate the rate of cooling of the earth's surface, on the hypothesis of an incandescent central mass. In other cases, again, we are not able to calculate the effects of our causes rigorously, but estimate them as well as we can, partly by physical reasonings, and partly by comparison with such analogous cases as we can find in the present state of things. Thus Mr. Lyell infers the change of climate which would result if land were transferred from the neighbourhood of the poles to that of the equator, by reasonings on the power of land and water to contain and communicate heat, supported by a reference to the different actual climates of places under the same latitude, but under different conditions as to the distribution of land and water.

Thus our Ætiology is constructed partly from calculation and reasoning, partly from phenomena. But we may observe that when we reason from phenomena to causes, we usually do so by various steps; often ascending from phenomena to mere laws of phenomena, before we can venture to connect the phenomenon confidently with its cause. Thus the law of subterranean heat, that it increases in descending below the surface, is now well established, although the doctrine which ascribes this effect to a central heat is not universally assented to.

14. *In the Geography of Plants and Animals.*—We may find in other subjects also, considerable contributions towards Ætiology, though not as yet a complete system of science. 1 The Ætiology of vegetables and animals, indeed, has been studied with great zeal in modern times,

as an essential preparative to geological theory; for how can we decide whether any assumed causes have produced the succession of species which we find in the earth's strata, except we know what effect of this kind given causes can produce? Accordingly, we find in Mr. Lyell's *Treatise on Geology* the most complete discussion of such questions as belong to these subjects:—for example, the question whether species can be transmuted into other species by the long continued influence of external causes, as climate, food, domestication, combined with internal causes, as habits, appetencies, progressive tendencies. We may observe, too, that ·as we have brought before us the inquiry what change difference of climate can produce in any species, we have also the inverse problem, how far a different development of the species, or a different collection of species, proves a difference of climate. In the same way, the geologist of the present day considers the question, whether, in virtue of causes now in action, species are from time to time extinguished; and in like manner the geologists of an earlier period discussed the question, now long completely decided whether fossil species in general are really extinct species.

15. *In Languages.*—Even with reference to the Ætiology of language, although this branch of science has hardly been considered separately from the glossological investigations in which it is employed or assumed to be employed, it might perhaps be possible to point out causes or conditions of change which, being general in their nature, must operate upon all languages alike. Changes made for the sake of euphony when words are modified and combined, occur in all dialects. Who can doubt that such changes of consonants as those by which the Greek roots become Gothic, and the Gothic, German, have for their cause some general principle in the pronunciation of each language? Again, we might attempt to

decide other questions of no small interest. Have the
terminations of verbs arisen from the accretion of pronouns; or, on the other hand, does the modification of a
verb imply a simpler mental process than the insulation
of a pronoun, as Adam Smith has maintained? Again,
when the language of a nation is changed by the invasion
and permanent mixture of an enemy of different speech,
is it generally true that it is changed from a synthetic to
an analytical structure? I will mention only one more
of these wide and general glossological inquiries. Is it
true, as Dr. Prichard has suggested*, that languages have
become more permanent as we come down towards later
times? May we justifiably suppose, with him, that in the
very earliest times, nations, when they had separated from
one stock, might lose all traces of this common origin
out of their languages, though retaining strong evidences
of it in their mythology, social forms, and arts, as appears
to be the case with the ancient Egyptians and the
Indians†?

Large questions of this nature cannot be treated pro-
fitably in any other way than by an assiduous study of
the most varied forms of living and dead languages. But
on the other hand, the study of languages should be prosecuted not only by a direct comparison of one with
another, but also with a view to the formation of a science
of causes and general principles, embracing such discussions as I have pointed out. It is only when such a
science has been formed, that we can hope to obtain any
solid and certain results in the Palætiology of language;
—to determine, with any degree of substantial proof,
what is the real evidence which the wonderful faculty of
speech, under its present developments and forms, bears to
the events which have taken place in its own history, and
in the history of man since his first origin.

* *Researches*, ii. 221.   † *Ib.*, ii. 192.

16. *Construction of Theories.*—When we have thus obtained, with reference to any such subject as those we have here spoken of, these two portions of science, a systematic description of the facts, and a rigorous analysis of the causes,—the *Phenomenology* and the *Ætiology* of the subject,—we are prepared for the third member which completes the science, the *Theory* of the actual facts. We can then take a view of the events which really have happened, discerning their connexion, interpreting their evidence, supplying from the context the parts which are unapparent. We can account for known facts by intelligible causes, we can infer latent facts from manifest effects, so as to obtain a distinct insight into the whole history of events up to the present time, and to see the last result of the whole in the present condition of things. The term Theory, when rigorously employed in such sciences as those which we here consider, bears nearly the sense which I have adopted : it implies a consistent and systematic view of the actual facts, combined with a true apprehension of their connexion and causes. Thus if we speak of a Theory of Mount Etna, or a Theory of the Paris Basin, we mean a connected and intelligible view of the events by which the rocks in these localities have come into their present condition. Undoubtedly the term Theory has often been used in a looser sense ; and men have put forth *Theories of the Earth*, which, instead of including the whole mass of actual geological facts and their causes, only assigned, in a vague manner, some causes by which some few phenomena might, it was conceived, be accounted for. Perhaps the portion of our Palætiological Sciences which we now wish to designate, would be more generally understood if we were to describe it as *Theoretical* or *Philosophical History;* as when we talk of the Theoretical History of Architecture, or the Philosophical History of Language. And in the same

manner we might speak of the Theoretical History of the Animal and Vegetable Kingdoms; meaning, a distinct account of the events which have produced the present distribution of species and families. But by whatever phrase we describe this portion of science, it is plain that such a Theory, such a Theoretical History, must result from the application of causes well understood to facts well ascertained. And if the term *Theory* be here employed we must recollect that it is to be understood, not in its narrower sense as opposed to facts, but in its wider signification, as including all known facts and differing from them only in introducing among them principles of intelligible connexion. The Theories of which we now speak are true Theories, precisely because they are identical with the total system of the Facts.

17. *No sound Palætiological Theory yet extant.*—It is not to disparage the present state of science to say that as yet no such theory exists on any subject. "Theories of the Earth" have been repeatedly published; but when we consider that even the facts of geology have been observed only on a small portion of the earth's surface, and even within those narrow bounds very imperfectly studied, we shall be able to judge how impossible it is that geologists should have yet obtained a well-established Theoretical History of the changes which have taken place in the crust of the terrestrial globe from its first origin. Accordingly, I have ventured in my History to designate the most prominent of the Theories which have hitherto prevailed as *premature* geological theories*: and we shall soon see that geological theory has not advanced beyond a few conjectures, and that its cultivators are at present mainly occupied with a controversy in which the two extreme hypotheses which first offer themselves to men's minds are opposed to each other.

Hist. Ind. Sci., iii. 603.

And if we have no theoretical history of the earth which merits any confidence, still less have we any theoretical History of Language, or of the Arts, which we can consider as satisfactory. The Theoretical History of the Vegetable and Animal Kingdoms is closely connected with that of the earth on which they subsist, and must follow the fortunes of geology. And thus we may venture to say that no Palætiological Science, as yet, possesses all its three members. Indeed most of them are very far from having completed and systematized their Phenomenology: in all, the cultivation of Ætiology is but just begun, or is not begun; in all, the Theory must reward the exertions of future, probably of distant, generations.

But in the mean time we may derive some instruction from the comparison of the two antagonist hypotheses of which I have spoken.

CHAPTER III.

OF THE DOCTRINE OF CATASTROPHES AND THE DOCTRINE OF UNIFORMITY.

1. *Doctrine of Catastrophes.*—I have already shown, in the History of Geology, that the attempts to frame a theory of the earth have brought into view two completely opposite opinions:—one, which represents the course of nature as *uniform* through all ages, the causes which produce change having had the same intensity in former times which they have at the present day;—the other opinion, which sees in the present condition of things evidences of *catastrophes*; changes of a more sweeping kind, and produced by more powerful agencies than those which occur in recent times. Geologists who held

the latter opinion, maintained that the forces which have elevated the Alps or the Andes to their present height could not have been any forces which are now in action: they pointed to vast masses of strata hundreds of miles long, thousands of feet thick, thrown into highly-inclined positions, fractured, dislocated, crushed: they remarked that upon the shattered edges of such strata they found enormous accumulations of fragments and rubbish, rounded by the action of water, so as to denote ages of violent aqueous action: they conceived that they saw instances in which whole mountains of rock in a state of igneous fusion, must have burst the earth's crust from below: they found that in the course of the revolutions by which one stratum of rock was placed upon another, the whole collection of animal species which tenanted the earth and the seas had been removed, and a new set of living things introduced in its place: finally, they found above all the strata vast masses of sand and gravel containing bones of animals, and apparently the work of a mighty deluge. With all these proofs before their eyes they thought it impossible not to judge that the agents of change by which the world was urged from one condition to another till it reached its present state must have been more violent, more powerful, than any which we see at work around us. They conceived that the evidence of " catastrophes " was irresistible.

2. *Doctrine of Uniformity.*—I need not here repeat the narrative (given in the History*) of the process by which this formidable array of proofs was, in the minds of some eminent geologists, weakened, and at last overcome. This was done by showing that the sudden breaks in the succession of strata were apparent only, the discontinuity of the series which occurred in one country being removed by terms interposed in another locality:

by urging that the total effect produced by existing causes, taking into account the accumulated result of long periods, is far greater than a casual speculator would think possible: by making it appear that there are in many parts of the world evidences of a slow and imperceptible rising of the land since it was the habitation of now existing species: by proving that it is not universally true that the strata separated in time by supposed catastrophes contain distinct species of animals: by pointing out the limited fields of the supposed diluvial action: and finally, by remarking that though the *creation* of species is a mystery, the *extinction* of them is going on in our own day. Hypotheses were suggested, too, by which it was conceived that the change of climate might be explained, which, as the consideration of the fossil remains seemed to show, must have taken place between the ancient and the modern times. In this manner the whole evidence of catastrophes was explained away: the notion of a series of paroxysms of violence in the causes of change was represented as a delusion arising from our contemplating short periods only in the action of present causes: length of time was called in to take the place of intensity of force: and it was declared that geology need not despair of accounting for the revolutions of the earth, as astronomy accounts for the revolutions of the heavens, by the universal action of causes which are close at hand to us, operating through time and space without variation or decay.

An antagonism of opinions, somewhat of the same kind as this, will be found to manifest itself in the other Palætiological Sciences as well as in Geology; and it will be instructive to endeavour to balance these opposite doctrines. I will mention some of the considerations which bear upon the subject.

3. *Is Uniformity probable à priori?*—The doctrine of

Uniformity in the course of nature has sometimes been represented by its adherents as possessing a great degree of *à priori* probability. It is highly unphilosophical, it has been urged, to assume that the causes of the geological events of former times were of a different kind from causes now in action, if causes of this latter kind can in any way be made to explain the facts. The analogy of all other sciences compels us, it was said, to explain phenomena by known, not by unknown, causes. And on these grounds the geological teacher recommended* "an earnest and patient endeavour to reconcile the indications of former change with the evidence of gradual mutations now in progress."

But on this we may remark, that if by *known* causes we mean causes acting with the same intensity which they have had during historical times, the restriction is altogether arbitrary and groundless. Let it be granted, for instance, that many parts of the earth's surface are now undergoing an imperceptible rise. It is not pretended that the rate of this elevation is rigorously uniform; what, then, are the limits of its velocity? Why may it not increase so as to assume that character of violence which we may term a *catastrophe* with reference to all changes hitherto recorded? Why may not the rate of elevation be such that we may conceive the strata to assume *suddenly* a position nearly vertical? and is it, in fact, easy to conceive a position of strata nearly vertical, a position which occurs so frequently, to be *gradually* assumed? In cases where the strata are nearly vertical, as in the Isle of Wight, and hundreds of other places, or where they are actually inverted, as sometimes occurs, are not the causes which have produced the effect as truly *known* causes, as those which have raised the coasts where we trace the former beach in an elevated terrace? If

* LYELL, b. iv. c. 1, p. 328.

the latter case proves *slow* elevation, does not the former case prove *rapid* elevation?   In neither case have we any measure of the time employed in the change; but does not the very nature of the results enable us to discern, that if one was gradual, the other was comparatively sudden?

The causes which are now elevating a portion of Scandinavia can be called known *causes,* only because we know the *effect.*   Are not the causes which have elevated the Alps and the Andes known causes in the same sense? We know nothing in either case which confines the intensity of the force within any limit, or prescribes to it any law of uniformity.   Why, then, should we make a merit of cramping our speculations by such assumptions? Whether the causes of change do act uniformly;— whether they oscillate only within narrow limits;— whether their intensity in former times was nearly the same as it now is;—these are precisely the questions which we wish Nature to answer to us impartially and truly: where is then the wisdom of "an earnest and patient endeavour" to secure an *affirmative* reply?

Thus I conceive that the assertion of an *à priori* claim to probability and philosophical spirit in favour of the doctrine of uniformity, is quite untenable.   We must learn from an examination of all the facts, and not from any assumption of our own, whether the course of nature be uniform.   The limit of intensity being really unknown, catastrophes are just as probable as uniformity.   If a volcano may repose for a thousand years, and then break out and destroy a city; why may not another volcano repose for ten thousand years, and then destroy a continent; or if a continent, why not the whole habitable surface of the earth?

4. *Cycle of Uniformity indefinite.*—But this argument may be put in another form.   When it is said that the

course of nature is uniform, the assertion is not intended to exclude certain smaller variations of violence and rest, such as we have just spoken of;—alternations of activity and repose in volcanos; or earthquakes, deluges, and storms, interposed in a more tranquil state of things. With regard to such occurrences, terrible as they appear at the time, they may not much affect the average rate of change; there may be a *cycle*, though an irregular one, of rapid and slow change; and if such cycles go on succeeding each other, we may still call the order of nature uniform, notwithstanding the periods of violence which it involves. The maximum and minimum intensities of the forces of mutation alternate with one another; and we may estimate the average course of nature as that which corresponds to something between the two extremes.

But if we thus attempt to maintain the uniformity of nature by representing it as a series of *cycles*, we find that we cannot discover, in this conception, any solid ground for excluding catastrophes. What is the length of that cycle, the repetition of which constitutes uniformity? What interval from the maximum to the minimum does it admit of? We may take for our cycle a hundred or a thousand years, but evidently such a proceeding is altogether arbitrary. We may mark our cycles by the greatest known paroxysms of volcanic and terremotive agency, but this procedure is no less indefinite and inconclusive than the other.

But further; since the cycle in which violence and repose alternate is thus indefinite in its length and in its range of activity, what ground have we for assuming more than *one* such cycle, extending from the origin of things to the present time? Why may we not suppose the maximum force of the causes of change to have taken place at the earliest period, and the tendency towards the minimum to have gone on ever since? Or instead of

only one cycle, there may have been several, but of such
length that our historical period forms a portion only of
the last;—the feeblest portion of the latest cycle. And
thus violence and repose may alternate upon a scale of
time and intensity so large, that man's experience supplies no evidence enabling him to estimate the amount.
The course of things is *uniform*, to an Intelligence which
can embrace the succession of several cycles, but it is
catastrophic to the contemplation of man, whose survey
can grasp a part only of one cycle. And thus the hypothesis of uniformity, since it cannot exclude degrees of
change, nor limit the range of these degrees, nor define
the interval of their recurrence, cannot possess any essential simplicity which, previous to inquiry, gives it a claim
upon our assent superior to that of the opposite catastrophic hypothesis.

5. *Uniformitarian Arguments are Negative only.*—
There is an opposite tendency in the mode of maintaining
the catastrophist and the uniformitarian opinions, which
depends upon their fundamental principles, and shows
itself in all the controversies between them. The Catastrophist is affirmative, the Uniformitarian is negative in
his assertions: the former is constantly attempting to
construct a theory; the latter delights in demolishing all
theories. The one is constantly bringing fresh evidence
of some great past event, or series of events, of a striking
and definite kind; his antagonist is at every step explaining away the evidence, and showing that it proves nothing.
One geologist adduces his proofs of a vast universal deluge;
but another endeavours to show that the proofs do not
establish either the universality or the vastness of such an
event. The inclined broken edges of a certain formation
covered with their own fragments beneath superjacent
horizontal deposits are at one time supposed to prove a
catastrophic breaking up of the earlier strata; but this

opinion is controverted by showing that the same formations, when pursued into other countries, exhibit a uniform gradation from the lower to the upper, with no
trace of violence. Extensive and lofty elevations of the
coast, continents of igneous rock, at first appear to indicate operations far more gigantic than those which now
occur; but attempts are soon made to show that time
only is wanting to enable the present age to rival the past
in the production of such changes. Each new fact adduced by the catastrophist is at first striking and apparently convincing; but as it becomes familiar, it strikes
the imagination less powerfully; and the uniformitarian,
constantly labouring to produce some imitation of it by
the machinery which he has so well studied, at last in
every case seems to himself to succeed, so far as to destroy
the effect of his opponent's evidence.

This is so with regard to more remote, as well as with
regard to immediate evidences of change. When it is
ascertained that in every part of the earth's crust the
temperature increases as we descend below the surface,
at first this fact seems to indicate a central heat: and a
central heat naturally suggests an earlier state of the
mass, in which it was incandescent, and from which it is
now cooling. But this original incandescence of the
globe of the earth is manifestly an entire violation of the
present course of things; it belongs to the catastrophist
view, and the advocates of uniformity have to explain it
away. Accordingly, one of them holds that this increase
of heat in descending below the surface may very possibly
not go on all the way to the centre. The heat which increases at first as we descend, may, he conceives, afterwards decrease; and he suggests causes which may have
produced such a succession of hotter and colder shells
within the mass of the earth. I have mentioned this
suggestion in the History of Geology; and have given

my reasons for believing it altogether untenable*. Other persons also, desirous of reconciling this subterraneous heat with the tenet of uniformity, have offered another suggestion:—that the warmth or incandescence of the interior parts of the earth does not arise out of an originally hot condition from which it is gradually cooling, but results from chemical action constantly going on among the materials of the earth's substance. And thus new attempts are perpetually making, to escape from the cogency of the reasonings which send us towards an original state of things different from the present. Those who theorize concerning an origin go on building up the fabric of their speculations, while those who think such theories unphilosophical, ever and anon dig away the foundation of this structure. As we have already said, the uniformitarian's doctrines are a collection of negatives.

This is so entirely the case, that the uniformitarian would for the most part shrink from maintaining as positive tenets the explanations which he so willingly uses as instruments of controversy. He puts forward his suggestions as difficulties, but he will not stand by them as doctrines. And this is in accordance with his general tendency; for any of his hypotheses, if insisted upon as positive theories, would be found inconsistent with the assertion of uniformity. For example, the nebular hypothesis appears to give to the history of the heavens an aspect which obliterates all special acts of creation, for, according to that hypothesis, new planetary systems are constantly forming; but when asserted as the origin of our own solar system, it brings with it an original incandescence, and an origin of the organic world. And if, instead of using the chemical theory of subterraneous heat to neutralize the evidence of original incandescence, we assert it as a positive tenet, we can no longer main-

* *Hist. Ind. Sci.*, iii. 562, and note.

tain the infinite past duration of the earth; for chemical
forces, as well as mechanical, tend to equilibrium; and
that condition once attained, their efficacy ceases.   Chemical affinities tend to form new compounds; and though,
when many and various elements are mingled together,
the play of synthesis and analysis may go on for a long
time, it must at last end.   If, for instance, a large portion of the earth's mass were originally pure potassium,
we can imagine violent igneous action to go on so long
as any part remained unoxidized; but when the oxidation
of the whole has once taken place, this action must be
at an end; for there is in the hypothesis no agency
which can reproduce the deoxidized metal.   Thus a perpetual motion is impossible in chemistry, as it is in
mechanics; and a theory of constant change continued
through infinite time, is untenable when asserted upon
chemical, no less than upon mechanical principles.   And
thus the scepticism of the uniformitarian is of force only
so long as it is employed against the dogmatism of the
catastrophist.   When the doubts are erected into dogmas,
they are no longer consistent with the tenet of uniformity.   When the negations become affirmations, the
negation of an origin vanishes also.

6. *Uniformity in the Organic World.*—In speaking of
the violent and sudden changes which constitute catastrophes, our thoughts naturally turn at first to great
mechanical and *physical* effects;—ruptures and displace-
ments of strata; extensive submersions and emersions of
land; rapid changes of temperature.   But the catastrophes
which we have to consider in geology affect the *organic* as
well as the inorganic world.   The sudden extinction of one
collection of species, and the introduction of another in
their place, is a catastrophe, even if unaccompanied by
mechanical violence.   Accordingly, the antagonism of
the catastrophist and uniformitarian school has shown

itself in this department of the subject, as well as in the other. When geologists had first discovered that the successive strata are each distinguished by appropriate organic fossils, they assumed at once that each of these collections of living things belonged to a separate creation. But this conclusion, as I have already said, Mr. Lyell has attempted to invalidate, by proving that in the existing order of things, some species become extinct; and by suggesting it as possible, that in the same order it may be true that new species are from time to time produced, even in the present course of nature. And in this, as in the other part of the subject, he calls in the aid of vast periods of time, in order that the violence of the changes may be softened down: and he appears disposed to believe that the actual extinction and creation of species may be so slow as to excite no more notice than it has hitherto obtained; and yet may be rapid enough, considering the immensity of geological periods, to produce such a succession of different collections of species as we find in the strata of the earth's surface.

7. *Origin of the present Organic World.*—The last great event in the history of the vegetable and animal kingdoms was that by which their various tribes were placed in their present seats. And we may form various hypotheses with regard to the sudden or gradual manner in which we may suppose this distribution to have taken place. We may assume that at the beginning of the present order of things, a stock of each species was placed in the vegetable or animal *province* to which it belongs, by some cause out of the common order of nature; or we may take a uniformitarian view of the subject, and suppose that the provinces of the organic world derived their population from some anterior state of things by the operation of natural causes.

Nothing has been pointed out in the existing order

of things which has any analogy or resemblance, of any valid kind, to that creative energy which must be exerted in the production of a new species. And to assume the introduction of new species as a part of the order of nature, without pointing out any natural fact with which such an event can be classed, would be to reject creation by an arbitrary act. Hence, even on natural grounds, the most intelligible view of the history of the animal and vegetable kingdoms seems to be, that each period which is marked by a distinct collection of species forms a cycle; and that at the beginning of each such cycle a creative power was exerted, of a kind to which there was nothing at all analogous in the succeeding part of the same cycle. If it be urged that in some cases the same species, or the same genus, runs through two geological formations, which must, on other grounds, be referred to different cycles of creative energy, we may reply that the creation of many new species does not imply the extinction of all the old ones.

Thus we are led by our reasonings to this view, that the present order of things was commenced by an act of creative power entirely different to any agency which has been exerted since. None of the influences which have modified the present races of animals and plants since they were placed in their habitations on the earth's surface can have had any efficacy in producing them at first. We are necessarily driven to assume, as the beginning of the present cycle of organic nature, an event not included in the course of nature. And we may remark that this necessity is the more cogent, precisely because other cycles have preceded the present.

8. *Nebular Origin of the Solar System.*—If we attempt to apply the same antithesis of opinion (the doctrines of catastrophe and uniformity,) to the other subjects of palætiological sciences, we shall be led to similar conclu-

sions. Thus if we turn our attention to astronomical palætiology, we perceive that the nebular hypothesis has a uniformitarian tendency. According to this hypothesis the formation of this our system of sun, planets, and satellites, was a process of the same kind as those which are still going on in the heavens. One after another, nebulæ condense into separate masses, which begin to revolve about each other by mechanical necessity, and form systems of which our solar system is a finished example. But we may remark, that the uniformitarian doctrine on this subject rests on most unstable foundations. We have as yet only very vague and imperfect reasonings to show that by such condensation a *material* system such as ours could result; and the introduction of *organized* beings into such a material system is utterly out of the reach of our philosophy. Here again, there- fore, we are led to regard the present order of the world as pointing towards an origin altogether of a different kind from anything which our material science can grasp.

9. *Origin of Languages.*—We may venture to say that we should be led to the same conclusion once more, if we were to take into our consideration those palætiolo- gical sciences which are beyond the domain of matter; for instance, the history of languages. We may explain many of the differences and changes which we become acquainted with, by referring to the action of causes of change which still operate. But what glossologist will venture to declare that the efficacy of such causes has been uniform; that the influences which mould a lan- guage, or make one language differ from others of the same stock, operated formerly with no more efficacy than they exercise now. "Where," as has elsewhere been asked, "do we now find a language in the process of formation, unfolding itself in inflexions, terminations, changes of vowels by grammatical relations, such as cha- racterize the oldest known languages?" Again, as another

proof how little the history of languages suggests to the
philosophical glossologist the persuasion of a uniform
action of the causes of change, I may refer to the
conjecture of Dr. Prichard, that the varieties of language
produced by the separation of one stock into several,
have been greater and greater as we go backwards in
history:—that* the formation of sister dialects from a
common language, (as the Scandinavian, German, and
Saxon dialects from the Teutonic, or the Gaelic, Erse and
Welsh from the Celtic,) belongs to the first millennium
before the Christian era; while the formation of cognate
languages of the same family, as the Sanskrit, Latin, Greek
and Gothic, must be placed at least two thousand years
before that era; and at a still earlier period took place
the separation of the great families themselves, the Indo-
European, Semitic, and others, in which it is now difficult to trace the features of a common origin. No
hypothesis except one of this kind will explain the existence of the families, groups, and dialects of languages,
which we find in existence. Yet this is an entirely different view from that which the hypothesis of the uniform
progress of change would give. And thus in the earliest
stages of man's career, the revolutions of language must
have been, even by the evidence of the theoretical history of language itself, of an order altogether different
from any which have taken place within the recent
history of man. And we may add, that as the early
stages of the progress of language must have been widely
different from those later ones of which we can in some
measure trace the natural causes, we cannot place the
origin of language in any point of view in which it comes
under the jurisdiction of natural causation at all.

10. *No Natural Origin discoverable.*—We are thus led
by a survey of several of the palætiological sciences to a
confirmation of the principle formerly asserted†, That

---

* *Researches,* ii., 224.   † *Hist. Ind. Sci.,* iii., 581.

in no palætiological science has man been able to arrive at a beginning which is homogeneous with the known course of events. We can in such sciences often go very far back ;—determine many of the remote circumstances of the past series of events ;—ascend to a point which seems to be near the origin ;—and limit the hypotheses respecting the origin itself :—but philosophers never have demonstrated, and, so far as we can judge, probably never will be able to demonstrate, what was that primitive state of things from which the progressive course of the world took its first departure. In all these paths of research, when we travel far backwards, the aspect of the earlier portions becomes very different from that of the advanced part on which we now stand ; but in all cases the path is lost in obscurity as it is traced backwards towards its starting point :—it becomes not only invisible, but unimaginable ; it is not only an interruption, but an abyss, which interposes itself between us and any intelligible beginning of things.

CHAPTER IV.

OF THE RELATION OF TRADITION TO PALÆ-TIOLOGY.

1. *Importance of Tradition.*—Since the Palætiological Sciences have it for their business to study the train of past events produced by natural causes down to the present time, the knowledge concerning such events which is supplied by the remembrance and records of man, in whatever form, must have an important bearing upon these sciences. All changes in the condition and extent of land and sea, which have taken place within man's observation, all effects of deluges, sea-waves, rivers, springs, volcanos, earthquakes, and the like, which come within

the reach of human history, have a strong interest for the palætiologist. Nor is he less concerned in all recorded instances of the modification of the forms and habits of plants and animals, by the operations of man, or by transfer from one land to another. And when we come to the Palætiology of Language, of Art, of Civilization, we find our subject still more closely connected with history; for in truth these are historical, no less than palætiological investigations. But, confining ourselves at present to the material sciences, we may observe that though the importance of the information which tradition gives us, in the sciences now under our consideration, as, for instance geology, has long been tacitly recognised; yet it is only recently that geologists have employed themselves in collecting their historical facts upon such a scale and with such comprehensive views as are required by the interest and use of collections of this kind. The Essay of Von Hoff*, *On the Natural Alterations in the Surface of the Earth which are proved by Tradition*, was the work which first opened the eyes of geologists to the extent and importance of this kind of investigation. Since that time the same path of research has been pursued with great perseverance by others, especially by Mr. Lyell; and is now justly considered as an essential portion of geology.

2. *Connexion of Tradition and Science.*—Events which we might naturally expect to have some bearing on geo- logy, are recorded in the historical writings which, even on mere human grounds, have the strongest claim to our respect as records of the early history of the world, and are confirmed by the traditions of various nations all over the globe, namely, the formation of the earth and its population, and a subsequent deluge. It has been made a matter of controversy how the narrative of these events

* Vol. i., 1822; vol. ii., 1824.

is to be understood, so as to make it agree with the facts which an examination of the earth's surface and of its vegetable and animal population discloses to us.   Such controversies, when they are considered as merely archæological, may occur in any of the palætiological sciences. We may have to compare and to reconcile the evidence of existing phenomena with that of historical tradition. But under some circumstances this process of conciliation may assume an interest of another kind, on which we will make a few remarks.

3. *Natural and Providential History of the World.*— We may contemplate the existence of man upon the earth, his origin and his progress, in the same manner as we contemplate the existence of any other race of animals; namely, in a purely palætiological view.   We may consider how far our knowledge of laws of causation enables us to explain his diffusion and migration, his differences and resemblances, his actions and works.   And this is the view of man as a member of the *natural* course of things.

But man, at the same time the contemplator and the subject of his own contemplation, endowed with faculties and powers which make him a being of a different nature from other animals, cannot help regarding his own actions and enjoyments, his recollections and his hopes, under an aspect quite different from any that we have yet had presented to us.   We have been endeavouring to place in a clear light the Fundamental Ideas, such as that of Cause, on which depends our knowledge of the natural course of things.   But there are other Ideas to which man necessarily refers his actions; he is led by his nature, not only to consider his own actions, and those of his fellow-men, as springing out of this or that cause, leading to this or that material result; but also as good or bad, as what they ought or ought not to be.

He has Ideas of *moral* relations as well as those Ideas of material relations with which we have hitherto been occupied. He is a moral as well as a natural agent.

Contemplating himself and the world around him by the light of his Moral Ideas, man is led to the conviction that his moral faculties were bestowed upon him by design and for a purpose; that he is the subject of a moral government; that the course of the world is directed by the Power which governs it, to the unfolding and perfecting of man's moral nature; that this guidance may be traced in the career of individuals and of the world; that there is a *providential* as well as a natural course of things.

Yet this view is beset by no small difficulties. The full development of man's moral faculties;—the perfection of his nature up to the measure of his own ideas;—the adaptation of his moral being to an ultimate destination, by its transit through a world full of moral evil, in which each has his share;—are effects for which the economy of the world appears to contain no adequate provision. Man, though aware of his moral nature, and ready to believe in an ultimate destination of purity and blessedness, is too feeble to resist the temptation of evil, and to restore his purity when once lost. He cannot but look for some confirmation of that providential order which he has begun to believe; some provision for those deficiencies in his moral condition which he has begun to feel.

He looks at the history of the world, and he finds that at a certain period it offers to him the promise of what he seeks. When the natural powers of man had been developed to their full extent, and were beginning to exhibit symptoms of decay; when the intellectual progress of the world appeared to have reached its limit, without supplying man's moral needs; we find the great Epoch in the Providential history of the world. We find the announcement of a Dispensation by which man's deficiencies shall be supplied and his aspirations fulfilled: we find a provision for the purification, the support, and the ultimate beatification of those who use the provided means. And thus the providential course of the world becomes consistent and intelligible.

4. *The Sacred Narrative.*—But with the new Dispensation, we receive, not only an account of its own scheme and history, but also a written narrative of the providential course of the world from the earliest times, and even from its first creation. This narrative is recognized and authorized by the new dispensation, and accredited by some of the same evidences as the dispensation itself. That the existence of such a sacred narrative should be a part of the providential order of things, cannot but appear natural; but naturally also, the study of it leads to some difficulties.

The sacred narrative in some of its earliest portions speaks of natural objects and occurrences respecting them. In the very beginning of the course of the world, we may readily believe (indeed as we have seen in the last chapter, our scientific researches lead us to believe) that such occurrences were very different from anything which now takes place;—different to an extent and in a manner which we cannot estimate. Now the narrative must speak of objects and occurrences in the words and phrases which have derived their meaning from their application to the existing natural state of things. When applied to an initial supernatural state therefore, these words and phrases cannot help being to us obscure and mysterious, perhaps ambiguous and seemingly contradictory.

5. *Difficulties in interpreting the Sacred Narrative.*—The moral and providential relations of man's condition are so much more important to him than mere natural

relations, that at first we may well suppose he will accept the Sacred Narrative, as not only unquestionable in its true import, but also as a guide in his views even of mere natural relations. He will try to modify the conceptions which he entertains of objects and their properties, so that the Sacred Narrative of the supernatural condition shall retain the first meaning which he had put upon it in virtue of his own habits in the usage of language.

But man is so constituted that he cannot persist in this procedure. The powers and tendencies of his intellect are such that he cannot help trying to attain true conceptions of objects and their properties by the study of things themselves. For instance, when he at first read of a firmament dividing the waters above from the waters below, he perhaps conceived a transparent floor in the skies, on which the superior waters rested which descend in rain; but as his observations and his reasonings satisfied him that such a floor could not exist, he became willing to allow (as St. Augustine allowed) that the waters above the firmament are in a state of vapour. And in like manner in other subjects, men, as their views of nature became more distinct and precise, modified, so far as it was necessary for consistency's sake, their first rude interpretations of the Sacred Narrative; so that, without in any degree losing its import as a view of the providential course of the world, it should be so conceived as not to contradict what they knew of the natural order of things.

But this accommodation was not always made without painful struggles and angry controversies. When men had conceived the occurrences of the Sacred Narrative in a particular manner, they could not readily and willingly adopt a new mode of conception; and they resisted all attempts to recommend it to them, as attacks upon the sacredness of the Narrative. They had clothed their belief of the workings of Providence in certain images; and they clung to those images with the persuasion that without them their belief could not subsist. Thus they imagined to themselves that the earth was a flat floor, solidly and broadly laid for the convenience of man, and they felt as if the kindness of Providence was disparaged, when it was maintained that the earth was a globe held together only by the mutual attraction of its parts.

The most memorable instance of a struggle of this kind is to be found in the circumstances which attended the introduction of the Heliocentric Theory of Copernicus to general acceptance. On this controversy I have already made some remarks in the History of Science*, and have attempted to draw from it some lessons which may be useful to us when any similar conflict of opinions may occur. I will here add a few reflections with a similar view.

6. *Such difficulties inevitable.*—In the first place, I remark that such modifications of the current interpretation of the words of Scripture appear to be an inevitable consequence of the progressive character of Natural Science. Science is constantly teaching us to describe known facts in new language, but the language of Scripture is always the same. And not only so, but the language of Scripture is necessarily adapted to the common state of man's intellectual development, in which he is supposed not to be possessed of science. Hence the phrases used by Scripture are precisely those which science soon teaches man to consider as inaccurate. Yet they are not on that account the less fitted for their proper purpose: for if any terms had been used, adapted to a more advanced state of knowledge, they must have been unintelligible among those to whom the Scripture was first

* i. 401.

addressed. If the Jews had been told that water existed in the clouds in small drops, they would have marvelled that it did not constantly descend; and to have explained the reason of this, would have been to teach Atmology in the sacred writings. If they had read in their Scripture that the earth was a sphere, when it appeared to be a plain, they would only have been disturbed in their thoughts or driven to some wild and baseless imaginations by a declaration to them so strange. If the Divine Speaker, instead of saying that he would set his bow in the clouds, had been made to declare that he would give to water the property of refracting different colours at different angles, how utterly unmeaning to the hearers would the words have been! And in these cases, the expressions, being unintelligible, startling, and bewildering, would have been such as tended to unfit the Sacred Narrative for its place in the providential dispensation of the world.

Accordingly, in the great controversy which took place in Galileo's time between the defenders of the then customary interpretations of Scripture, and the assertors of the Copernican system of the universe, when the innovators were upbraided with maintaining opinions contrary to Scripture, they replied that Scripture was not intended to teach men astronomy, and that it expressed the acts of divine power in images which were suited to the ideas of unscientific men. To speak of the rising and setting and travelling of the sun, of the fixity and of the foundations of the earth, was to use the only language which would have made the Sacred Narrative intelligible. To extract from these and the like expressions doctrines of science, was, they declared, in the highest degree unjustifiable; and such a course could lead, they held, to no result but a weakening of the authority of Scripture in proportion as its credit was identified with that of these modes of applying it.   And this judgment has since been generally assented to by those who most reverence and value the study of the designs of Providence as well as that of the works of nature.

7. *Science tells us nothing concerning Creation.*—Other apparent difficulties arise from the accounts given in the Scripture of the first origin of the world in which we live: for example, light is represented as created before the sun.   With regard to difficulties of this kind, it appears that we may derive some instruction from the result to which we were led in the last chapter;—namely, that in the sciences which trace the progress of natural occurrences, we can in no case go back to an origin, but in every instance appear to find ourselves separated from it by a state of things, and an order of events, of a kind altogether different from those which come under our experience.   The thread of induction respecting the natural course of the world snaps in our fingers, when we try to ascertain where its beginning is.   Since, then, science can teach us nothing positive respecting the beginning of things, she can neither contradict nor confirm what is taught by Scripture on that subject; and thus, as it is unworthy timidity to fear contradiction, so is it ungrounded presumption to look for confirmation in such cases.   The providential history of the world has its own beginning, and its own evidence; and we can only render the system insecure, by making it lean on our material sciences.   If any one were to suggest that the nebular hypothesis countenances the Scripture history of the formation of this system, by showing how the luminous matter of the sun might exist previous to the sun itself, we should act wisely in rejecting such an attempt to weave together these two heterogeneous threads;— the one a part of a providential scheme, the other a fragment of physical speculation.

We shall best learn those lessons of the true philosophy of science which it is our object to collect, by attending to portions of science which have gone through such
crises as we are now considering; nor is it requisite, for
this purpose, to bring forwards any subjects which are
still under discussion. It may, however, be mentioned
that such maxims as we are now endeavouring to establish,
and the one before us in particular, bear with a peculiar
force upon those Palætiological Sciences of which we have
been treating in the present Book.

8. *Scientific views, when familiar, do not disturb the
authority of Scripture.*—There is another reflection which
may serve to console and encourage us in the painful
struggles which thus take place, between those who maintain interpretations of Scripture already prevalent and
those who contend for such new ones as the new discoveries of science require. It is this;—that though the
new opinion is resisted by one party as something
destructive of the credit of Scripture and the reverence
which is its due, yet, in fact, when the new interpretation
has been generally established and incorporated with
men's current thoughts, it ceases to disturb their views
of the authority of the Scripture or of the truth of its
teaching. When the language of Scripture, invested
with its new meaning, has become familiar to men, it is
found that the ideas which it calls up are quite as recon-
cileable as the former ones were with the most entire
acceptance of the providential dispensation. And when
this has been found to be the case, all cultivated persons
look back with surprise at the mistake of those who
thought that the essence of the revelation was involved
in their own arbitrary version of some collateral circumstance in the revealed narrative. At the present day, we
can hardly conceive how reasonable men could ever have
imagined that religious reflections on the stability of the

earth, and the beauty and use of the luminaries which revolve round it, would be interfered with by an acknowledgment that this rest and motion are apparent only*. And thus the authority of revelation is not shaken by any changes introduced by the progress of science in the mode of interpreting expressions which describe physical objects and occurrences; provided the new interpretation is admitted at a proper season, and in a proper spirit; so as to soften, as much as possible, both the public controversies and the private scruples which almost inevitably accompany such an alteration.

9. *When should old Interpretations be given up?*—But the question then occurs, What is the proper season for a religious and enlightened commentator to make such a change in the current interpretation of sacred Scripture? At what period ought the established exposition of a passage to be given up, and a new mode of understanding the passage, such as is, or seems to be, required by new discoveries respecting the laws of nature, accepted in its place? It is plain, that to introduce such an alteration lightly and hastily would be a procedure fraught with inconvenience; for if the change were made in such a manner, it might be afterwards discovered that it had been adopted without sufficient reason, and that it was necessary to reinstate the old exposition. And the minds of the readers of Scripture, always to a certain extent and for a time disturbed by the subversion of their long-established notions, would be distressed without any need, and might be seriously unsettled. While, on the other hand, a too protracted and obstinate resistance to the innovation, on the part of the scriptural expositors, would tend to identify, at least in the minds of many, the authority of the Scripture with the truth of the exposition; and therefore would bring discredit upon the

---

* I have here borrowed a sentence or two from my own History.

revealed word, when the established interpretation was finally proved to be untenable.

A rule on this subject, propounded by some of the most enlightened dignitaries of the Roman Catholic church, on the occasion of the great Copernican controversy begun by Galileo, seems well worthy of our attention. The following was the opinion given by Cardinal Bellarmine at the time:—" When a *demonstration* shall be found to establish the earth's motion, it will be proper to interpret the sacred Scriptures otherwise than they have hitherto been interpreted in those passages where mention is made of the stability of the earth and movement of the heavens." This appears to be a judicious and reasonable maxim for such cases in general.   So long as the supposed scientific discovery is doubtful, the exposition of the meaning of Scripture given by commentators of established credit is not wantonly to be disturbed: but when a scientific theory, irreconcileable with this ancient interpretation, is clearly proved, we must give up the interpretation, and seek some new mode of understanding the passage in question, by means of which it may be consistent with what we know; for if it be not, our conception of the things so described is no longer consistent with itself.

It may be said that this rule is indefinite, for who shall decide when a new theory is completely demonstrated, and the old interpretation become untenable? But to this we may reply, that if the rule be assented to, its application will not be very difficult.   For when men have admitted as a general rule, that the current interpretations of scriptural expressions respecting natural objects and events may possibly require, and in some cases certainly will require, to be abandoned, and new ones admitted, they will hardly allow themselves to contend for such interpretations as if they were essential parts of revelation; and will look upon the change of exposition, whether it come sooner or later, without alarm or anger.   And when men lend themselves to the progress of truth in this spirit, it is not of any material importance at what period a new and satisfactory interpretation of the scriptural difficulty is found; since a scientific exactness in our apprehension of the meaning of such passages as are now referred to is very far from being essential to our full acceptance of revelation.

10. *In what Spirit should the Change be accepted?*— Still these revolutions in scriptural interpretation must always have in them something which distresses and disturbs religious communities.   And such uneasy feelings will take a different shape, according as the community acknowledges or rejects a paramount interpretative authority in its religious leaders.   In the case in which the interpretation of the Church is binding upon all its members, the more placid minds rest in peace upon the ancient exposition, till the spiritual authorities announce that the time for the adoption of a new view has arrived; but in these circumstances, the more stirring and inquisitive minds, which cannot refrain from the pursuit of new truths and exact conceptions, are led to opinions which, being contrary to those of the Church, are held to be sinful.   On the other hand, if the religious constitution of the community allow and encourage each man to study and interpret for himself the Sacred Writings, we are met by evils of another kind.   In this case, although, by the unforced influence of admired commentators, there may prevail a general agreement in the usual interpretation of difficult passages, yet as each reader of the Scripture looks upon the sense which he has adopted as being his own interpretation, he maintains it, not with the tranquil acquiescence of one who has deposited his judgment in the hands of his Church, but with the keenness

and strenuousness of self-love.   In such a state of things, though no judicial severities can be employed against the innovators, there may arise more angry controversies than in the other case.

It is impossible to overlook the lesson which here offers itself, that it is in the highest degree unwise in the friends of religion, whether individuals or communities, unnecessarily to embark their credit in expositions of Scripture on matters which appertain to natural science.   By delivering physical doctrines as the teaching of revelation, religion may lose much, but cannot gain anything.   This maxim of practical wisdom has often been urged by Christian writers.   Thus St. Augustin says*: " In obscure matters and things far removed from our senses, if we read anything, even in the divine Scripture, which may produce diverse opinions without damaging the faith which we cherish, let us not rush headlong by positive assertion to either the one opinion or the other ; lest, when a more thorough discussion has shown the opinion which we had adopted to be false, our faith may fall with it : and we should be found contending, not for the doctrine of the sacred Scriptures, but for our own ; endeavouring to make our doctrine to be that of the Scriptures, instead of taking the doctrine of the Scriptures to be ours."   And in nearly the same spirit, at the time of the Copernican controversy, it was thought proper to append to the work of Copernicus a postil, to say that the work was written to account for the phenomena, and that people must not run on blindly and condemn either of the opposite opinions.   Even when the Inquisition, in 1616, thought itself compelled to pronounce a decision upon this subject, the verdict was delivered in very moderate language ;—that " the doctrine of the earth's motion appeared to be contrary to Scrip-

* Lib. i. *de Genesi*, cap. 18.

ture:" and yet, moderate as this expression is, it has been
blamed by judicious members of the Roman church as
deciding a point such as religious authorities ought not
to pretend to decide; and has brought upon that church
no ordinary weight of general condemnation.   Kepler
pointed out, in his lively manner, the imprudence of
employing the force of religious authorities on such subjects: *Acies dolabræ in ferrum illisa, postea nec in lignum
valet amplius. Capiat hoc cujus interest.*   " If you *will* try
to chop iron, the axe becomes unable to cut even wood."

11. *In what Spirit should the Change be urged?*—But
while we thus endeavour to show in what manner the
interpreters of Scripture may most safely and most properly accept the discoveries of science, we must not
forget that there may be errors committed on the other
side also; and that men of science, in bringing forward
views which may for a time disturb the minds of lovers
of Scripture, should consider themselves as bound by
strict rules of candour, moderation, and prudence.   In-
tentionally to make their supposed discoveries a means
of discrediting, contradicting, or slighting the sacred
Scriptures, or the authority of religion, is in them unpar-
donable.   As men who make the science of Truth the
business of their lives, and are persuaded of her genuine
superiority, and certain of her ultimate triumph, they are
peculiarly bound to urge her claims in a calm and temperate spirit; not forgetting that there are other kinds
of truth besides that which they peculiarly study.   They
may properly reject authority in matters of science; but
they are to leave it its proper office in matters of religion.
I may here again quote Kepler's expressions: " In Theology we balance authorities, in Philosophy we weigh
reasons.   A holy man was Lactantius who denied that
the earth was round; a holy man was Augustin, who
granted the rotundity, but denied the antipodes; a holy

thing to me is the Inquisition, which allows the smallness of the earth, but denies its motion; but more holy to me is Truth; and hence I prove, from philosophy, that the earth is round, and inhabited on every side, of small size, and in motion among the stars,—and this I do with no disrespect to the Doctors." I the more willingly quote such a passage from Kepler, because the entire ingenuousness and sincere piety of his character does not allow us to suspect in him anything of hypocrisy or latent irony. That similar professions of respect may be made ironically, we have a noted example in the celebrated Introduction to *Galileo's Dialogue on the Copernican System;* probably the part which was most offensive to the authorities. "Some years ago," he begins, "a wholesome edict was promulgated at Rome, which, in order to check the perilous scandals of the present age, imposed silence upon the Pythagorean opinion of the mobility of the earth. There were not wanting," he proceeds, "persons who rashly asserted that this decree was the result, not of a judicious inquiry, but of passion ill-informed; and complaints were heard that counsellors, utterly unacquainted with astronomical observation, ought not to be allowed, with their sudden prohibitions, to clip the wings of speculative intellects. *At the hearing of rash lamentations like these, my zeal could not keep silence.*" And he then goes on to say, that he wishes, in his *Dialogue,* to show that the subject had been fully examined at Rome. Here the irony is quite transparent, and the sarcasm glaringly obvious. I think we may venture to say that this is not the temper in which scientific questions should be treated; although by some, perhaps, the prohibition of public discussion may be considered as justifying any evasion which is likely to pass unpunished.

12. *Duty of Mutual Forbearance.*—We may add, as a further reason for mutual forbearance in such cases, that

the true interests of both parties are the same. The man of science is concerned, no less than any other person, in the truth and import of the divine dispensation; the religious man, no less than the man of science, is, by the nature of his intellect, incapable of believing two contradictory declarations. Hence they have both alike a need for understanding the Scripture in some way in which it shall be consistent with their understanding of nature. It is for their common advantage to conciliate, as Kepler says, the finger and the tongue of God, his works and his word. And they may find abundant reason to bear with each other, even if they should adopt for this purpose different interpretations, each finding one satisfactory to himself; or if any one should decline employing his thoughts on such subjects at all. I have elsewhere* quoted a passage from Kepler† which appears to me written in a most suitable spirit: "I beseech my reader that, not unmindful of the Divine goodness bestowed upon man, he do with me praise and celebrate the wisdom of the Creator, which I open to him from a more inward explication of the form of the world, from a searching of causes, from a detection of the errors of vision; and that thus not only in the firmness and stability of the earth may we perceive with gratitude the preservation of all living things in nature as the gift of God: but also that in its motion, so recondite, so admirable, we may acknowledge the wisdom of the Creator. But whoever is too dull to receive this science, or too weak to believe the Copernican system without harm to his piety, him, I say, I advise that, leaving the school of astronomy, and condemning, if so he please, any doctrines of the philosophers, he follow his own path, and desist from this wandering through the universe; and that, lifting up his natural eyes with which alone he can see, he pour himself out

* *Bridgewater Tr.*, b. 314.   † *Com. Stell. Mart.*, Introd.

from his own heart in worship of God the Creator, being certain that he gives no less worship to God than the astronomer, to whom God has given to see more clearly with his inward eyes, and who, from what he has himself discovered, both can and will glorify God."

13. *Case of Galileo.*—I may perhaps venture here to make a remark or two upon this subject with reference to a charge brought against a certain portion of the *History of the Inductive Sciences.* Complaint has been made* that the character of the Roman church, as shown in its behaviour towards Galileo, is misrepresented in the account given of it in the History of Astronomy. It is asserted that Galileo provoked the condemnation he incurred; first, by pertinaciously demanding the assent of the ecclesiastical authorities to his opinion of the consistency of the Copernican doctrine with Scripture; and afterwards by contumaciously, and, as we have seen, contumeliously violating the silence which the Church had enjoined upon him. It is further declared that the statement which represents it as the habit of the Roman church to dogmatize on points of natural science is unfounded; as well as the opinion that in consequence of this habit, new scientific truths were promulgated less boldly in Italy than in other countries. I shall reply very briefly on these subjects; for the decision of them is by no means requisite in order to establish the doctrines to which I have been led in the present chapter, nor, I hope, to satisfy my reader that my views have been collected from an impartial consideration of scientific history.

With regard to Galileo, I do not think it can be denied that he obtruded his opinions upon the ecclesiastical authorities in an unnecessary and imprudent manner. He was of an ardent character, strongly convinced himself, and urged on still more by the conviction which he

* *Dublin Review,* No. ix., July, 1838, p. 72.

produced among his disciples, and thus he became impatient for the triumph of truth.   This judgment of him has recently been delivered by various independent authorities, and has undoubtedly considerable foundation*.   As to the question whether authority in matters of natural science were habitually claimed by the authorities of the Church of Rome, I have to allow that I cannot produce instances which establish such a habit.   We who have been accustomed to have daily before our eyes the Monition which the Romish editors of Newton thought it necessary to prefix—*Cæterum latis a summo Pontifice contra telluris motum Decretis, nos obsequi profitemur*—were not likely to conjecture that this was a solitary instance of the interposition of the Papal authority on such subjects.   But although it would be easy to find declarations of heresy delivered by Romish Universities, and writers of great authority, against tenets belonging to the natural sciences, I am not aware that any other case can be adduced in which the Church or the Pope can be shown to have pronounced such a sentence.   I am well contented to acknowledge this; for I should be far more gratified by finding myself compelled to hold up the seventeenth century as a model for the nineteenth in this respect, than by having to sow enmity between the admirers of the past and the present through any disparaging contrast†.

* Besides the *Dublin Review*, I may quote the *Edinburgh Review*, which I suppose will not be thought likely to have a bias in favour of the exercise of ecclesiastical authority in matters of science; though certainly there is a puerility in the critic's phraseology which does not add to the weight of his judgment.   "Galileo contrived to surround the truth with every variety of obstruction.   The tide of knowledge, which had hitherto advanced in peace, he crested with angry breakers, and he involved in its surf both his friends and his foes."—*Ed. Rev.*, No. cxxiii. p. 126.

† I may add that the most candid of the adherents of the Church of Rome condemn the assumption of authority in matters of science, made, in this one instance at least, by the ecclesiastical tribunals.   The

With respect to the attempt made in my History to characterize the intellectual habits of Italy as produced by her religious condition,—certainly it would ill become any student of the history of science to speak slightingly of that country, always the mother of sciences, always ready to catch the dawn and hail the rising of any new light of knowledge. But I think our admiration of this activity and acuteness of mind is by no means inconsistent with the opinion, that new truths were promulgated more boldly beyond the Alps, and that the subtilty of the Italian intellect loved to insinuate what the rough German bluntly asserted. Of the decent duplicity with which forbidden opinions were handled, the reviewer himself gives us instances, when he boasts of the liberality with which Copernican professors were placed in important stations by the ecclesiastical authorities, soon after the doctrine of the motion of the earth had been declared by the same authorities contrary to Scripture. And in the same spirit is the process of demanding from Galileo a public and official recantation of opinions which he had repeatedly been told by his ecclesiastical superiors he might hold as much as he pleased. I think it is easy to believe that among persons so little careful to reconcile public profession with private conviction, official decorum was all that was demanded. When Galileo had made his renunciation of the earth's motion on his knees, he rose and said, as we are told, *E pur si muove*—"and yet it *does* move." This is sometimes represented as the heroic soliloquy of a mind cherishing its conviction of the truth, in spite of persecution; I think we may more naturally conceive it uttered as a playful epigram in the ear of a cardinal's secretary, with a full knowledge that it would be immediately repeated to his master.

author of the *Ages of Faith* (Book viii. p. 248), says, "A Congregation, it is to be lamented, declared the new system to be opposed to Scripture, and therefore heretical."

Besides the Ideas involved in the material sciences, of which we have already examined the principal ones, there is one Idea or Conception which our Sciences do not indeed include, but to which they not obscurely point; and the importance of this Idea will make it proper to speak of it, though this must be done very briefly.

CHAPTER V.

OF THE CONCEPTION OF A FIRST CAUSE.

1. AT the end of the last chapter but one, we were led to this result,—that we cannot, in any of the Palætiological Sciences, ascend to a beginning which is of the same nature as the existing cause of events, and which depends upon causes that are still in operation. Philosophers never have demonstrated, and probably never will be able to demonstrate, what was the original condition of the solar system, of the earth, of the vegetable and animal worlds, of languages, of arts. On all these subjects the course of investigation, followed backwards as far as our materials allow us to pursue it, ends at last in an impenetrable gloom. We strain our eyes in vain when we try, by our natural faculties, to discern an Origin.

2. Yet speculative men have been constantly employed in attempts to arrive at that which thus seems to be placed out of their reach. The Origin of Languages, the Origin of the present Distribution of Plants and Animals, the Origin of the Earth, have been common subjects of diligent and persevering inquiry. Indeed inquiries respecting such subjects have been, at least till lately, the usual form which Palætiotogical researches have assumed. *Cosmogony*, the origin of the world, of which, in such speculations, the earth was considered as a principal part,

has been a favourite study both of ancient and of modern times: and most of the attempts at Geology previous to the present period have been *Cosmogonies* or *Geogonies*, rather than that more genuine science which we have endeavoured to delineate. Glossology, though now an extensive body of solid knowledge, was mainly brought into being by inquiries concerning the original language spoken by men; and the nature of the first separation and diffusion of languages, the first peopling of the earth by man and by animals, were long sought after with ardent curiosity, although of course with reference to the authority of the Scriptures, as well as the evidence of natural phenomena. Indeed the interest of such inquiries even yet is far from being extinguished. The disposition to explore the past in the hope of finding, by the light of natural reasoning as well as by the aid of revelation, the origin of the present course of things, appears to be unconquerable. What was the beginning? is a question which the human race cannot desist from perpetually asking. And no failure in obtaining a satisfactory answer can prevent inquisitive spirits from again and again repeating the inquiry, although the blank abyss into which it is uttered does not even return an echo.

3. What, then, is the reason of an attempt so pertinacious yet so fruitless? By what motive are we impelled thus constantly to seek what we can never find? Why are the error of our conjectures, the futility of our reasonings, the precariousness of our interpretations, over and over again proved to us in vain? Why is it impossible for us to acquiesce in our ignorance and to relinquish the inquiry? Why cannot we content ourselves with examining those links of the chain of causes which are nearest to us;—those in which the connexion is intelligible and clear; instead of fixing our attention upon those remote portions where we can no longer estimate its co-

herence? In short, why did not men from the first take for the subject of their speculations the Course of Nature rather than the Origin of Things?

To this we reply, that in doing what they have thus done, in seeking what they have sought, men are impelled by an intellectual necessity. They cannot conceive a series of connected occurrences without a commencement; they cannot help supposing a cause for the whole, as well as a cause for each part; they cannot be satisfied with a succession of causes without assuming a First Cause. Such an assumption is necessarily impressed upon our minds by our contemplation of a series of causes and effects; that *there must be a First Cause*, is accepted by all intelligent reasoners as an Axiom: and like other Axioms, its truth is necessarily implied in the Idea which it involves.

4. The evidence of this axiom may be illustrated in several ways. In the first place, the axiom is assumed in the argument usually offered to prove the existence of the Deity. Since, it is said, the world now exists, and since nothing cannot produce something, something must have existed from eternity. This Something is the First Cause: it is God.

Now what I have to remark here is this: the conclusiveness of this argument, as a proof of the existence of one independent, immutable Deity, depends entirely upon the assumption of the axiom above stated. The world, a series of causes and effects, exists: therefore there must be, not only this series of causes and effects, but also a First Cause. It will be easily seen, that without the axiom, that in every series of causes and effects there must be a First Cause, the reasoning is altogether inconclusive.

5. Or to put the matter otherwise: The argument for the existence of the Deity was stated thus: Something

exists, therefore something must have existed from eternity. Granted, the opponent might say; but this *something* which has existed from eternity, why may it not be this very series of causes and effects which is now going on, and which appears to contain in itself no indication of beginning or end? And thus, without the assumption of the necessity of a First Cause, the force of the argument may be resisted.

6. But, it may be asked, how do those who have written to prove the existence of the Deity reply to such an objection as the one just stated? It is natural to suppose that, on a subject so interesting and so long discussed, all the obvious arguments, with their replies, have been fully brought into view. What is the result in this case?

The principal modes of replying to the above objection, that the series of causes and effects which now exists, may have existed from eternity, appear to be these.

In the first place, our minds cannot be satisfied with a series of successive, dependent, causes and effects, without something first and independent. We pass from effect to cause, and from that to a higher cause, in search of something on which the mind can rest; but if we can do nothing but repeat this process, there is no use in it. We move our limbs, but make no advance. Our question is not answered, but evaded. The mind cannot acquiesce in the destiny thus presented to it, of being referred from event to event, from object to object, along an interminable vista of causation and time. Now this mode of stating the reply,—to say that the mind *cannot thus be satisfied*, appears to be equivalent to saying that the mind is conscious of a principle in virtue of which such a view as this must be rejected;—the mind takes refuge in the assumption of a First Cause, from an employment inconsistent with its own nature.

7. Or again, we may avoid the objection, by putting the argument for the existence of a Deity in this form : The series of causes and effects which we call the *world*, or the *course of nature*, may be considered as a *whole*, and this whole must have a cause of its existence. The whole collection of objects and events may be comprehended as a single effect, and of this effect there must be a cause. This Cause of the Universe must be superior to, and independent of the special events, which, happening in time, make up the universe of which He is the cause. He must exist and exercise causation, before these events can begin: He must be the First Cause.

Although the argument is here somewhat modified in form, the substance is the same as before. For the assumption that we may consider the whole series of causes and effects as a *single effect*, is equivalent to the assumption that besides partial causes, we must have a First Cause. And thus the Idea of a First Cause, and the axiom which asserts its necessity, are recognized in the usual argumentation on this subject.

8. This Idea of a First Cause, and the principle involved in the Idea, have been the subject of discussion in another manner. As we have already said, we assume as an axiom that a First Cause must exist; and we assert that God, the First Cause, exists eternal and immutable, by the necessity which the axiom implies. Hence God is said to exist necessarily;—to be a necessarily existing being. And when this *necessary existence* of God had been spoken of, it soon began to be contemplated as a sufficient reason, and as an absolute demonstration of His existence; without any need of referring to the world as an effect, in order to arrive at God as the cause. And thus men conceived that they had obtained a proof of the existence of the Deity, *à priori*, from ideas, as well as *à posteriori*, from effects.

9. Thus, Thomas Aquinas employs this reasoning to prove the *eternity* of God[*]. "Oportet ponere aliquod primum necessarium quod est per se ipsum necessarium; et hoc est Deus, cum sit prima causa ut dictum est : igitur Deus æternus est, cum omne necessarium per se sit æternum." It is true that the schoolmen never professed to be able to prove the *existence* of the Deity *à priori :* but they made use of this conception of necessary existence in a manner which approached very near to such an attempt. Thus Suarez[†] discusses the question, "Utrum aliquo modo possit *à priori* demonstrari Deum esse." And resolves the question in this manner : "Ad hunc ergo modum dicendum est : Demonstrato *à posteriori* Deum esse ens necessarium et a se, ex hoc attributo posse *à priori* demonstrari præter illud non posse esse aliud ens necessarium et a se, et consequenter demonstrari Deum esse."

But in modern times attempts were made by Descartes and Samuel Clarke, to prove the Divine existence at once *à priori,* from the conception of necessary existence; which, it was argued, could not subsist without actual existence. This argumentation was acutely and severely criticized by Dr. Waterland.

10. Without dwelling upon a subject, the discussion of which does not enter into the design of the present work, I may remark that the question whether an *à priori* proof of the existence of a First Cause be possible, is a question concerning the nature of our Ideas, and the evidence of the axioms which they involve, of the same kind as many questions which we have already had to discuss. Is our Conception or Idea of a First Cause gathered from the effects we see around us? It is plain that we must answer, here as in other cases, that the Idea is not

---

[*] AQUIN. *Contr. Gentil.* lib. i. c. 14, p. 21.
[†] *Metaphys.*, tom. ii. disp. 29, sect. 3, p. 28.

extracted from the phenomena, but assumed in order that the phenomena may become intelligible to the mind;—that the Idea is a necessary one, inasmuch as it does not depend upon observation for its evidence; but that it depends upon observation for its development, since without some observation, we cannot conceive the mind to be cognisant of the relation of causation at all. In this respect, however, the Idea of a First Cause is no less necessary than the ideas of Space, or Time, or Cause in general. And whether we call the reasoning derived from such a necessity an argument *à priori* or *à posteriori*, in either case it possesses the genuine character of demonstration, being founded upon axioms which command universal assent.

11. I have, however, spoken of our *Conception* rather than of our *Idea* of a First Cause; for the notion of a First Cause appears to be rather a modification of the Fundamental Idea of Cause, which was formerly discussed, than a separate and peculiar Idea. And the Axiom, *that there must be a First Cause*, is recognised by most persons as an application of the general Axiom of Causation, *that every effect must have a cause;* this latter Axiom being applied to the world, considered in its totality, as a single effect. This distinction, however, between an Idea and a Conception, is of no material consequence to our argument; provided we allow the maxim, that there must be a First Cause, to be necessarily and evidently true; whether it be thought better to speak of it as an independent Axiom, or to consider it as derived from the general Axiom of Causation.

12. Thus we necessarily infer a First Cause, although the Palætiological Sciences only point *towards* it, and do not lead us *to* it. But I must observe further; that in each of the series of events which form the subject of Palætiological research, the First Cause is the *same*.

Without here resting upon reasoning founded upon our
Conception of a First Cause, I may remark that this
identity is proved by the close connexion of all the
branches of natural science, and the way in which the
causes and the events of each are interwoven with those
which belong to the others. We must needs believe
that the First Cause which produced the earth and its
atmosphere is also the Cause of the plants which clothe
its surface; that the First Cause of the vegetable and of
the animal world are the same; that the First Cause
which produced light produced also eyes; that the First
Cause which produced air and organs of articulation produced also language and the faculties by which language
is rendered possible: and if *those* faculties, then also all
man's other faculties;—the powers by which, as we have
said, he discerns right and wrong, and recognizes a providential as well as a natural course of things. Nor can
we think otherwise than that the Being who gave these
faculties, bestowed them for some purpose;—bestowed
them for that purpose which alone is compatible with
their nature:—the purpose, namely, of guiding and elevating man in his present career, and of preparing him
for another state of being to which they irresistibly direct
his hopes. And thus, although, as we have said, no one
of the Palætiological Sciences can be traced continuously
to an origin, yet they not only each point to an origin,
but all to the same origin. Their lines are broken indeed,
as they run backwards into the early periods of the world,
but yet they all appear to converge to the same invisible
point. And this point, thus indicated by the natural
course of things, can be no other than that which is
disclosed to us as the starting point of the providential
course of the world; for we are persuaded by such reasons
as have just been hinted, that the Creator of the natural
world can be no other than the Author and Governor
and Judge of the moral and spiritual world.

13. Thus we are led, by our material sciences, and especially by the Palætiological class of them, to the borders of a higher region, and to a point of view from which we have a prospect of other provinces of knowledge, in which other faculties of man are concerned besides his intellectual, other interests involved besides those of speculation. On these it does not belong to our present plan to dwell: but even such a brief glance as we have taken of the connexion of material with moral speculations may not be useless, since it may serve to show that the principles of truth which we are now laboriously collecting among the results of the physical sciences, may possibly find some application in those parts of knowledge towards which men most naturally look with deeper interest and more serious reverence.

We have been employed up to the present stage of this work in examining the materials of knowledge, namely, Facts and Ideas; and we have dwelt particularly upon the latter element; inasmuch as the consideration of it is, on various accounts, and especially at the present time, by far the most important. We have now to proceed to the remainder of our task;—to determine the processes by which those materials may actually be made to constitute knowledge. We have surveyed the stones of our building: we have found them exactly squared, and often curiously covered with significant imagery and important inscriptions. We have now to discover how they may best be fitted into their places, and cemented together, so that rising stage above stage, they may grow at last into that fair and lofty temple of Truth for which we cannot doubt that they were intended by the Great Architect.

THE

PHILOSOPHY

OF THE

INDUCTIVE SCIENCES.

———————

PART II.

OF KNOWLEDGE.

De Scientiis tum demum bene sperandum est, quando por SCALAM veram et per gradus continuos, et non intermissos aut hiulcos, a particularibus ascendetur ad Axiomatur minora, et deinde ad media, alia aliis superiora, et postremò demum ad generalissima.

In constituendo autem Axiomate, Forma INDUCTIONIS alia quam adhuc in usu fuit, excogitanda est; et quæ non ad Principia tantum (quæ vocant) probanda et invenienda, sed etiam ad Axiomata minora, et media, denique omnia.

BACON, Nov. Org., Aph. civ. cv.

BOOK XI.

OF THE CONSTRUCTION OF SCIENCE.

CHAPTER I.

OF TWO PRINCIPAL PROCESSES BY WHICH
SCIENCE IS CONSTRUCTED.

To the subject of the present Book all that has preceded is subordinate and preparatory. The First Part of
this work treated of Ideas: we now enter upon the
Second Part, in which we have to consider the Knowledge which arises from them. It has already been stated
that knowledge requires us to possess both Facts and
Ideas;—that every step in our knowledge consists in
applying the ideas and conceptions furnished by our
minds to the facts which observation and experiment
offer to us. When our conceptions are clear and distinct,
when our facts are certain and sufficiently numerous, and
when the conceptions, being suited to the nature of the
facts, are applied to them so as to produce an exact and
universal accordance, we attain knowledge of a precise
and comprehensive kind, which we may term *Science.*
And we apply this term to our knowledge still more
decidedly when, facts being thus included in exact and
general propositions, such propositions are, in the same
manner, included with equal rigour in propositions of a
higher degree of generality; and these again in others of
a still wider nature, so as to form a large and systematic
whole.

But after thus stating, in a general way, the nature of science, and the elements of which it consists, we have been examining with a more close and extensive scrutiny, some of those elements ; and we must now return to our main subject, and apply to it the results of our long investigation. We have been exploring the realm of Ideas; we have been passing in review the difficulties in which the workings of our own minds involve us when we would make our conceptions consistent with themselves : and we have endeavoured to get a sight of the true solutions of these difficulties. We have now to inquire how the results of these long and laborious efforts of thought find their due place in the formation of our knowledge. What do we gain by these attempts to make our notions distinct and consistent ; and in what manner is the gain of which we thus become possessed, carried to the general treasure-house of our permanent and indestructible knowledge? After all this battling in the world of ideas, all this struggling with the shadowy and changing forms of intellectual perplexity, how do we secure to ourselves the fruits of our warfare, and assure ourselves that we have really pushed forwards the frontier of the empire of Science? It is by such an appropriation that the task which we have had in our hands during the last nine Books of this work, must acquire its real value and true place in our design.

In order to do this, we must reconsider, in a more definite and precise shape, the doctrine which has already been laid down;—that our knowledge consists in applying Ideas to Facts; and that the conditions of real knowledge are that the ideas be distinct and appropriate, and exactly applied to clear and certain facts. The steps by which our knowledge is advanced are those by which one or the other of these two processes is rendered more complete ;—by which *conceptions* are *made more clear* in themselves, or by which the conceptions more strictly *bind together the facts*. These two processes may be considered as together constituting the whole formation of our knowledge; and the principles which have been established in the preceding Books, bear principally upon the former of these two operations;—upon the business of elevating our conceptions to the highest possible point of precision and generality. But these two portions of the progress of knowledge are so clearly connected with each other, that we shall consider them in immediate succession. And having now to consider these operations in a more exact and formal manner than it was before possible to do, we shall designate them by certain constant and technical phrases. We shall speak of the two processes by which we arrive at science, as *the Explication of Conceptions* and *the Colligation of Facts:* we shall show how the discussions in which we have been engaged have been necessary in order to promote the former of these offices; and we shall endeavour to point out modes, maxims, and principles by which the second of the two tasks may also be furthered.

CHAPTER II.

OF THE EXPLICATION OF CONCEPTIONS.

1. WE have given the appellation of *Ideas* to certain comprehensive forms of thought,—as *space, number, cause, composition, resemblance,*—which we apply to the phenomena which we contemplate. But the special modifications of these ideas which are exemplified in particular facts, we have termed *Conceptions;* as *a circle, a square number, an accelerating force, a neutral combination* of elements, a *genus.* Such Conceptions involve in themselves

certain necessary and universal relations derived from the
ideas just enumerated; and these relations are an indispensable portion of the texture of our knowledge. But
to determine the contents and limits of this portion of
our knowledge, requires an examination of the ideas and
conceptions from which it proceeds. The conceptions
must be, as it were, carefully *unfolded*, so as to bring
into clear view the elements of truth with which they are
marked from their ideal origin. This is one of the processes by which our knowledge is extended and made
more exact; and this I shall describe as the *Explication
of Conceptions.*

In the preceding Books we have discussed a great
many of the Fundamental Ideas of the most important
existing sciences. We have, in those Books, abundant
exemplifications of the process now under our consideration. We shall here add a few general remarks, suggested by the survey which we have thus made.

2. (I.) Such discussions as those in which we have been
engaged concerning our fundamental ideas, have been the
course by which, historically speaking, those conceptions
which the existing sciences involve have been rendered
so clear as to be fit elements of exact knowledge. The
disputes concerning the various kinds and measures of
Force were an important part of the progress of the science of mechanics. The struggles by which philosophers
attained a right general conception of plane, of circular,
of elliptical Polarization, were some of the most difficult
steps in the modern discoveries of optics. A conception
of the Atomic Constitution of bodies, such as shall include
what we know, and assume nothing more, is even now a
matter of conflict among chemists. The debates by which,
in recent times, the conceptions of Species and Genera
have been rendered more exact, have improved the science
of botany: the imperfection of the science of mineralogy

arises in a great measure from the circumstance, that in that subject, the conception of a Species is not yet fixed. In physiology, what a vast advance would that philosopher make, who should establish a precise, tenable, and consistent conception of Life!

Thus discussions and speculations concerning the import of very abstract and general terms and notions, may be, and in reality have been, far from useless and barren. Such discussions arose from the desire of men to impress their opinions on others, but they had the effect of making the opinions much more clear and distinct. In trying to make others understand them, they learnt to understand themselves. Their speculations were begun in twilight, and ended in the full brilliance of day. It was not easily and at once, without expenditure of labour or time, that men arrived at those notions which now form the elements of our knowledge; on the contrary, we have, in the history of science, seen how hard discoverers, and the forerunners of discoverers, have had to struggle with the indistinctness and obscurity of the intellect, before they could advance to the critical point at which truth became clearly visible. And so long as, in this advance, some speculators were more forward than others, there was a natural and inevitable ground of difference of opinion, of argumentation, of wrangling. But the tendency of all such controversy is to diffuse truth and to dispel error. Truth is consistent, and can bear the tug of war; error is incoherent, and falls to pieces in the struggle. True conceptions can endure the sun, and become clearer as a fuller light is obtained; confused and inconsistent notions vanish like visionary spectres at the break of a brighter day. And thus all the controversies concerning such conceptions as science involves have ever ended in the establishment of the side on which the truth was found.

3. Indeed, so complete has been the victory of truth

in most of these instances, that at present we can hardly imagine the struggle to have been necessary. The very essence of these triumphs is that they lead us to regard the views we reject as not only false, but inconceivable. And hence we are led rather to look back upon the vanquished with contempt than upon the victors with gratitude. We now despise those who in the Copernican controversy could not conceive the apparent motion of the sun on the heliocentric hypothesis; or those who, in opposition to Galileo, thought that a uniform force might be that which generated a velocity proportional to the space; or those who held there was something absurd in Newton's doctrine of the different refrangibility of differently coloured rays; or those who imagined that when elements combine, their sensible qualities must be manifest in the compound; or those who were reluctant to give up the distinction of vegetables into herbs, shrubs, and trees. We cannot help thinking that men must have been singularly dull of comprehension to find a difficulty in admitting what is to us so plain and simple. We have a latent persuasion that we in their place should have been wiser and more clear-sighted;—that we should have taken the right side, and given our assent at once to the truth.

4. Yet in reality such a persuasion is a mere delusion. The persons who, in such instances as the above, were on the losing side, were very far, in most cases, from being persons more prejudiced, or stupid, or narrow-minded, than the greater part of mankind now are; and the cause for which they fought was far from being a manifestly bad one, till it had been so decided by the result of the war. It is the peculiar character of scientific contests, that what is only an epigram with regard to other warfare is a truth in this; and they who are defeated are really in the wrong. But they may, never-

theless, be men of great subtilty, sagacity, and genius; and we nourish a very foolish self-complacency when we suppose that we are their superiors. That this is so, is proved by recollecting that many of those who have made very great discoveries have laboured under the imperfection of thought which was the obstacle to the next step in knowledge. Though Kepler detected with great acuteness the numerical laws of the solar system, he laboured in vain to conceive the very simplest of the laws of motion by which the paths of the planets are governed. Though Priestley made some important steps in chemistry, he could not bring his mind to admit the doctrine of a general principle of oxidation. How many ingenious men in the last century rejected the Newtonian attraction as an impossible chimera! How many more, equally intelligent, have, in the same manner, in our own time, rejected, I do not now mean as false, but as inconceivable, the doctrine of luminiferous undulations! To err in this way is the lot, not only of men in general, but of men of great endowments, and very sincere love of truth.

5. And those who liberate themselves from such perplexities, and who thus go on in advance of their age in such matters, owe their superiority in no small degree to such discussions and controversies as those to which we now refer. In such controversies, the conceptions in question are turned in all directions, examined on all sides; the strength and the weakness of the maxims which men apply to them are fully tested; the light of the brightest minds is diffused to others. Inconsistency is unfolded into self-contradiction; axioms are built up into a system of necessary truths; and ready exemplifications are accumulated of that which is to be proved or disproved concerning the ideas which are the basis of the controversy.

The History of Mechanics from the time of Kepler

to that of Lagrange, is perhaps the best exemplification of the mode in which the progress of a science depends upon such disputes and speculations as give clearness and generality to its elementary conceptions. This, it is to be recollected, is the kind of progress of which we are now speaking; and this is the principal feature in the portion of scientific history which we have mentioned. For almost all that was to be done by reference to observation, was executed by Galileo and his disciples. What remained was the task of generalization and simplification. And this was promoted in no small degree by the various controversies which took place within that period concerning mechanical conceptions:—as, for example, the question concerning the measure of the force of percussion ;—the war of the *vis viva* ;—the controversy of the centre of oscillation ;—of the independence of statics and dynamics ;—of the principle of least action ;—of the evidence of the laws of motion ;—and of the number of laws really distinct. None of these discussions was without its influence in giving generality and clearness to the mechanical ideas of mathematicians: and therefore, though remote from general apprehension, and dealing with very abstract notions, they were of eminent use in the perfecting the science of mechanics. Similar controversies concerning fundamental notions, those, for example, which Galileo himself had to maintain, were no less useful in the formation of the science of hydrostatics. And the like struggles and conflicts, whether they take the form of controversies between several persons, or only operate in the efforts and fluctuations of the discoverer's mind, are always requisite before the conceptions acquire that clearness which makes them fit to appear in the enunciation of scientific truth.

This, then, is one object of the preceding Books ;—to bring under the reader's notice the main elements of the controversies which have thus had so important a share in the formation of the existing body of science, and the decisions on the controverted points to which the mature examination of the subject has led; and thus to give an abundant exhibition of that step which we term the Explication of Conceptions.

6. (II.) The result of such controversies as we have been speaking of, often appears to be summed up in a *Definition;* and the controversy itself has often assumed the form of a battle of definitions. For example, the inquiry concerning the laws of falling bodies led to the question whether the proper definition of a *uniform force* is, that it generates a velocity proportional to the *space* from rest, or to the *time*. The controversy of the *vis viva* was, what was the proper definition of the *measure of force*. A principal question in the classification of minerals is, what is the definition of a *mineral species*. Physiologists have endeavoured to throw light on their subject, by defining *organization*, or some similar term.

7. It is very important for us to observe, that these controversies have never been questions of insulated and *arbitrary* definitions, as men seem often tempted to suppose them to have been. In all cases there is a tacit assumption of some proposition which is to be expressed by means of the definition, and which gives it its importance. The dispute concerning the definition thus acquires a real value, and becomes a question concerning true and false. Thus in the discussion of the question, What is a uniform force? it was taken for granted that gravity is a uniform force:—in the debate of the *vis viva*, it was assumed that in the mutual action of bodies the whole effect of the force is unchanged:—in the zoological definition of species, (that it consists of individuals which have, or may have, sprung from the same parents,) it is presumed that individuals so related resemble each other

more than those which are excluded by such a definition; or perhaps, that species so defined have permanent and definite differences. A definition of organization, or of any other term, which was not employed to express some principle, would be of no value.

The establishment, therefore, of a right definition of a term may be a useful step in the explication of our conceptions; but this will be the case then only when we have under our consideration some proposition in which the term is employed. For then the question really is, how the conception shall be understood and defined in order that the proposition may be true.

8. The establishment of a proposition requires an attention to observed facts, and can never be rightly derived from our conceptions alone. We must hereafter consider the necessity which exists that the facts should be rightly bound together, as well as that our conceptions should be clearly employed, in order to lead us to real knowledge. But we may observe here that, in such cases at least as we are now considering, the two processes are co-ordinate. To unfold our conceptions by the means of definitions has never been serviceable to science, except when it has been associated with an immediate use of the definitions. The endeavour to define a Uniform Force was combined with the assertion that gravity is a uniform force: the attempt to define Accelerating Force was immediately followed by the doctrine that accelerating forces may be compounded: the process of defining Momentum was connected with the principle that momenta gained and lost are equal: naturalists would have given in vain the definition of Species which we have quoted, if they had not also given the characters of species so separated. Definition and Proposition are the two handles of the instrument by which we apprehend truth; the former is of no use without the latter. Defi-

nition may be the best mode of explaining our conception, but that which alone makes it worth while to explain it in any mode, is the opportunity of using it in the expression of truth. When a definition is propounded to us as a useful step in knowledge, we are always entitled to ask what principle it serves to enunciate. If there be no answer to this inquiry, we define and give clearness to our conceptions in vain. While we labour at such a task, we do but light up a vacant room; we sharpen a knife with which we have nothing to cut; we take exact aim, while we load our artillery with blank cartridge; we apply strict rules of grammar to sentences which have no meaning.

If, on the other hand, we have under our consideration a proposition probably established, every step which we can make in giving distinctness and exactness to the terms which this proposition involves, is an important step towards scientific truth. In such cases, any improvement in our definition is a real advance in the explication of our conception. The clearness of our expressions casts a light upon the ideas which we contemplate and convey to others.

9. (III.) But though *definition* may be subservient to a right explication of our conceptions, it is *not essential* to that process. It is absolutely necessary to every advance in our knowledge, that those by whom such advances are made should possess clearly the conceptions which they employ: but it is by no means necessary that they should unfold these conceptions in the words of a formal definition. It is easily seen, by examining the course of Galileo's discoveries, that he had a distinct conception of the *moving force* which urges bodies downwards upon an inclined plane, while he still hesitated whether to call it Momentum, Energy, Impetus, or Force, and did not venture to offer a definition of the thing which was the subject of his thoughts. The con-

ception of *polarization* was clear in the minds of many
optical speculators, from the time of Huyghens and
Newton to that of Young and Fresnel. This conception
we have defined to be "opposite properties depending
upon opposite positions;" but this notion was, by the
discoverers, though constantly assumed and expressed by
means of superfluous hypotheses, never clothed in definite
language. And in the mean time, it was the custom,
among subordinate writers on the same subjects, to say
that the term polarization had no definite meaning, and
was merely an expression of our ignorance. The definition which was offered by Haüy and others of a *mineralogical species;*—"The same elements combined in the
same proportions, with the same fundamental form;"—
was false, inasmuch as it was incapable of being rigorously applied to any one case; but this defect did not
prevent the philosophers who propounded such a definition from making many valuable additions to mineralogical knowledge, in the way of identifying some species
and distinguishing others. The right conception which
they possessed in their minds prevented their being
misled by their own very erroneous definition. The want
of any precise definitions of *strata,* and *formations,* and
epochs, among geologists, has not prevented the discussions which they have carried on upon such subjects from
being highly serviceable in the promotion of geological
knowledge. For however much the apparent vagueness
of these terms might leave their arguments open to cavil,
there was a general understanding prevalent among the
most intelligent cultivators of the science, as to what was
meant in such expressions; and this common understanding sufficed to determine what evidence should be considered conclusive and what inconclusive in these inquiries.
And thus the distinctness of conception, which is a real
requisite of scientific progress, existed in the minds of the

inquirers, although definitions, which are a partial and accidental evidence of this distinctness, had not yet been hit upon. The idea had been developed in men's minds, although a clothing of words had not been contrived for it, nor, perhaps, the necessity of such a vehicle felt: and thus that essential condition of the progress of knowledge of which we are here speaking existed; while it was left to the succeeding speculators to put this unwritten rule in the form of a verbal statute.

10. (IV.) Men are often prone to consider it as a thoughtless *omission* of an essential circumstance, and as a *neglect* which involves some blame, when knowledge thus assumes a form in which definitions, or rather conceptions, are implied but are not expressed. But in such a judgment, they assume *that* to be a matter of choice requiring attention only, which is in fact as difficult and precarious as any other portion of the task of discovery. To define, so that our definition shall have any scientific value, requires no small portion of that sagacity by which truth is detected. As we have already said, definitions and propositions are co-ordinate in their use and in their origin. In many cases, perhaps in most, the proposition which contains a scientific truth, is apprehended with confidence, but with some vagueness and vacillation, before it is put in a positive, distinct, and definite form. It is thus known to be true, before it can be enunciated in terms each of which is rigorously defined. The business of definition is part of the business of discovery. When it has been clearly seen what ought to be our definition, it must be pretty well known what truth we have to state. The definition, as well as the discovery, supposes a decided step in our knowledge to have been made. The writers on Logic in the middle ages, made *Definition* the last stage in the progress of knowledge; and in this arrangement at least, the history of science, and

the philosophy derived from the history, confirm their speculative views. If the explication of our conceptions ever assume the form of a definition, it will not be as an arbitrary process, or as a matter of course, but as the mark of one of those happy efforts of sagacity to which all the successive advances of our knowledge are owing.

11. (V.) Our conceptions, then, even when they become so clear as the progress of knowledge requires, are not adequately expressed, or necessarily expressed at all, by means of definitions. We may ask, then, whether there is any *other mode* of expression in which we may look for the evidence and exposition of that peculiar exactness of thought which the formation of science demands. And in answer to this inquiry, we may refer to the previous discussions respecting many of the fundamental ideas of the sciences. It has there been seen that these ideas involve many elementary truths which enter into the texture of our knowledge, introducing into it connexions and relations of the most important kind, although these elementary truths cannot be deduced from any verbal definition of the idea. It has been seen that these elementary truths may often be enunciated by means of *Axioms,* stated in addition to, or in preference to, Definitions. For example, the Idea of Cause, which forms the basis of the science of mechanics, makes its appearance in our elementary mechanical reasonings, not as a definition, but by means of the axioms that causes are measured by their effects, and that reaction is equal and opposite to action. Such axioms, tacitly assumed or occasionally stated as maxims of acknowledged validity, belong to all the ideas which form the foundations of the sciences, and are constantly employed in the reasoning and speculations of those who think clearly on such subjects. It may often be a task of some difficulty to detect and enunciate in words the principles which are thus, perhaps silently

and unconsciously, taken for granted by those who have a share in the establishment of scientific truth : but inasmuch as these principles are an essential element in our knowledge, it is very important to our present purpose to separate them from the associated materials and to trace them to their origin. This accordingly I have attempted to do, with regard to a considerable number of the most prominent of such ideas, in the preceding Books. The reader will there find many of these ideas resolved into axioms and principles by means of which their effect upon the elementary reasonings of the various sciences may be expressed. That part of the Work is intended to form, in some measure, a representation of the ideal side of our physical knowledge ;—a table of those contents of our conceptions which are not received directly from facts; —an exhibition of rules to which we know that truth must conform.

12. In order, however, that we may see the necessary cogency of these rules, we must possess, clearly and steadily, the ideas from which the rules flow. In order to perceive the necessary relations of the circles of the sphere, we must possess clearly the Idea of solid Space : —in order that we may see the demonstration of the composition of forces, we must have the Idea of Cause moulded into a distinct conception of statical force. This is that *Clearness of Ideas* which we stipulate for in any one's mind, as the first essential condition of his making any new step in the discovery of truth. And we now see what answer we are able to give, if we are asked for a Criterion of this Clearness of Idea. The Criterion is, that the person shall *see* the necessity of the Axioms belonging to each Idea ;—shall accept them in such a manner as to perceive the cogency of the reasonings founded upon them. Thus a person has a clear idea of space who follows the reasonings of geometry and fully apprehends

their conclusiveness. The explication of conceptions, which we are speaking of as an essential part of real knowledge, is the process by which we bring the clearness of our ideas to bear upon the formation of our knowledge. And this is done, as we have now seen, not always, nor generally, nor principally, by laying down a definition of the conception; but by acquiring such a possession of it in our minds as enables, indeed compels us, to admit along with the conception, all the axioms and principles which it necessarily implies, and by which it produces its effect upon our reasonings.

13. (VI.) But in order that we may make any real advance in the discovery of truth, our ideas must not only be clear, they must also be *appropriate*. Each science has for its basis a different class of ideas; and the steps which constitute the progress of one science can never be made by employing the ideas of another kind of science. No genuine advance could ever be obtained in mechanics by applying to the subject the ideas of space and time merely:—no advance in chemistry by the use of mere mechanical conceptions:—no discovery in physiology, by referring facts to mere chemical and mechanical principles. Mechanics must involve the conception of force; —chemistry, the conception of elementary composition; —physiology, the conception of vital powers. Each science must advance by means of its appropriate conceptions. Each has its own field, which extends as far as its principles can be applied. I have already noted the separation of several of these fields by the divisions of the preceding Books. The Mechanical, the Secondary Mechanical, the Chemical, the Classificatory, the Biological Sciences form so many great provinces in the kingdom of knowledge, each in a great measure possessing its own peculiar fundamental principles. Every attempt to build up a new science by the application of principles which

belong to an old one, will lead to frivolous and barren speculations.

This truth has been exemplified in all the instances in which subtle speculative men have failed in their attempts to frame new sciences, and especially in the essays of the ancient schools of philosophy in Greece, as has already been stated in the History of Science. Aristotle and his followers endeavoured in vain to account for the mechanical relation of forces in the lever by applying the *inappropriate* geometrical conceptions of the properties of the circle:—they failed in explaining the *form* of the luminous spot made by the sun shining through a hole, because they applied the *inappropriate* conception of a circular *quality* in the sun's light:—they speculated to no purpose about the elementary composition of bodies, because they assumed the *inappropriate* conception of *likeness* between the elements and the compound, instead of the genuine notion of elements merely *determining* the qualities of the compound. And in like manner, in modern times, we have seen, in the history of the fundamental ideas of the physiological sciences, how all the *inappropriate* mechanical and chemical and other ideas which were applied in succession to the subject failed in bringing into view any genuine physiological truth.

14. That the real cause of the failure in the instances above mentioned lay in the *conceptions*, is plain. It was not ignorance of the facts which in these cases prevented the discovery of the truth. Aristotle was as well acquainted with the fact of the proportion of the weights which balance on a lever as Archimedes was, although Archimedes alone gave the true mechanical reason for the proportion. Aristotle knew that the rays of light are straight lines, although he was not a sufficiently just thinker to apply the conception of rays in explaining the

round image which the sun produces by shining through a triangular hole*.

With regard to the doctrine of the four elements indeed, the inapplicability of the conception of composition of qualities, required, perhaps, to be proved by some reference to facts. But this conception was devised at first, and accepted by succeeding times, in a blind and gratuitous manner, which could hardly have happened if men had been awake to the necessary condition of our knowledge;—that the conceptions which we introduce into our doctrines are not arbitrary or accidental notions, but certain peculiar modes of apprehension strictly determined by the subject of our speculations.

15. (VII.) It may, however, be said that this injunction that we are to employ *appropriate* conceptions only in the formation of our knowledge, cannot be of practical use, because we can only determine what ideas *are* appropriate, by finding that they truly combine the facts. And this is to a certain extent true. Scientific discovery must ever depend upon some happy thought, of which we cannot trace the origin; some fortunate cast of intellect, rising above all rules. No maxims can be given which inevitably lead to discovery. No precepts will elevate a man of ordinary endowments to the level of a man of genius: nor will an inquirer of truly inventive mind need to come to the teacher of inductive philosophy to learn how to exercise the faculties which nature has given him. Such persons as Kepler or Fresnel, or Brewster, will have

---

* The Edinburgh Reviewer of my History (*Ed. Rev.* No. cxxxiii., p. 118,) thinks that Aristotle committed this error because he did not try the experiment under various forms. Yet he cannot but allow that the result follows by mere geometrical reasoning, without any experiment, from the fact which Aristotle did know, that the rays of light are straight lines. Is the Reviewer's mind in that stage of speculation in which the truths of geometry are considered as best proved experimentally?

their powers of discovering truth little augmented by any injunctions respecting distinct and appropriate ideas; and such men may very naturally question the utility of rules altogether.

16. But yet the opinions which such persons may entertain, will not lead us to doubt concerning the value of the attempts to analyse and methodize the process of discovery. Who would attend to Kepler if he had maintained that the speculations of Francis Bacon were worthless? Notwithstanding what has been said, we may venture to assert that the maxim which points out the necessity of ideas appropriate as well as clear, for the purpose of discovering truth, is not without its use. It may, at least, have a value as a caution or prohibition, and may thus turn us away from labours certain to be fruitless. We have already seen that this maxim, if duly attended to, would have at once condemned as wrongly directed the speculations of physiologists of the mathematical, mechanical, chemical, and vital-fluid schools; since the ideas which the teachers of these schools introduce, cannot suffice for the purposes of physiology, which seeks truths respecting the vital powers. Again, it is clear from similar considerations that no definition of a mineralogical species by chemical characters alone can answer the end of science, since we seek to make mineralogy, not an analytical but a classificatory science*. Even before the appropriate conception is matured in men's minds so that they see clearly what it is, they may still have light enough to see what it is not.

17. (VIII.) Another result of this view of the necessity of appropriate ideas, combined with a survey of the

---

* This agrees with what M. Necker has well observed in his "*Règne Mineral*," that those who have treated mineralogy as a merely chemical science, have substituted the analysis of substances for the classification of individuals. See above, b. viii. chap. 3. p. **506**.

history of science is, that though for the most part, as we shall see, the progress of science consists in accumulating and combining facts rather than in debating concerning definitions; there are still certain periods when the *discussion* of Definitions may be the most useful mode of cultivating some special branch of science. This discussion is of course always to be conducted by the light of facts; and as has already been said, along with the settlement of every good Definition will occur the corresponding establishment of some Proposition. But still at particular periods, the want of a Definition, or of the clear conceptions which Definition supposes, may be peculiarly felt. A good and tenable Definition of Species in mineralogy would at present be perhaps the most important step which the science could make. A just conception of the nature of life, (and if expressed by means of a Definition so much the better,) can hardly fail to give its possessor an immense advantage in the speculations which now come under the consideration of physiologists. And controversies respecting Definitions, in these cases and such as these, may be very far from idle and unprofitable.

Thus the knowledge that clear and appropriate ideas are requisite for discovery, although it does not lead to any very precise precepts or supersede the value of natural sagacity and inventiveness, may still be of use to us in our pursuit after truth. It may show us what course of research is, in each stage of science, recommended by the general analogy of the history of knowledge; and it may both save us from hopeless and barren paths of speculation, and make us advance with more courage and confidence, to know that we are looking for discoveries in the manner in which they have always hitherto been made.

18. (IX.) Another consequence follows from the views presented in this Chapter, and it is the last I shall

at present mention. *No scientific discovery* can, with any justice, be considered *due to accident.* In whatever manner facts may be presented to the notice of a discoverer, they can never become the materials of exact knowledge, except they find his mind already provided with precise and suitable conceptions by which they may be analysed and connected. Indeed, as we have already seen, facts cannot be observed as facts, except in virtue of the conceptions which the observer* himself unconsciously supplies; and they are not facts of observation for any purpose of discovery, except these familiar and unconscious acts of thought be themselves of a just and precise kind. But supposing the facts to be adequately observed, they can never be combined into any new truth, except by means of some new conceptions, clear and appropriate, such as I have endeavoured to characterize. When the observer's mind is prepared with such instruments, a very few facts, or it may be a single one, may bring the process of discovery into action. But in such cases, this previous condition of the intellect, and not the single fact, is really the main and peculiar cause of the success. The fact is merely the occasion by which the engine of discovery is brought into play sooner or later. It is, as I have elsewhere said, only the spark which discharges a gun already loaded and pointed; and there is little propriety in speaking of such an accident as the cause why the bullet hits the mark. If it were true that the fall of an apple was the occasion of Newton's pursuing the train of thought which led to the doctrine of universal gravitation, the habits and constitution of Newton's intellect, and not the apple, were the real source of this great event in the progress of knowledge. The common love of the marvellous, and the vulgar desire to bring down the greatest achievements of genius to our

* B. i. c. 2.

own level, may lead men to ascribe such results to any casual circumstances which accompany them; but no one who fairly considers the real nature of great discoveries, and the intellectual processes which they involve, can seriously hold the opinion of their being the effect of accident.

19. Such accidents never happen to common men. Thousands of men, even of the most inquiring and speculative men, had seen bodies fall; but who, except Newton, ever followed the accident to such consequences? And in fact, how little of his train of thought was contained in, or even directly suggested by, the fall of the apple! If the apple fall, said the discoverer, why should not the moon, the planets, the satellites fall? But how much previous thought, what a steady conception of the universality of the laws of motion gathered from other sources, were requisite, that the inquirer should see any connexion in these cases! Was it by accident that he saw in the apple an image of the moon, and of every body in the solar system?

20. The same observations may be made with regard to the other cases which are sometimes adduced as examples of accidental discovery. It has been said*, "By the accidental placing of a rhomb of calcareous spar upon a book or line Bartholinus discovered the property of the *Double Refraction* of light." But Bartholinus could have seen no such consequence in the accident if he had not previously had a clear conception of *single refraction*. A lady, in describing an optical experiment which had been shown her, said of her teacher, "He told me to *increase and diminish the angle of refraction*, and at last I found that he only meant me to move my head up and down." At any rate, till the lady had acquired the notions which the technical terms convey, she could not have made

* *Ed. Rev.*, No. cxxxiii., p. 121.

Bartholinus's discovery by means of his accident. "By accidentally combining two rhombs in different positions," it is added*, "Huyghens discovered the *Polarization* of Light." Supposing that this experiment had been made without design, what Huyghens really observed, was that the images appeared and disappeared alternately as he turned the rhombs round. But was it an easy or an obvious business to analyse this curious alternation into the circumstances of the rays of light having *sides*, as Newton expressed it, and into the additional hypotheses which are implied in the term polarization? Those will be able to answer this question who have found how far from easy it is to understand clearly what is meant by polarization in this case, now that the property is fully established. Huyghens's success depended on his clearness of thought, for this enabled him to perform the intellectual analysis, which never would have occurred to most men, however often they had "accidentally combined two rhombs in different positions." "By accidentally looking through a prism of the same substance, and turning it round, Malus discovered the polarization of light by reflection." Malus saw that, in some positions, the light reflected from the windows of the Louvre thus seen through the prism, became dim. Another man would have supposed this dimness the result of accident; but his mind was differently constituted and disciplined. He considered the position of the window, and of the prism; repeated the experiment over and over; and in virtue of the eminently distinct conceptions of space which he possessed, resolved the phenomena into its geometrical conditions. A believer in accident would not have sought them; a person of less clear ideas would not have found them. A person must have a strange confidence in the virtue of chance, and the worthlessness of intellect,

* *Ed. Rev.*, No. cxxxiii., p. 121.

who can say*, even in the heat of debate, or the reckless-
ness of anonymous criticism, that "in all these fundamental discoveries appropriate ideas had no share," and
that the discoveries "might have been made by the most
ordinary observers."

21. I have now, I trust, shown in various ways, how
the *Explication of Conceptions*, including in this term
their clear development from Fundamental Ideas in the
discoverer's mind, as well as their precise expression in
the form of Definitions or Axioms, when that can be
done, is an essential part in the establishment of all exact
and general physical truths. In doing this, I havee ndea-
voured to explain in what sense the possession of clear
and appropriate ideas is a main requisite for every step
in scientific discovery. That it is far from being the only
step, I shall soon have to show; and if any obscurity
remain on the subject treated of in the present chapter,
it will, I hope, be removed when we have examined the
other elements which enter into the constitution of our
knowledge.

Chapter III.

OF FACTS AS THE MATERIALS OF SCIENCE.

We have now to examine how science is built up
by the combination of facts. In doing this, we suppose
that we have already obtained a supply of definite and
certain facts, free from obscurity and doubt. We must,
therefore, first consider under what conditions facts can
assume this character.

---

* This is said by the Edinburgh Reviewer of my History, in an
attempt to disprove the necessity of clear and appropriate ideas for
discovery. *Ed. Rev.*, No. cxxxiii., p. 122.

When we inquire what facts are to be made the materials of science, perhaps the answer which we should most commonly receive would be, that they must be *true facts*, as distinguished from any mere inferences or opinions of our own. We should probably be told that we must be careful in such a case to consider as facts only what we really observe: that we must assert only what we see; and believe nothing except upon the testimony of our senses.

But such maxims are far from being easy to apply, as a little examination will convince us.

1. It has been explained in the preceding part of this work that all perception of external objects and occurrences involves an active as well as a passive process of the mind;—includes not only sensations, but also ideas by which sensations are bound together, and have a unity given to them. From this it follows that there is a difficulty in separating in our perceptions what we receive from without, and what we ourselves contribute from within;—what we perceive, and what we infer. In many cases, this difficulty is obvious to all: as, for example, when we witness the performances of a juggler or a ventriloquist. In these instances we imagine ourselves to see and to hear what certainly we do not see and hear. The performer takes advantage of the habits by which our minds supply interruptions and infer connexions; and by giving us fallacious indications, he leads us to perceive as an actual fact, what does not happen at all. In these cases it is evident that we ourselves assist in making the fact; for we make one which does not really exist. In other cases, though the fact which we perceive be true, we can easily see that a large portion of the perception is our own act; as when from the sight of a bird of prey we infer a carcass, or when we

read a half-obliterated inscription. In the latter case, the mind supplies the meaning, and perhaps half the letters; yet we do not hesitate to say that we actually read the inscription. Thus, in many cases, our own inferences and interpretations enter into our facts. But this happens in many instances in which it is at first sight less obvious. When any one has seen an oak-tree blown down by a strong gust of wind, he does not think of the occurrence any otherwise than as a fact of which he is assured by his senses. Yet by what sense does he perceive the force which he thus supposes the wind to exert? By what sense does he distinguish an oak-tree from all other trees? It is clear upon reflection, that in such a case, his own mind supplies the conception of extraneous impulse and pressure, by which he thus interprets the motions observed, and the distinction of different kinds of trees according to which he thus names the one under his notice. The idea of force, and the idea of definite resemblances and differences, are thus combined with the impressions on our senses, and form an undistinguished portion of that which we consider as the fact. And it is evident that we can in no other way perceive force, than by seeing motion; and cannot give a name to any object without not only seeing a difference of single objects, but supposing a difference of classes of objects. When we speak as if we saw impulse and attraction, things and classes, we really see only objects of various forms and colours, more or less numerous, variously combined. But do we really perceive so much as this? When we see the form, the size, the number, the motion of objects, are these really mere impressions on our senses, unmodified by any contribution or operation of the mind itself? A very little attention will suffice to convince us that this is not the case. When we see a windmill turning, it may happen, as we have elsewhere noticed*, that we mistake the direction in which the sails turn: when we look at certain diagrams, they may appear either convex or concave: when we see the moon first in the horizon and afterwards high up in the sky, we judge her to be much larger in the former than in the latter position, although to the eye she subtends the same angle. And in these cases and the like, it has been seen that the error and confusion which we thus incur arise from the mixture of acts of the mind itself with impressions on the senses. But such acts are, as we have also seen, inseparable portions of the process of perception. A certain activity of the mind is involved, not only in seeing objects erroneously, but in seeing them at all. With regard to solid objects, this is generally acknowledged. When we seem to see an edifice occupying space in all dimensions, we really see only a representation of it as it appears referred by perspective to a surface. The inference of the solid form is an operation of our own, alike when we look at a reality and when we look at a picture. But we may go further. Is plane figure really a mere sensation? If we look at a decagon, do we see at once that it has ten sides, or is it not necessary for us to count them: and is not counting an act of the mind? All objects are seen in space; all objects are seen as one or many: but are not the idea of space and the idea of number requisite in order that we may thus apprehend what we see? That these ideas of space and number involve a connexion derived from the mind and not from the senses, appears, as we have already seen, from this, that those ideas afford us the materials of universally and necessary truths: such truths as the senses cannot possibly supply. And thus even the perception of such facts as the size, shape, and number of objects, cannot be said to be impressions of sense, distinct from all acts of mind, and cannot be expected to be free from error on the ground of their being mere observed facts.

Thus the difficulty which we have been illustrating, of distinguishing facts from inferences and from interpretations of facts, is not only great, but amounts to an impossibility. The separation at which we aimed in the outset of this discussion, and which was supposed to be necessary in order to obtain a firm groundwork for science, is found to be unattainable. We cannot obtain a sure basis of facts by rejecting all inferences and judgments of our own, for such inferences and judgments form an unavoidable element in all facts. We cannot exclude our ideas from our perceptions, for our perceptions involve our ideas.

2. But still it cannot be doubted that in selecting the facts which are to form the foundation of science, we must reduce them to their most simple and certain form; and must reject everything from which doubt or error may arise. Now since this, it appears, cannot be done, by rejecting the ideas which all facts involve, in what manner are we to conform to the obvious maxim that the facts which form the basis of science must be perfectly definite and certain?

The analysis of facts into Ideas and Sensations, which we have so often referred to, suggests the answer to this inquiry. We are not able, nor need we endeavour, to exclude ideas from our facts, but we may be able to discern, with perfect distinctness, the ideas which we include. We cannot observe any phenomena without applying to them such ideas as space and number, cause and resemblance, and usually several others; but we may avoid applying these ideas in a wavering or obscure manner, and confounding ideas with one another. We cannot read any of the inscriptions which nature presents to us, without interpreting them by means of some language which we ourselves are accustomed to speak, but we may make it our business to acquaint ourselves perfectly with the language which we thus employ, and to interpret it according to the rigorous rules of grammar and analogy.

This maxim, that when facts are employed as the basis of science, we must distinguish clearly the ideas which they involve, and must apply these in a distinct and rigorous manner, will be found to be a more precise guide than we might perhaps at first expect. We may notice one or two Rules which flow from it.

3. In the first place, facts, when used as the materials of physical science, must be *referred to conceptions of the intellect only*, all emotions of fear, admiration, and the like, being rejected or subdued. Thus the observations of phenomena which are related as portents and prodigies, striking terror and boding evil, are of no value for purposes of science. The tales of armies seen warring in the sky, the sound of arms heard from the clouds, fiery dragons, chariots, swords seen in the air, may refer to meteorological phenomena; but the records of phenomena observed in the state of mind which these descriptions imply can be of no scientific value. We cannot make the poets our observers.

> Armorum sonitum toto Germania cœlo
> Audiit; insolitis tremuerunt motibus Alpes.
> Vox quoque per lucos vulgo exaudita silentes
> Ingens, et simulacra modis pallentia miris
> Visa sub obscurum noctis: pecudesque locutæ.

The mixture of fancy and emotion with the observation of facts has often disfigured them to an extent which is too familiar to all to need illustration. We have an example of this result in the manner in which Comets are described in the treatises of the middle ages. In such works, these bodies are regularly distributed into several classes, accordingly as they assume the form of a sword, of a spear, of a cross, and so on. When such resemblances had become matters of interest, the impressions of the senses were governed, not by the rigorous conceptions of form and colour, but by these assumed images; and under these circumstances we can attach little value to the statement of what was seen.

In all such phenomena, the reference of the objects to the exact ideas of space, number, position, motion, and the like, is the first step of science: and accordingly, this reference was made at an early period in those sciences which made an early progress, as, for instance, astronomy. Yet even in astronomy there appears to have been a period when the predominant conceptions of men in regarding the heavens and the stars pointed to mythical story and supernatural influence, rather than to mere relations of space, time, and motion: and of this primeval condition of those who gazed at the stars, we seem to have remnants in the constellations, in the mythological names of the planets, and in the early prevalence of astrology. It was only at a later period, when men had begun to measure the places, or at least to count the revolutions of the stars, that astronomy had its birth.

4. And thus we are led to another Rule :—that in collecting facts which are to be made the basis of science, the facts are to be observed, as far as possible, *with reference to place, figure, number, motion,* and the like conceptions; which depending upon the ideas of space and time, are the most universal, exact, and simple of our conceptions. It was by early attention to these relations in the case of the heavenly bodies, that the ancients formed the science of astronomy: it was by not making precise observations of this kind in the case of terrestrial bodies, that they failed in framing a science of the mechanics of motion. They succeeded in optics as far as they made observations of this nature; but when they ceased to

trace the geometrical paths of rays in the actual experiment, they ceased to go forwards in the knowledge of this subject.

5. But we may state a further Rule:—that though these relations of time and space are highly important in almost all facts, we are not to confine ourselves to these: but are to consider the phenomena *with reference to other conceptions also:* it being always understood that these conceptions are to be made as exact and rigorous as those of geometry and number. Thus the science of Harmonics arose from considering sounds with reference to *concords* and *discords;* the science of Mechanics arose from not only observing motions as they take place in time and space, but further, referring them to *force* as their *cause.* And in like manner other sciences depend upon other Ideas, which, as I have endeavoured to show, are not less fundamental than those of time and space; and like them, capable of leading to rigorous consequences.

6. Thus the facts which we assume as the basis of science are to be freed from all the mists which imagination and passion throw round them; and to be separated into those elementary facts which exhibit simple and evident relations of Time, or Space, or Cause, or some other ideas equally clear. We resolve the complex appearances which nature offers to us, and the mixed and manifold modes of looking at these appearances which rise in our thoughts, into limited, definite, and clearly-understood portions. This process we may term the *Decomposition of Facts.* It is the beginning of exact knowledge,—the first step in the formation of all science. This decomposition of facts into elementary facts, clearly understood and surely ascertained, must precede all discovery of the laws of nature.

7. But though this step is necessary, it is not infallibly

sufficient. It by no means follows that when we have thus decomposed facts into elementary truths of observation, we shall soon be able to combine these so as to obtain truths of a higher and more speculative kind. We have examples which show us how far this is from being a necessary consequence of the former step. Observations of the weather, made and recorded for many years, have not led to any general truths, forming a science of Meteorology: and although great numerical precision has been given to such observations by means of barometers, thermometers, and other instruments, still no general laws regulating the cycles of change of such phenomena have yet been discovered. In like manner the faces of crystals, and the sides of the polygons which these crystals form, were counted, and thus numerical facts were obtained, perfectly true and definite, but still of no value for purposes of science. And when it was discovered what element of the form of crystals it was important to observe and measure, namely, the angle made by two faces with each other, this discovery was a step of a higher order, and did not belong to that department, of mere exact observation of manifest facts, with which we are here concerned.

8. When the complex facts which nature offers to us are thus decomposed into simple facts, the decomposition, in general, leads to the introduction of *terms* and phrases, more or less technical, by which these simple facts are described. Thus when astronomy was thus made a science of measurement, the things measured were soon described as *hours*, and *days*, and *cycles*, *altitude* and *declination*, *phases* and *aspects*. In the same manner in music, the concords had names assigned them, as *diapente, diatessaron, diapason;* in studying optics, the *rays* of light were spoken of as having their course altered by *reflection* and *refraction;* and when useful observations began to be

made in mechanics, the observers spoke of *force, pressure, momentum, inertia,* and the like.

When we take phenomena in which the leading idea is Resemblance, and resolve them into precise component facts, we obtain some kind of classification; as, for instance, when we lay down certain rules by which particular trees, or particular animals are to be known. This is the earliest form of natural history; and the classification which it involves is that which corresponds, nearly or exactly, with the usual names of the objects thus classified.

9. Thus the first attempts to render observation certain and exact, lead to a decomposition of the obvious facts into elementary facts, connected by the ideas of space, time, number, cause, likeness, and others: and into a classification of the simple facts, more or less just, and marked by names either common or technical. Elementary facts, and individual objects, thus observed and classified, form the materials of science; and any improvement in classification or nomenclature, or any discovery of a connexion among the materials thus accumulated, leads us fairly within the precincts of science. We must now, therefore, consider the manner in which science is built up of such materials ;—the process by which they are brought into their places, and the texture of the bond which unites and cements them.

CHAPTER IV.

OF THE COLLIGATION OF FACTS.

FACTS such as the last chapter speaks of are, by means of such conceptions as are described in the preceding chapter, bound together so as to give rise to those general propositions of which science consists. Thus the facts that the planets revolve about the sun in certain periodic times and at certain distances, are included and connected in Kepler's law by means of such conceptions as the *squares of numbers*, the *cubes of distances*, and the *proportionality* of these quantities. Again the existence of this proportion in the motions of any two planets, forms a set of facts which may all be combined by means of the conception of a certain *central accelerating force*, as was proved by Newton. The whole of our physical knowledge consists in the establishment of such propositions; and in all such cases facts are bound together by the aid of suitable conceptions. This part of the formation of our knowledge we have called the *Colligation of Facts*: and we may apply this term to every case in which, by an act of the intellect, we establish a precise connexion among the phenomena which are presented to our senses. The knowledge of such connexions, accumulated and systematized, is Science. On the steps by which science is thus collected from phenomena we shall proceed now to make a few remarks.

1. Science begins with *common* observation of facts, in which we are not conscious of any peculiar discipline or habit of thought exercised in observing. Thus the common perceptions of the appearances and recurrences of the celestial luminaries, were the first steps of astronomy: the obvious cases in which bodies fall or are supported were the beginning of mechanics; the familiar aspects of visible things were the origin of optics; the usual distinctions of well-known plants first gave rise to botany. Facts belonging to such parts of our knowledge are noticed by us, and accumulated in our memories in the common course of our habits, almost without our being aware that we are observing and collecting facts. Yet such facts may lead to many scientific truths; for

instance, in the first stages of Astronomy (as we have shown in the History) such facts lead to Methods of Intercalation and Rules of the Recurrence of Eclipses. In succeeding stages of science, more especial attention and preparation on the part of the observer, and a selection of certain kinds of facts, becomes necessary ; but there is an early period in the progress of knowledge at which man is a physical philosopher without seeking to be so, or being aware that he is so.

2. But in all stages of the progress, even in that early one of which we have just spoken, it is necessary, in order that the facts may be fit materials of any knowledge, that they should be decomposed into elementary facts, and these observed with precision. Thus in the first infancy of astronomy, the recurrence of phases of the moon, of places of the sun's rising and setting, of planets, of eclipses, was observed to take place at intervals of certain definite numbers of days and in a certain exact order; and thus it was that the observations became portions of astronomical science. In other cases, although the facts were equally numerous, and their general aspect equally familiar, they led to no science, because their exact circumstances were not apprehended. A vague and loose mode of looking at facts very easily observable, left men for a long time under the belief that a body ten times as heavy as another falls ten times as fast;—that objects immersed in water are always magnified, without regard to the form of the surface;—that the magnet exerts an irresistible force;—that crystal is always found associated with ice;—and the like. These and many others are examples how blind and careless man can be, even in observation of the plainest and commonest appearances; and they show us that the mere faculties of perception, although constantly exercised upon innumerable objects, may long fail in leading to any exact knowledge.

3. If we further inquire what was the favourable condition through which some special classes of facts were, from the first, fitted to become portions of science, we shall find it to have been principally this;—that these facts were considered with reference to the ideas of time, number, and space, which are ideas possessing peculiar definiteness and precision; so that with regard to them confusion and indistinctness are hardly possible. The interval from new moon to new moon was always a particular number of days: the sun in his yearly course rose and set near to a known succession of distant objects: the moon's path passed among the stars in a certain order: —these are observations in which mistake and obscurity are not likely to occur, if the smallest degree of attention is bestowed upon the task. To count a number is, from the first opening of man's mental faculties, an operation which no science can render more precise. The relations of space are nearest to those of number in obvious and universal evidence. Sciences depending upon these ideas arise with the first dawn of intellectual civilization. But few of the other ideas which man employs in the acquisition of knowledge possess this clearness in their common use. The idea of *resemblance* may be noticed as coming next to those of space and number in original precision; and the idea of *cause*, in a certain more vague and general mode of application, sufficient for the purposes of common life, but not for the ends of science, exercises a very extensive influence over men's thoughts. But the other ideas on which science depends, with the conceptions which arise out of them, are not unfolded till a much later period of intellectual progress; and therefore, except in such limited cases as I have noticed, the observations of common spectators and uncultivated nations, however numerous or varied, are of little or no effect in giving rise to science.

4. Let us now suppose that, besides common every-day perception of facts, we turn our attention to some other occurrences and appearances, with a design of obtaining from them speculative knowledge. This process is more peculiarly called *observation*, or, when we ourselves occasion the facts, *experiment*. But the same remark which we have already made, still holds good here. These facts can be of no value except they are resolved into those exact conceptions which contain the essential circumstances of the case. They must be determined, not indeed necessarily, as has sometimes been said, "according to Number, Weight, and Measure;" for, as we have endeavoured to show in the preceding Books *, there are many other conceptions to which phenomena may be subordinated quite different from these, and yet not at all less definite and precise. But in order that the facts obtained by observation and experiment may be capable of being used in furtherance of our exact and solid knowledge, they must be apprehended and analysed according to some conceptions which, applied for this purpose, give distinct and definite results, such as can be steadily taken hold of and reasoned from; that is, they must be referred to clear and appropriate ideas, according to the manner in which we have already explained this condition of the derivation of our knowledge. The phenomena of light, when they are such as to indicate sides in the ray, must be referred to the conception of *polarization*; the phenomena of mixture, when there is an alteration of qualities as well as quantities, must be combined by a conception of *elementary composition*. And thus when mere position, and number, and resemblance, will no longer answer the purpose of enabling us to connect the facts, we call in other ideas, in such cases more efficacious, though less obvious.

* Books v., vi., vii., viii., ix., x.

5. But how are we in these cases to discover such ideas, and to judge which will be efficacious, in leading to a scientific combination of our experimental data? To this question we must in the first place answer, that the first and great instrument by which facts, so observed with a view to the formation of exact knowledge, are combined into important and permanent truths, is that peculiar sagacity which belongs to the genius of a discoverer; and which, while it supplies those distinct and appropriate conceptions which lead to its success, cannot be limited by rules, or expressed in definitions. It would be difficult or impossible to describe in words the habits of thought which led Archimedes to refer the conditions of equilibrium on the lever to the conception of pressure, while Aristotle could not see in them anything more than the results of the strangeness of the properties of the circle;—or which impelled Pascal to explain by means of the conception of the weight of air the fact which his predecessors had connected by the notion of nature's horror of a vacuum;—or which caused Vitellio and Roger Bacon to refer the magnifying power of a convex lens to the bending of the rays of light towards the perpendicular by refraction, while others conceived the effect to result from the matter of medium, with no consideration of its form. These are what are commonly spoken of as felicitous and inexplicable strokes of inventive talent; and such, no doubt, they are. No rules can ensure to us similar success in new cases; or can enable men who do not possess similar endowments to make like advances in knowledge.

6. Yet still we may do something in tracing the process by which such discoveries are made; and this it is here our business to do. We may observe that these, and the like discoveries, are not improperly described as happy *guesses*; and that guesses, in these as in other

instances, imply various suppositions made, of which some one turns out to be the right one. We may, in such cases, conceive the discoverer as inventing and trying many conjectures, till he finds one which answers the purpose of combining the scattered facts into a single rule. The discovery of general truths from special facts is performed, commonly at least, and more commonly than at first appears, by the use of a series of suppositions, or *hypotheses*, which are looked at in quick succession, and of which the one which really leads to truth is rapidly detected, and when caught sight of, firmly held, verified, and followed to its consequences. In the minds of most discoverers, this process of invention, trial, and acceptance or rejection of the hypothesis, goes on so rapidly that we cannot trace it in its successive steps. But in some instances we can do so; and we can also see that the other examples of discovery do not differ essentially from these. The same intellectual operations take place in other cases, although this often happens so instantaneously that we lose the trace of the progression. In the discoveries made by Kepler, we have a curious and memorable exhibition of this process in its details. Thanks to his communicative disposition, we know that he made nineteen hypotheses with regard to the motion of Mars, and calculated the results of each, before he established the true doctrine, that the planet's path is an ellipse. We know, in like manner, that Galileo made wrong suppositions respecting the laws of falling bodies, and Mariotte, concerning the motion of water in a siphon, before they hit upon the correct view of these cases.

7. But it has very often happened in the history of science, that the erroneous hypotheses which preceded the discovery of the truth have been made, not by the discoverer himself, but by his precursors; to whom he thus owed the service, often an important one in such

cases, of exhausting the most tempting forms of error. Thus the various fruitless suppositions by which Kepler endeavoured to discover the law of refraction, led the way to its real detection by Snell; Kepler's numerous imaginations concerning the forces by which the celestial motions are produced,—his " physical reasonings " as he termed them,—were a natural prelude to the truer physical reasonings of Newton. The various hypotheses by which the suspension of vapour in air had been explained, and their failure, left the field open for Dalton with his doctrine of the mechanical mixture of gases. In most cases, if we could truly analyse the operation of the thoughts of those who make, or who endeavour to make discoveries in science, we should find that many more suppositions pass through their minds than those which are expressed in words; many a possible combination of conceptions is formed and soon rejected. There is a constant invention and activity, a perpetual creating and selecting power at work, of which the last results only are exhibited to us. Trains of hypotheses are called up and pass rapidly in review; and the judgment makes its choice from the varied group.

8. It would, however, be a great mistake to suppose that the hypotheses, among which our choice thus lies, are constructed by an enumeration of obvious cases, or by a wanton alteration of relations which occur in some first hypothesis. It may, indeed, sometimes happen that the proposition which is finally established is such as may be formed, by some slight alteration, from those which are justly rejected. Thus Kepler's elliptical theory of Mars's motions involved relations of lines and angles much of the same nature as his previous false suppositions : and the true law of refraction so much resembles those erroneous ones which he tried, that we cannot help wondering how he chanced to miss it. But it more frequently

happens that new truths are brought into view by the application of new ideas, not by new modifications of old ones. The cause of the properties of the lever was learnt, not by introducing any new geometrical combination of lines and circles, but by referring the properties to genuine mechanical conceptions. When the motions of the planets were to be explained, this was done, not by merely improving the previous notions, of cycles of time, but by introducing the new conception of epicycles in space. The doctrine of the four simple elements was expelled, not by forming any new scheme of elements which should impart, according to new rules, their sensible qualities to their compounds, but by considering the elements of bodies as neutralizing each other. The fringes of shadows could not be explained by ascribing new properties to the single rays of light, but were reduced to law by referring them to the interference of several rays.

Since the true supposition is thus very frequently something altogether diverse from all the obvious conjectures and combinations, we see here how far we are from being able to reduce discovery to rule, or to give any precepts by which the want of real invention and sagacity shall be supplied. We may warn and encourage these faculties when they exist, but we cannot create them, or make great discoveries when they are absent.

9. The conceptions which a true theory requires are very often clothed in a *hypothesis* which connects with them several superfluous and irrelevant circumstances. Thus the conception of the polarization of light was originally represented under the image of *particles* of light having their poles all turned in the same direction. The laws of heat may be made out perhaps most conveniently by conceiving heat to be a *fluid*. The attraction of gravitation might have been successfully applied to the explanation of facts, if Newton had throughout treated attraction as the

result of an *ether* diffused through space; a supposition which he has noticed as a possibility. The doctrine of definite and multiple proportions may be conveniently expressed by the hypothesis of *atoms*. In such cases the hypothesis may serve at first to facilitate the introduction of a new conception. Thus a pervading ether might for a time remove a difficulty, which some persons find considerable, of imagining a body to exert force at a distance. A particle with poles is more easily conceived than polarization in the abstract. And if hypotheses thus employed will really explain the facts by means of a few simple assumptions, the laws may afterwards be reduced to a simpler form than that in which they were first suggested. The general laws of heat, of attraction, of polarization, of multiple proportions, are now certain, whatever image we may form to ourselves of their ultimate causes.

10. In order, then, to discover scientific truths, suppositions consisting either of new conceptions, or of new combinations of old ones, are to be made, till we find one which succeeds in binding together the facts. But how are we to find this? How is the trial to be made? What is meant by success in these cases? To this we reply, that our inquiry must be, whether the facts have the same relation in the hypothesis which they have in reality;—whether the results of our suppositions agree with the phenomena which nature presents to us. For this purpose we must both carefully observe the phenomena, and steadily trace the consequences of our assumptions, till we can bring the two into comparison. The Conceptions which our hypotheses involve, being derived from certain Fundamental Ideas, afford a basis of rigorous reasoning, as we have shown in the Books respecting those Ideas. And the results to which this reasoning leads will be susceptible of being verified or contradicted by observation of the facts.

Thus the Epicyclical Theory of the Moon, once assumed, determined what the moon's place among the stars ought to be at any given time, and could therefore be tested by actually observing the moon's places. The doctrine that musical strings of the same length stretched with weights of 1, 4, 9, 16, would give the musical intervals of an octave, a fifth, a fourth, in succession, could be put to the trial by any one whose ear was capable of appreciating those intervals: and the inference which follows from this doctrine by numerical reasoning,—that there must be certain imperfections in the concords of every musical scale,—could in like manner be confirmed by trying various modes of *Temperament.* In like manner all received theories in science, up to the present time, have been established by taking up some supposition, and comparing it, directly or by means of its remoter consequences, with the facts it was intended to embrace. Its agreement, under certain cautions and conditions, of which we may hereafter speak, is held to be the evidence of its truth. It answers its genuine purpose, the colligation of facts.

11. When we have, in any subject, succeeded in one attempt of this kind, and obtained some true bond of unity by which the phenomena are held together, the subject is open to further prosecution; which ulterior process may, for the most part, be conducted in a more formal and technical manner. The first great outline of the subject is drawn; and the finishing of the resemblance of nature demands a more minute pencilling, but perhaps requires less of genius in the master. In the pursuance of this task, rules and precepts may be given, and features and leading circumstances pointed out, of which it may often be useful to the inquirer to be aware.

Before proceeding further, I shall speak of some characteristic marks which belong to such scientific processes as are now the subject of our consideration, and which may sometimes aid us in determining when the task has been rightly executed.

CHAPTER V.

OF CERTAIN CHARACTERISTICS OF SCIENTIFIC INDUCTION.

1. THE two operations spoken of in the preceding chapters,—the Explication of the Conceptions of our own minds, and the Colligation of observed Facts by the aid of such Conceptions,—are, as we have just said, inseparably connected with each other. When united, and employed in collecting knowledge from the phenomena which the world presents to us, they constitute the mental process of *Induction;* which is usually and justly spoken of as the genuine source of all our *real general knowledge* respecting the external world. And we see, from the preceding analysis of this process into its two constituents, from what origin it derives each of its characters. It is *real*, because it arises from the combination of real facts, but it is *general*, because it implies the possession of general ideas. Without the former, it would not be knowledge of the external world; without the latter, it would not be knowledge at all. When Ideas and Facts are separated from each other, the neglect of facts gives rise to empty speculations, idle subtleties, visionary inventions, false opinions concerning the laws of phenomena, disregard of the true aspect of nature: while the want of ideas leaves the mind overwhelmed, bewildered, and stupified by particular sensations, with no means of connecting the past with the future, the absent with the present, the example with the rule; open to the impression of all

appearances, but capable of appropriating none. Ideas are the *Form*, facts the *Material*, of our structure. Knowledge does not consist in the empty mould, or in the brute mass of matter, but in the rightly-moulded substance. Induction gathers general truths from particular facts;— and in her harvest, the corn and the reaper, the solid ears and the binding band, are alike requisite. All our know-ledge of nature is obtained by Induction; the term being understood according to the explanation we have now given. And our knowledge is then most complete, then most truly deserves the name of Science, when both its elements are most perfect;—when the Ideas which have been concerned in its formation have, at every step, been clear and consistent;—and when they have, at every step also, been employed in binding together real and certain Facts. Of such Induction I have already given so many examples and illustrations in the two preceding chapters, that I need not now dwell further upon the subject.

2. Induction is familiarly spoken of as the process by which we collect a *general proposition* from a number of *particular cases:* and it appears to be frequently imagined that the general proposition results from a mere juxta-position of the cases, or at most, from merely conjoining and extending them. But if we consider the process more closely, as exhibited in the cases lately spoken of, we shall perceive that this is an inadequate account of the matter. The particular facts are not merely brought together, but there is a new element added to the combi-nation by the very act of thought by which they are com-bined. There is a conception of the mind introduced in the general proposition, which did not exist in any of the observed facts. When the Greeks, after long observing the motions of the planets, saw that these motions might be rightly considered as produced by the motion of one wheel revolving in the inside of another wheel, these

wheels were creations of their minds, added to the facts which they perceived by sense. And even if the wheels were no longer supposed to be material, but were reduced to mere geometrical spheres or circles, they were not the less products of the mind alone,—something additional to the facts observed. The same is the case in all other discoveries. The facts are known, but they are insulated and unconnected, till the discoverer supplies from his own stores a principle of connexion. The pearls are there, but they will not hang together till some one provides the string. The distances and periods of the planets were all so many separate facts; by Kepler's Third Law they are connected into a single truth: but the conceptions which this law involves were supplied by Kepler's mind, and without these, the facts were of no avail. The planets described ellipses round the sun, in the contemplation of others as well as of Newton; but Newton conceived the deflection from the tangent in these elliptical motions in a new light,—as the effect of a central force following a certain law; and then it was that such a force was discovered truly to exist.

Thus* in each inference made by Induction, there is introduced some general conception, which is given, not by the phenomena, but by the mind. The conclusion is not contained in the premises, but includes them by the introduction of a new generality. In order to obtain our inference, we travel beyond the cases which we have before us; we consider them as mere exemplifications of some ideal case in which the relations are complete and intelligible. We take a standard, and measure the facts by it; and this standard is constructed by us, not offered by Nature. We assert, for example, that a body left to itself will move on with unaltered velocity;

* I repeat here remarks made at the end of the *Mechanical Euclid*, p. 178.

not because our senses ever disclosed to us a body doing this, but because (taking this as our ideal case) we find that all actual cases are intelligible and explicable by means of the Conception of *Forces*, causing change and motion, and exerted by surrounding bodies. In like manner, we see bodies striking each other, and thus moving and stopping, accelerating and retarding each other: but in all this, we do not perceive by our senses that abstract quantity *Momentum*, which is always lost by one body as it is gained by another. This Momentum is a creation of the mind, brought in among the facts, in order to convert their apparent confusion into order, their seeming chance into certainty, their perplexing variety into simplicity. This the Conception of *Momentum gained and lost* does: and in like manner, in any other case in which a truth is established by Induction, some Conception is introduced, some Idea is applied, as the means of binding together the facts, and thus producing the truth.

3. Hence in every inference by Induction there is some Conception *superinduced* upon the Facts: and we may henceforth conceive this to be the peculiar import of the term *Induction*. I am not to be understood as asserting that the term was originally or anciently employed with this notion of its meaning; for the peculiar feature just pointed out in Induction has generally been overlooked. This appears by the accounts generally given of Induction. "Induction," says Aristotle*, "is when by means of one extreme term† we infer the other extreme term to be true of the middle term." Thus, (to take such exemplifications as belong to our subject,) from

---

* *Analyt. Prior.*, lib. ii., c. 23. Περὶ τῆς ἐπαγωγῆς.

† The syllogism here alluded to would be this:—
> Mercury, Venus, Mars, describe ellipses about the Sun;
> All Planets do what Mercury, Venus, Mars, do;
> Therefore all Planets describe ellipses about the Sun.

knowing that Mercury, Venus, Mars, describe ellipses about the Sun, we infer that all Planets describe ellipses about the Sun. In making this inference syllogistically, we assume that the evident proposition, " Mercury, Venus, Mars, do what all Planets do," may be taken *conversely,* " All Planets do what Mercury, Venus, Mars, do." But we remark that, in this passage, Aristotle (as was natural in his line of discussion) turns his attention entirely to the *evidence* of the inference; and overlooks a step which is of far more importance to our knowledge, namely, the *invention* of the second extreme term. In the above instance, the particular luminaries, Mercury, Venus, Mars, are one logical *extreme;* the general designation Planets is the *middle term;* but having these before us, how do we come to think of *description of ellipses,* which is the other extreme of the syllogism? When we have once invented this " second extreme term," we may, or may not, be satisfied with the evidence of the syllogism; we may, or may not, be convinced that, so far as this property goes, the extremes are co-extensive with the middle term*; but the *statement* of the syllogism is the important step in science. We know how long Kepler laboured, how hard he fought, how many devices he tried, before he hit upon this *term,* the elliptical motion. He rejected, as we know, many other "second extreme terms," for example, various combinations of epicyclical constructions, because they did not represent with sufficient accuracy the special facts of observation. When he had established his pre-miss, that " Mars does describe an ellipse about the Sun," he does not hesitate to *guess* at least that, in this respect, he might *convert* the other premiss, and assert that " All the Planets do what Mars does." But the main business was, the inventing and verifying the proposition respect-

---

* Εἰ οὖν ἀντιστρέφει τὸ Γ τῷ Β καὶ μὴ ὑπερτείνει τὸ μέσον.—ARISTOT. *Ibid.*

ing the ellipse.   The Invention of the Conception was the great step in the *discovery;* the Verification of the Proposition was the great step in the *proof* of the discovery.   If Logic consists in pointing out the conditions of proof, the Logic of Induction must consist in showing what are the conditions of proof in such inferences as this: but this subject must be pursued in the next chapter; I now speak principally of the act of *invention* which is requisite in every inductive inference.

4. Although in every inductive inference an act of invention is requisite, the act soon slips out of notice. Although we bind together facts by superinducing upon them a new conception, this conception, once introduced and applied, is looked upon as inseparably connected with the facts, and necessarily implied in them.   Having once had the phenomena bound together in their minds in virtue of the conception, men can no longer easily restore them back to the detached and incoherent condition in which they were before they were thus combined.   The pearls once strung, they seem to form a chain by their nature.   Induction has given them a unity which it is so far from costing us an effort to preserve, that it requires an effort to imagine it dissolved.   For instance, we usually represent to ourselves the earth as round, the earth and the planets as revolving about the sun, and as drawn to the sun by a central force; we can hardly understand how it could cost the Greeks, and Copernicus, and Newton so much pains and trouble to arrive at a view which is to us so familiar.   These are no longer to us conceptions caught hold of and kept hold of by a severe struggle; they are the simplest modes of conceiving the facts: they are really facts.   We are willing to *own* our obligation to those discoverers, but we hardly *feel* it: for in what other manner (we ask in our thoughts,) could we represent the facts to ourselves?

Thus we see why it is that this step of which we now speak, the invention of a new Conception in every inductive inference, is so generally overlooked that it has hardly been noticed by preceding philosophers. When once performed by the discoverer, it takes a fixed and permanent place in the understanding of every one. It is a thought which, once breathed forth, permeates all men's minds. All fancy they nearly or quite knew it before. It oft was thought, or almost thought, though never till now expressed. Men accept it and retain it, and know it cannot be taken from them, and look upon it as their own. They will not and cannot part with it, even though they may deem it trivial and obvious. It is a secret, which once uttered, cannot be recalled, even though it be despised by those to whom it is imparted. As soon as the leading term of a new theory has been pronounced and understood, all the phenomena change their aspect. There is a standard to which we cannot help referring them. We cannot fall back into the helpless and bewildered state in which we gazed at them when we possessed no principle which gave them unity. Eclipses arrive in mysterious confusion: the notion of a *Cycle* dispels the mystery. The Planets perform a tangled and mazy dance; but *Epicycles* reduce the maze to order. The Epicycles themselves run into confusion; the conception of an *Ellipse* makes all clear and simple. And thus from stage to stage, new elements of intelligible order are introduced. But this intelligible order is so completely adopted by the human understanding, as to seem part of its texture. Men ask whether Eclipses follow a Cycle; whether the Planets describe Ellipses; and they imagine that so long as they do not *answer* such questions rashly, they take nothing for granted. They do not recollect how much they assume in *asking* the question:—how far the conceptions of Cycles and of Ellipses

are beyond the visible surface of the celestial phenomena:
—how many ages elapsed, how much thought, how much
observation, were needed, before men's thoughts were
fashioned into the words which they now so familiarly
use. And thus they treat the subject, as we have seen
Aristotle treating it; as if it were a question, not of
invention, but of proof; not of substance, but of form:
as if the main thing were not *what* we assert, but *how* we
assert it. But for our purpose it is requisite to bear in
mind the feature which we have thus attempted to mark;
and to recollect that in every inference by induction,
there is a Conception supplied by the mind and superinduced upon the Facts.

5. In collecting scientific truths by Induction we
often find (as has already been observed,) a Definition
and a Proposition established at the same time,—introduced together and mutually dependent on each other.
The combination of the two constitutes the Inductive act;
and we may consider the Definition as representing the
superinduced Conception, and the Proposition as exhibiting the Colligation of Facts.

6. To discover a conception of the mind which will
justly represent a train of observed facts is, in some measure, a process of conjecture, as I have stated already;
and as I then observed, the business of conjecture is
commonly conducted by calling up before our minds
several suppositions, and selecting that one which most
agrees with what we know of the observed facts. Hence
he who has to discover the laws of nature may have to
invent many suppositions before he hits upon the right
one; and among the endowments which lead to his success, we must reckon that fertility of invention which
ministers to him such imaginary schemes, till at last he
finds the one which conforms to the true order of
nature. A facility in devising hypotheses, therefore, is

so far from being a fault in the intellectual character of a discoverer, that it is, in truth, a faculty indispensable to his task. It is, for his purposes, much better that he should be too ready in contriving, too eager in pursuing systems which promise to introduce law and order among a mass of unarranged facts, than that he qhould be barren of such inventions and hopeless of such success. Accordingly, as we have already noticed, great discoverers have often invented hypotheses which would not answer to all the facts, as well as those which would; and have fancied themselves to have discovered laws, which a more careful examination of the facts overturned.

The tendencies of our speculative nature*, carrying us onwards in pursuit of symmetry and rule, and thus producing all true theories, perpetually show their vigour by overshooting the mark. They obtain something, by aiming at much more. They detect the order and connexion which exist, by conceiving imaginary relations of order and connexion which have no existence. Real discoveries are thus mixed with baseless assumptions; profound sagacity is combined with fanciful conjecture; not rarely, or in peculiar instances, but commonly, and in most cases; probably in all, if we could read the thoughts of discoverers as we read the books of Kepler. To try wrong guesses is, with most persons, the only way to hit upon right ones. The character of the true philosopher is, not that he never conjectures hazardously, but that his conjectures are clearly conceived, and brought into rigid contact with facts. He sees and compares distinctly the ideas and

---

* I here take the liberty of characterizing inventive minds in general in the same phraseology which, in the History of Science, I have employed in reference to particular examples. These expressions are what I have used in speaking of the discoveries of Copernicus.—*Hist. Ind. Sci.*, vol. i. p. 373.

the things;—the relations. of his notions to each other and to phenomena. Under these conditions it is not only excusable, but necessary for him, to snatch at every semblance of general rule,—to try all promising forms of simplicity and symmetry.

Hence advances in knowledge* are not commonly made without the previous exercise of some boldness and license in guessing. The discovery of new truths requires, undoubtedly, minds careful and scrupulous in examining what is suggested; but it requires, no less, such as are quick and fertile in suggesting. What is invention, except the talent of rapidly calling before us the many possibilities, and selecting the appropriate one? It is true that when we have rejected all the inadmissible suppositions, they are often quickly forgotten; and few think it necessary to dwell on these discarded hypotheses, and on the process by which they were condemned. But all who discover truths must have reasoned upon many errors to obtain each truth; every accepted doctrine must have been one chosen out of many candidates. If many of the guesses of philosophers of bygone times now appear fanciful and absurd because time and observation have refuted them, others, which were at the time equally gratuitous, have been confirmed in a manner which makes them appear marvellously sagacious. To form hypotheses, and then to employ much labour and skill in refuting, if they do not succeed in establishing them, is a part of the usual process of inventive minds. Such a proceeding belongs to the *rule* of the genius of discovery, rather than (as has often been taught in modern times) to the *exception*.

7. But if it be an advantage for the discoverer of

* These observations are made on occasion of Kepler's speculations, and are illustrated by reference to his discoveries —*Hist. Ind. Sci.*, vol. i. p. 411—414.

truth that he be ingenious and fertile in inventing hypotheses which may connect the phenomena of nature, it is indispensably requisite that he be diligent and careful in comparing his hypotheses with the facts, and ready to abandon his invention as soon as it appears that it does not agree with the course of actual occurrences. This constant comparison of his own conceptions and supposition with observed facts under all aspects, forms the leading employment of the discoverer: this candid and simple love of truth, which makes him willing to suppress the most favourite production of his own ingenuity as soon as it appears to be at variance with realities, constitutes the first characteristic of his temper. He must have neither the blindness which cannot, nor the obstinacy which will not, perceive the discrepancy of his fancies and his facts. He must allow no indolence, or partial views, or self-complacency, or delight in seeming demonstration, to make him tenacious of the schemes which he devises, any further than they are confirmed by their accordance with nature. The framing of hypotheses is, for the inquirer after truth, not the end, but the beginning of his work. Each of his systems is invented, not that he may admire it and follow it into all its consistent consequences, but that he may make it the occasion of a course of active experiment and observation. And if the results of this process contradict his fundamental assumptions, however ingenious, however symmetrical, however elegant his system may be, he rejects it without hesitation. He allows no natural yearning for the offspring of his own mind to draw him aside from the higher duty of loyalty to his sovereign, Truth: to her he not only gives his affections and his wishes, but strenuous labour and scrupulous minuteness of attention.

We may refer to what we have said of Kepler, Newton, and other eminent philosophers, for illustrations of

this character.   In Kepler we have remarked* the
courage and perseverance with which he undertook and
executed the task of computing his own hypotheses:
and, as a still more admirable characteristic, that he never
allowed the labour he had spent upon any conjecture to
produce any reluctance in abandoning the hypothesis, as
soon as he had evidence of its inaccuracy.   And in the
history of Newton's discovery that the moon is retained
in her orbit by the force of gravity, we have noticed the
same moderation in maintaining the hypothesis, after it
had once occurred to the author's mind.   The hypothesis
required that the moon should fall from the tangent of
her orbit every second through a space of sixteen feet;
but according to his first calculations it appeared that in
fact she only fell through a space of thirteen feet in that
time.   The difference seems small, the approximation
encouraging, the theory plausible; a man in love with
his own fancies would readily have discovered or invented
some probable cause of the difference.   But Newton
acquiesced in it as a disproof of his conjecture, and "laid
aside at that time any further thoughts of this matter†."

8. It has often happened that those who have undertaken to instruct mankind have not possessed this pure
love of truth and comparative indifference to the main-
tenance of their own inventions.   Men have frequently
adhered with great tenacity and vehemence to the hypotheses which they have once framed; and in their affection for these, have been prone to overlook, to distort,
and to misinterpret facts.   In this manner hypotheses
have so often been prejudicial to the genuine pursuit of
truth, that they have fallen into a kind of obloquy; and
have been considered as dangerous temptations and fallacious guides.   Many warnings have been uttered against
the fabrication of hypotheses by those who profess to

---

* *Hist. Ind. Sci.*, i., 414.   † *Ib.*, ii., 160.

teach philosophy; many disclaimers of such a course by those who cultivate science.

Thus we shall find Bacon frequently discommending this habit, under the name of "anticipation of the mind," and Newton thinks it necessary to say emphatically "hypotheses non fingo." It has been constantly urged that the inductions by which sciences are formed must be *cautious* and *rigorous;* and the various imaginations which passed through Kepler's brain, and to which he has given utterance, have been blamed or pitied as lamentable instances of an unphilosophical frame of mind. Yet it has appeared in the preceding remarks that hypotheses rightly used are among the helps, far more than the dangers, of science;—that scientific induction is not a "cautious" or a "rigorous" process in the sense of *abstaining from* such suppositions, but in *not adhering to* them till they are confirmed by fact, and in carefully seeking from facts confirmation or refutation. Kepler's character was, not that he was peculiarly given to the construction of hypotheses, but that he narrated with extraordinary copiousness and candour the course of his thoughts, his labours, and his feelings. In the minds of most persons, as we have said, the inadmissible suppositions, when rejected, are soon forgotten : and thus the trace of them vanishes from the thoughts, and the successful hypothesis alone holds its place in our memory. But in reality, many other transient suppositions must have been made by all discoverers;—hypotheses which are not afterwards asserted as true systems, but entertained for an instant;—" tentative hypotheses," as they have been called. Each of these hypotheses is followed by its corresponding train of observations, from which it derives its power of leading to truth. The hypothesis is like the captain, and the observations like the soldiers of an army : while he appears to command them, and in this way to work his own will, he does in fact derive all his power of conquest from their obedience, and becomes helpless and useless if they mutiny.

Since the discoverer has thus constantly to work his way onwards by means of hypotheses, false and true, it is highly important for him to possess talents and means for rapidly *testing* each supposition as it offers itself. In this as in other parts of the work of discovery, success has in general been mainly owing to the native ingenuity and sagacity of the discoverer's mind. Yet some rules tending to further this object have been delivered by eminent philosophers, and some others may perhaps be suggested. Of these we shall here notice only some of the most general, leaving for a future chapter the consideration of some more limited and detailed processes by which, in certain cases, the discovery of the laws of nature may be materially assisted.

9. A maxim which it may be useful to recollect is this;—that hypotheses may often be of service to science, when they involve a certain portion of incompleteness, and even of error. The object of such inventions is to bind together facts which without them are loose and detached; and if they do this, they may lead the way to a perception of the true rule by which the phenomena are associated together, even if they themselves somewhat misstate the matter. The imagined arrangement enables us to contemplate as a whole a collection of special cases which perplex and overload our minds when they are considered in succession ; and if our scheme has so much of truth in it as to conjoin what is really connected, we may afterwards duly correct or limit the mechanism of this connexion. If our hypothesis renders a reason for the agreement of cases really similar, we may afterwards find this reason to be false, but we shall be able to translate it into the language of truth.

A conspicuous example of such an hypothesis, one which was of the highest value to science, though very incomplete, and as a representation of nature altogether false, is seen in the *doctrine of epicycles* by which the ancient astronomers explained the motions of the sun, moon, and planets. This doctrine connected the places and velocities of these bodies at particular times in a manner which was, in its general features, agreeable to nature. Yet this doctrine was erroneous in its assertion of the circular nature of all the celestial motions, and in making the heavenly bodies revolve round the earth. It was, however, of immense value to the progress of astronomical science; for it enabled men to express and reason upon many important truths which they discovered respecting the motion of the stars, up to the time of Kepler. Indeed we can hardly imagine that astronomy could, in its outset, have made so great a progress under any other form, as it did in consequence of being cultivated in this shape of the incomplete and false epicyclical hypothesis.

We may notice another instance of an exploded hypothesis, which is generally mentioned only to be ridiculed, and which undoubtedly is both false in the extent of its assertion, and unphilosophical in its expression; but which still, in its day, was not without merit. I mean the doctrine of *Nature's horror of a vacuum* (*fuga vacui,*) by which the action of siphons and pumps and many other phenomena were explained, till Mersenne and Pascal taught a truer doctrine. This hypothesis was of real service; for it brought together many facts which really belong to the same class, although they are very different in their first aspect. A scientific writer of modern times* appears to wonder that men did not at once divine the weight of the air from which the phenomena formerly

---

* DELUC, *Modifications de l'Atmosphere*, partie i.

ascribed to the *fuga vacui* really result.   "Loaded, com-
pressed by the atmosphere," he says, "they did not recognize its action.   In vain all nature testified that air was
elastic and heavy; they shut their eyes to her testimony.
The water rose in pumps and flowed in siphons at that
time as it does at this day.   They could not separate the
boards of a pair of bellows of which the holes were
stopped; and they could not bring together the same
boards without difficulty if they were at first separated.
Infants sucked the milk of their mothers; air entered
rapidly into the lungs of animals at every inspiration;
cupping-glasses produced tumours on the skin; and in
spite of all these striking proofs of the weight and elasticity of the air, the ancient philosophers maintained
resolutely that air was light, and explained all these
phenomena by the horror which they said nature had for
a vacuum."   It is curious that it should not have occurred
to the author while writing this, that if these facts, so
numerous and various, can all be accounted for by *one*
principle, there is a strong presumption that the principle
is not altogether baseless.   And in reality is it not true
that nature *does* abhor a vacuum, and do all she can to
avoid it?   No doubt this power is not unlimited; and
we can trace it to a mechanical cause, the pressure of the
circumambient air.   But the tendency, arising from this
pressure, which the bodies surrounding a space void of
air have to rush into it, may be expressed, in no extravagant or unintelligible manner, by saying that nature
has a repugnance to a vacuum.

That imperfect and false hypotheses, though they
may thus explain *some* phenomena, and may be useful in
the progress of science, cannot explain *all* phenomena;
—and that we are never to rest in our labours or acquiesce in our results, till we have found some view of the
subject which *is* consistent with *all* the observed facts:—

will of course be understood.   We shall afterwards have
to speak of the other steps of such a progress.

10. The hypotheses which we accept ought to explain
phenomena which we have observed.   But they ought to
do more than this: they ought to *foretel* phenomena
which have not yet been observed;—at least all of the
same kind as those which the hypothesis was invented to
explain.   For our assent to the hypothesis implies that
it is held to be true of all particular instances.   That
these cases belong to past or to future times, that they
have or have not already occurred, makes no difference
in the applicability of the rule to them.   Because the
rule prevails, it includes all cases; and will determine
them all, if we can only calculate its real consequences.
Hence it will predict the results of new combinations, as
well as explain the appearances which have occurred in
old ones.   And that it does this with certainty and cor-
rectness, is one mode in which the hypothesis is to be
verified as right and useful.

The scientific doctrines which have at various periods
been established have been verified in this manner.   For
example, the Epicyclical Theory of the heavens was confirmed by its *predicting* truly eclipses of the sun and
moon, configurations of the planets, and other celestial
phenomena; and by its leading to the construction of
Tables by which the places of the heavenly bodies were
given at every moment of time.   The truth and accuracy
of these predictions were a proof that the hypothesis was
valuable and, at least to a great extent, true; although,
as was afterwards found, it involved a false representation
of the structure of the heavens.   In like manner, the
discovery of the Laws of Refraction enabled mathematicians to *predict*, by calculation, what would be the
effect of any new form or combination of transparent
lenses.   Newton's hypothesis of Fits of Easy Transmis-

sion and Easy Reflection in the particles of light, although not confirmed by other kinds of facts, involved a true statement of the law of the phenomena which it was framed to include and served to *predict* the forms and colours of thin plates for a wide range of given cases. The hypothesis that Light operates by Undulations and Interferences, afforded the means of *predicting* results under a still larger extent of conditions.   In like manner in the progress of chemical knowledge, the doctrine of Phlogiston supplied the means of *foreseeing* the consequence of many combinations of elements, even before they were tried; but the Oxygen Theory, besides affording predictions, at least equally exact, with regard to the general results of chemical operations, included all the facts concerning the relations of weight of the elements and their compounds, and enabled chemists to *foresee* such facts in untried cases.   And the Theory of Electromagnetic Forces, as soon as it was rightly understood, enabled those who had mastered it to *predict* motions such as had not been before observed, which were accordingly found to take place.

Men cannot help believing that the laws laid down by discoverers must be in a great measure identical with the real laws of nature, when the discoverers thus determine effects beforehand in the same manner in which nature herself determines them when the occasion occurs.   Those who can do this, must, to a considerable extent, have detected nature's secret;—must have fixed upon the conditions to which she attends, and must have seized the rules by which she applies them.   Such a coincidence of untried facts with speculative assertions cannot be the work of chance, but implies some large portion of truth in the principles on which the reasoning is founded.   To trace order and law in that which has been observed, may be considered as interpreting what nature has written down for us, and will commonly prove that we understand her alphabet. But to predict what has not been observed, is to attempt ourselves to use the legislative phrases of nature; and when she responds plainly and precisely to that which we thus utter, we cannot but suppose that we have in a great measure made ourselves masters of the meaning and structure of her language. The prediction of results, even of the same kind as those which have been observed, in new cases, is a proof of real success in our inductive processes.

11. We have here spoken of the prediction of facts *of the same kind* as those from which our rule was collected. But the evidence in favour of our induction is of a much higher and more forcible character when it enables us to explain and determine cases of a *kind different* from those which were contemplated in the formation of our hypothesis. The instances in which this has occurred, indeed, impress us with a conviction that the truth of our hypothesis is certain. No accident could give rise to such an extraordinary coincidence. No false supposition could, after being adjusted to one class of phenomena, so exactly represent a different class, when the agreement was unforeseen and uncontemplated. That rules springing from remote and unconnected quarters should thus leap to the same point, can only arise from *that* being the point where truth resides.

Accordingly the cases in which inductions from classes of facts altogether different have thus *jumped together*, belong only to the best established theories which the history of science contains. And as I shall have occasion to refer to this peculiar feature in their evidence, I will take the liberty of describing it by a particular phrase; and will term it the *Consilience of Inductions.*

It is exemplified principally in some of the greatest discoveries. Thus it was found by Newton that the doctrine of the attraction of the sun varying according to the inverse square of this distance, which explained Kepler's *third law* of the proportionality of the cubes of the distances to the squares of the periodic times of the planets, explained also his *first* and *second laws* of the elliptical motion of each planet; although no connexion of these laws had been visible before.   Again, it appeared that the force of universal gravitation, which had been inferred from the *perturbations* of the moon and planets by the sun and by each other, also accounted for the fact, apparently altogether dissimilar and remote, of the *precession of the equinoxes.*   Here was a most striking and surprising coincidence, which gave to the theory a stamp of truth beyond the power of ingenuity to counterfeit. In like manner in optics; the hypothesis of alternate fits of easy transmission and reflection would explain the colours of thin plates, and indeed was devised and adjusted for that very purpose; but it could give no account of the phenomena of the fringes of shadows.   But the doctrine of interferences, constructed at first with reference to phenomena of the nature of the *fringes,* explained also the *colours of thin plates* better than the supposition of the fits invented for that very purpose.   And we have in physical optics another example of the same kind, which is quite as striking as the explanation of precession by inferences from the facts of perturbation.   The doctrine of undulations propagated in a spheroidal form was contrived at first by Huyghens, with a view to explain the laws of *double refraction* in calc-spar; and was pursued with the same view by Fresnel.   But in the course of the investigation it appeared, in a most unexpected and wonderful manner, that this same doctrine of spheroidal undulations, when it was so modified as to account for the directions of the two refracted rays, accounted also for the positions of their *planes of polarization**; a

------

* *Hist. Ind. Sci.,* ii. 420.

phenomenon which, taken by itself, it had perplexed previous mathematicians, even to represent.

The theory of universal gravitation, and of the undulatory theory of light, are, indeed, full of examples of this Consilience of Inductions. With regard to the latter, it has been justly asserted by Herschel, that the history of the undulatory theory was a succession of *felicities**. And it is precisely the unexpected coincidences of results drawn from distant parts of the subject which are properly thus described. Thus the laws of the *modification of polarization* to which Fresnel was led by his general views, accounted for the rule respecting the *angle at which light is polarized*, discovered by Brewster†. The conceptions of the theory pointed out peculiar *modifications* of the phenomena when *Newton's rings* were produced by polarized light, which were ascertained to take place in fact, by Arago and Airy‡. When the beautiful phenomena of *dipolarized light* were discovered by Arago and Biot, Young was able to declare that they were reducible to the general laws of *interference* which he had already established§. And what was no less striking a confirmation of the truth of the theory, *measures* of the same element deduced from various classes of facts were found to coincide. Thus the *length* of a luminiferous undulation, calculated by Young from the measurement of *fringes* of shadows, was found to agree very nearly with the previous calculation from the colours of *thin plates*‖.

No example can be pointed out, in the whole history of science, so far as I am aware, in which this Consilience of Inductions has given testimony in favour of an hypothesis afterwards discovered to be false. If we take one class of facts only, knowing the law which they follow, we may construct an hypotheses, or perhaps several, which may represent them: and as new circumstances are

---

* See *Hist. Ind. Sci.*, ii. 435.   † *Ib.*, 423.

‡ *Ib.*, 450.   § *Ib.*, 426.   ‖ *Ib.*, 406.

discovered, we may often adjust the hypothesis so as to correspond to these also. But when the hypothesis, of itself and without adjustment for the purpose, gives us the rule and reason of a class of facts not contemplated in its construction, we have a criterion of its reality, which has never yet been produced in favour of falsehood.

12. In the preceding section I have spoken of the hypothesis with which we compare our facts as being framed *all at once*, each of its parts being included in the original scheme. In reality, however, it often happens that the various suppositions which our system contains are *added* upon occasion of different researches. Thus in the Ptolemaic doctrine of the heavens, new epicycles and eccentrics were added as new inequalities of the motions of the heavenly bodies were discovered; and in the Newtonian doctrine of material rays of light, the supposition that these rays had "fits," was added to explain the colours of thin plates; and the supposition that they had "sides" was introduced on occasion of the phenomena of polarization. In like manner other theories have been built up of parts devised at different times.

This being the mode in which theories are often framed, we have to notice a distinction which is found to prevail in the progress of true and of false theories. In the former class all the additional suppositions *tend to simplicity* and harmony; the new suppositions resolve themselves into the old ones, or at least require only some easy modification of the hypothesis first assumed: the system becomes more coherent as it is further extended. The elements which we require for explaining a new class of facts are already contained in our system. Different members of the theory run together, and we have thus a constant convergence to unity. In false theories, the contrary is the case. The new suppositions are something altogether additional;—not suggested by the original scheme; perhaps difficult to reconcile with it. Every such addition adds to the complexity of the hypothetical system, which at last becomes unmanageable, and is compelled to surrender its place to some simpler explanation.

Such a false theory, for example, was the ancient doctrine of eccentrics and epicycles. It explained the general succession of the places of the Sun, Moon, and Planets; it would not have explained the proportion of their magnitudes at different times, if these could have been accurately observed; but this the ancient astronomers were unable to do. When, however, Tycho and other astronomers came to be able to observe the planets accurately in all positions, it was found that *no* combination of *equable* circular motions would exactly represent all the observations. We may see, in Kepler's works, the many new modifications of the epicyclical hypothesis which offered themselves to him; some of which would have agreed with the phenomena with a certain degree of accuracy, but not so great a degree as Kepler, fortunately for the progress of science, insisted upon obtaining. After these epicycles had been thus accumulated, they all disappeared and gave way to the simpler conception of an *elliptical* motion. In like manner, the discovery of new inequalities in the moon's motions encumbered her system more and more with new machinery, which was at last rejected all at once in favour of the *elliptical* theory. Astronomers could not but suppose themselves in a wrong path when the prospect grew darker and more entangled at every step.

Again; the Cartesian system of vortices might be said to explain the primary phenomena of the revolutions of planets about the sun, and satellites about planets. But the elliptical form of the orbits required new suppositions.   Bernoulli ascribed this curve to the shape of the planet, operating on the stream of the vortex in a manner similar to the rudder of a boat.   But then the motions of the aphelia, and of the nodes,—the perturbations,—even the action of gravity to the earth,—could not be accounted for without new and independent suppositions   Here was none of the simplicity of truth.   The theory of gravitation on the other hand became more simple as the facts to be explained became more numerous.   The attraction of the sun accounted for the motions of the planets; the attraction of the planets was the cause of the motion of the satellites.   But this being assumed, the perturbations, the motions of the nodes and aphelia, only made it requisite to extend the attraction of the sun to the satellites, and that of the planets to each other:—the tides, the spheroidal form of the earth, the precession, still required nothing more than that the moon and sun should attract the parts of the earth, and that these should attract each other;—so that all the suppositions resolved themselves into the single one, of the universal gravitation of all matter.   It is difficult to imagine a more convincing manifestation of simplicity and unity.

Again, to take an example from another science;— the doctrine of phlogiston brought together many facts in a very plausible manner,—combustion, acidification, and others,—and very naturally prevailed for a while.   But the balance came to be used in chemical operations, and the facts of weight as well as of combination were to be accounted for.   On the phlogistic theory, it appeared that this could not be done without a new supposition, and *that* a very strange one;—that phlogiston was an element not only not heavy, but absolutely light, so that it diminished the weight of the compounds into which it entered. Some chemists for a time adopted this extravagant view; but the wiser of them saw, in the necessity of such a supposition to the defence of the theory, an evidence that the hypothesis of an element *phlogiston* was erroneous. And the opposite hypothesis, which taught that oxygen was subtracted, and not phlogiston added, was accepted because it required no such novel and inadmissible assumption.

Again, we find the same evidence of truth in the progress of the undulatory theory of light, in the course of its application from one class of facts to another. Thus we explain reflection and refraction by undulations; when we come to thin plates, the requisite "fits" are already involved in our fundamental hypothesis, for they are the length of an undulation: the phenomena of diffraction also require such intervals; and the intervals thus required agree exactly with the others in magnitude, so that no new property is needed. Polarization for a moment appears to require some new hypothesis; yet this is hardly the case; for the direction of our vibrations is hitherto arbitrary:—we allow polarization to decide it, and we suppose the undulations to be transverse. Having done this for the sake of polarization, we turn to the phenomena of double refraction, and inquire what new hypothesis they require. But the answer is, that they require none: the supposition of transverse vibrations, which we have made in order to explain polarization, gives us also the law of double refraction. Truth may give rise to such a coincidence; falsehood cannot. Again, the facts of dipolarization come into view. But they hardly require any new assumption; for the difference of optical elasticity of crystals in different directions, which is already assumed in uniaxal crystals*, is extended to biaxal exactly according to the law of symmetry; and this being done, the laws of the phenomena, curious and complex as they are, are fully explained. The phenomena

* *Hist. Ind. Sci.*, ii. 427.

of circular polarization by internal reflection, instead of requiring a new hypothesis, are found to be given by an interpretation of an apparently inexplicable result of an old hypothesis. The circular polarization of quartz and its double refraction does indeed appear to require a new assumption, but still not one which at all disturbs the form of the theory; and in short, the whole history of this theory is a progress, constant and steady, often striking and startling, from one degree of evidence and consistence to another of higher order.

In the emission theory, on the other hand, as in the theory of solid epicycles, we see what we may consider as the natural course of things in the career of a false theory. Such a theory may, to a certain extent, explain the phenomena which it was at first contrived to meet; but every new class of facts requires a new supposition— an addition to the machinery: and as observation goes on, these incoherent appendages accumulate, till they overwhelm and upset the original frame-work. Such has been the hypothesis of the material emission of light. In its original form it explained reflection and refraction: but the colours of thin plates added to it the fits of easy transmission and reflection; the phenomena of diffraction further invested the emitted particles with complex laws of attraction and repulsion; polarization gave them sides: double refraction subjected them to peculiar forces emanating from the axes of the crystal: finally, dipolarization loaded them with the complex and unconnected contrivance of moveable polarization: and even when all this had been done, additional mechanism was wanting. There is here no unexpected success, no happy coincidence, no convergence of principles from remote quarters. The philosopher builds the machine, but its parts do not fit. They hold together only while he presses them. This is not the character of truth.

As another example of the application of the maxim

now under consideration, I may perhaps be allowed to refer to the judgment which, in the History of Thermotics, I have ventured to give respecting Laplace's Theory of Gases. I have stated*, that we cannot help forming an unfavourable judgment of this theory, by looking for that great characteristic of true theory; namely, that the hypotheses which were assumed to account for *one class* of facts are found to explain *another class* of a different nature. Thus Laplace's first suppositions explain the connexion of compression with density, (the law of Boyle and Mariotte,) and the connexion of elasticity with heat, (the law of Dalton and Gay Lussac.) But the theory requires other assumptions when we come to latent heat; and yet these new assumptions produce no effect upon the calculations in any application of the theory. When the hypothesis, constructed with reference to the elasticity and temperature, is applied to another class of facts, those of latent heat, we have no Simplification of the Hypothesis, and therefore no evidence of the truth of the theory.

13. The two last sections of this chapter direct our attention to two circumstances, which tend to prove, in a manner which we may term irresistible, the truth of the theories which they characterize:—the *Consilience of Inductions* from different and separate classes of facts; —and the progressive *Simplification of the Theory* as it is extended to new cases. These two Characters are, in fact, hardly different; they are exemplified by the same cases. For if these Inductions, collected from one class of facts, supply an unexpected explanation of a new class, which is the case first spoken of, there will be no need for new machinery in the hypothesis to apply it to the newly-contemplated facts; and thus we have a case in which the system does not become more complex when its application is extended to a wider field, which was

* *Hist. Ind. Sci.*, ii. 530.

the character of true theory in its second aspect.   The Consiliences of our Inductions give rise to a constant Convergence of our Theory towards Simplicity and Unity.

But, moreover, both these cases of the extension of the theory, without difficulty or new suppositions, to a wider range and to new classes of phenomena, may be conveniently considered in yet another point of view; namely, as successive steps by which we gradually ascend in our speculative views to a higher and higher point of generality.   For when the theory, either by the concurrence of two indications, or by an extension without complication, has included a new range of phenomena, we have, in fact, a new induction of a more general kind, to which the inductions formerly obtained are subordinate, as particular cases to a general proposition.   We have in such examples, in short, an instance of *successive generalization.* This is a subject of great importance, and deserving of being well illustrated; it will come under our notice in the next chapter.

Chapter VI.

OF THE LOGIC OF INDUCTION.

1. The subject to which the present chapter refers is described by phrases which are at the present day familiarly used in speaking of the progress of knowledge. We hear very frequent mention of *ascending from particular to general* propositions, and from these to propositions still more general;—of truths *included* in other truths of a higher degree of generality;—of different *stages of generalization;*—and of the *highest step* of the process of discovery, to which all others are subordinate and preparatory.   As these expressions, so familiar to our

ears, especially since the time of Francis Bacon, denote, very significantly, processes and relations which are of great importance in the formation of science, it is necessary for us to give a clear account of them, illustrated with general exemplifications, and this we shall endeavour to do.

We have, indeed, already explained that science consists of propositions which include the facts from which they were collected; and other wider propositions, collected in like manner from the former, and including them. Thus, that the stars, the moon, the sun, rise, culminate, and set, are facts *included* in the proposition that the heavens, carrying with them all the celestial bodies, have a diurnal revolution about the axis of the earth. Again, the observed monthly motions of the moon, and the annual motions of the sun, are *included* in certain propositions concerning the movements of those luminaries with respect to the stars. But all these propositions are really *included* in the doctrine that the earth, revolving on its axis, moves round the sun, and the moon round the earth. These movements, again, considered as facts, are explained and *included* in the statement of the forces which the earth exerts upon the moon, and the sun upon the earth. Again, this doctrine of the forces of these two bodies is *included* in the assertion, that all the bodies of the solar system, and all parts of matter, exert forces, each upon each. And we might easily show that all the leading facts in astronomy are comprehended in the same generalization. In like manner with regard to any other science, so far as its truths have been well established and fully developed, we might show that it consists of a gradation of propositions, proceeding from the most special facts to the most general theoretical assertions. We shall exhibit this gradation in some of the principal branches of science.

2. This gradation of truths, successively included in other truths, may be conveniently represented by *Tables* resembling the genealogical tables by which the derivation of descendants from a common ancestor is exhibited; except that it is proper in this case to invert the form of the Table, and to make it converge to unity downwards instead of upwards, since it has for its purpose to express, not the derivation of many from one, but the collection of one truth from many things. Two or more co-ordinate facts or propositions may be ranged side by side, and joined by some mark of connexion, (a bracket, as ⌣ or └────┘,) beneath which may be placed the more general proposition which is collected by induction from the former. Again, propositions co-ordinate with this more general one may be placed on a level with it; and the combination of these, and the result of the combination, may be indicated by brackets in the same manner; and so on, through any number of gradations. By this means the streams of knowledge from various classes of facts will constantly run together into a smaller and smaller number of channels; like the confluent rivulets of a great river, coming together from many sources, uniting their ramifications so as to form larger branches, these again uniting in a single trunk. The *genealogical tree* of each great portion of science, thus formed, will contain all the leading truths of the science arranged in their due co-ordination and subordination. Such Tables, constructed for the sciences of Astronomy and of Optics, will be given at the end of this chapter.

3. The union of co-ordinate propositions into a proposition of a higher order, which occurs in this Tree of Science wherever two twigs unite in one branch, is, in each case, an example of *Induction*. The single proposition is collected by the process of induction from its several members. But here we may observe, that the

image of a mere *union* of the parts at each of these points, which the figure of a tree or a river presents, is very inadequate to convey the true state of the case; for in Induction, as we have seen, besides mere collection of particulars, there is always a *new conception*, a principle of connexion and unity, supplied by the mind, and super-induced upon the particulars. There is not merely a juxtaposition of materials, by which the new proposition contains all that its component parts contained; but also a formative act exerted by the understanding, so that these materials are contained in a new shape. We must remember, therefore, that our Inductive Tables, although they represent the elements and the order of these inductive steps, do not fully represent the whole signification of the process in each case.

4. The principal features of the progress of science spoken of in the last chapter are clearly exhibited in these Tables; namely, the *Consilience of Inductions*, and the constant Tendency to Simplicity observable in true theories. Indeed in all cases in which from propositions of considerable generality, propositions of a still higher degree are obtained, there is a convergence of inductions; and if in one of the lines which thus converge, the steps be rapidly and suddenly made in order to meet the other line, we may consider that we have an example of Consilience. Thus when Newton had collected from Kepler's Laws the Central Force of the sun, and from these, combined with other facts, the Universal Force of all the heavenly bodies, he suddenly turned round to include in his generalization the Precession of the Equinoxes, which he declared to arise from the attraction of the sun and moon upon the protuberant part of the terrestrial spheroid. The apparent remoteness of this fact, in its nature, from the others with which he thus associated it, causes this part of his reasoning to strike us

as a remarkable example of *Consilience.* Accordingly, in the Table of Astronomy we find that the columns which contain the facts and theories relative to the *sun* and *planets,* after exhibiting several stages of induction within themselves, are at length suddenly connected with a column till then quite distinct, containing the *precession of the equinoxes.* In like manner, in the Table of Optics, the columns which contain the facts and theories relative to *double refraction,* and those which include *polarization by crystals,* each go separately through several stages of induction; and then these two sets of columns are suddenly connected by Fresnel's mathematical induction that double refraction and polarization arise from the same cause: thus exhibiting a remarkable *Consilience.*

5. The constant *Tendency to Simplicity* in the sciences of which the progress is thus represented, appears from the form of the Table itself; for the single trunk into which all the branches converge, contains in itself the substance of all the propositions by means of which this last generalization was arrived at. It is true, that this ultimate result is sometimes not so simple as in the Table it appears: for instance, the ultimate generalization of the Table exhibiting the progress of Physical Optics,— namely, that Light consists in Undulations,—must be understood as including some other hypotheses; as, that the undulations are transverse, that the ether through which they are propagated has its elasticity in crystals and other transparent bodies regulated by certain laws; and the like. Yet still, even acknowledging all the complication thus implied, the Table in question evidences clearly enough the constant advance towards unity, consistency, and simplicity, which have marked the progress of this Theory. The same is the case in the Inductive Table of Astronomy in a still greater degree.

6. These Tables naturally afford the opportunity of

assigning to each of the distinct steps of which the progress of science consists, the name of the *Discoverer* to
whom it is due.   Every one of the inductive processes
which the brackets of our Tables mark, directs our attention to some person by whom the induction was first
distinctly made.   These names I have endeavoured to
put in their due places in the Tables; and the Inductive
Tree of our knowledge in each science becomes, in this
way, an exhibition of the claims of each discoverer to
distinction, and, as it were, a Genealogical Tree of scientific nobility.   It is by no means pretended that such a
tree ·includes the names of all the meritorious labourers
in each department of science.   Many persons are most
usefully employed in collecting and verifying truths, who
do not advance to any new truths.   The labours of a
number of such are included in each stage of our ascent.
But such Tables as we have now before us will present
to us the names of all the most eminent discoverers: for
the main steps of which the progress of science consists,
are transitions from more particular to more general
truths, and must therefore be rightly given by these
Tables; and those must be the greatest names in science
to whom the principal events of its advance are thus due.

7. The Tables, as we have presented them, exhibit
the course by which we pass from particular to general
through various gradations, and so to the most general.
They display the order of *discovery*.   But by reading
them in an inverted manner, beginning at the single
comprehensive truths with which the Tables end, and
tracing these back into the more partial truths, and these
again into special facts, they answer another purpose;—
they exhibit the process of *verification* of discoveries once
made.   For each of our general propositions is true in
virtue of the truth of the narrower propositions which it
involves; and we cannot satisfy ourselves of its truth in

any other way than by ascertaining that these its constituent elements are true. To assure ourselves that the sun attracts the planets with forces varying inversely as the square of the distance, we must analyse by geometry the motion in an ellipse about the focus, so as to see that it does imply such a force. We must also verify those calculations by which the observed places of each planet are stated to be included in an ellipse. These calculations involve assumptions respecting the path which the earth describes about the sun, which assumptions must again be verified by reference to observation. And thus, proceeding from step to step, we resolve the most general truths into their constituent parts; and these again into their parts; and by testing, at each step, both the reality of the asserted ingredients and the propriety of the conjunction, we establish the whole system of truths, however wide and various it may be.

8. It is a very great advantage, in such a mode of exhibiting scientific truths, that it resolves the verification of the most complex and comprehensive theories, into a number of small steps, of which almost any one falls within the reach of common talents and industry. That *if* the particulars of any one step be true, the generalization also is true, any person with a mind properly disciplined may satisfy himself by a little study. That each of these particular propositions *is* true, may be ascertained, by the same kind of attention, when this proposition is resolved into *its* constituent and more special propositions. And thus we may proceed, till the most general truth is broken up into small and manageable portions. Of these portions, each may appear by itself narrow and easy; and yet they are so woven together, by hypothesis and conjunction, that the truth of the parts necessarily assures us of the truth of the whole. The verification is of the same nature as the verification of a

large and complex statement of great sums received by a mercantile office on various accounts from many quarters. The statement is separated into certain comprehensive heads, and these into others less extensive; and these again into smaller collections of separate articles, each of which can be inquired into and reported on by separate persons. And thus at last, the mere addition of numbers performed by these various persons, and the summation of the results which they obtain, executed by other accountants, is a complete and entire security that there is no error in the whole of the process.

9. This comparison of the process by which we verify scientific truth to the process of book-keeping in a large commercial establishment, may appear to some persons not sufficiently dignified for the subject. But, in fact, the possibility of giving this formal and business-like aspect to the evidence of science, as involved in the process of successive generalization, is an inestimable advantage. For if no one could pronounce concerning a wide and profound theory except he who could at once embrace in his mind the whole range of inference, extending from the special facts up to the most general principles, none but the greatest geniuses would be entitled to judge concerning the truth or error of scientific discoveries. But, in reality, we seldom need to verify more than one or two steps of such discoveries at one time; and this may commonly be done (when the discoveries have been fully established and developed,) by any one who brings to the task clear conceptions and steady attention. The progress of science is gradual: the discoveries which are successively made, are also verified successively. We have never any very large collections of them on our hands at once. The doubts and uncertainties of any one who has studied science with care and perseverance are generally confined to a

few points. If he can satisfy himself upon these, he has no misgivings respecting the rest of the structure; which has indeed been repeatedly verified by other persons in like manner. The fact that science is capable of being resolved into separate processes of verification, is that which renders it possible to form a great body of scientific truth, by adding together a vast number of truths, of which many men, at various times and by multiplied efforts, have satisfied themselves. The treasury of Science is constantly rich and abundant, because it accumulates the wealth which is thus gathered by so many, and reckoned over by so many more : and the dignity of Knowledge is no more lowered by the multiplicity of the tasks on which her servants are employed, and the narrow field of labour to which some confine themselves, than the rich merchant is degraded by the number of offices which it is necessary for him to maintain, and the minute articles of which he requires an exact statement from his accountants.

10. The analysis of doctrines inductively obtained, into their constituent facts, and the arrangement of them in such a form that the conclusiveness of the induction may be distinctly seen, may be termed *the Logic of Induction*. By Logic has generally been meant a system which teaches us so to arrange our reasonings that their truth or falsehood shall be evident in their form. In *deductive* reasonings, in which the general principles are assumed, and the question is concerning their application and combination in particular cases, the device which thus enables us to judge whether our reasonings are conclusive, is the *Syllogism;* and this *form,* along with the rules which belong to it, does in fact supply us with a criterion of deductive or demonstrative reasoning. The *Inductive Table,* such as it is presented in the present chapter, in like manner supplies the means of ascertaining the truth of our *inductive* inferences, so far as the *form* in which

our reasoning may be stated can afford such a criterion.
Of course some care is requisite in order to reduce a
train of demonstration into the form of a series of syllogisms; and certainly not less thought and attention
are required for resolving all the main doctrines of any
great department of science into a graduated table of
co-ordinate and subordinate inductions. But in each
case, when this task is once executed, the evidence or
want of evidence of our conclusions appears at once in a
most luminous manner. In each step of induction, our
Table enumerates the particular facts, and states the general theoretical truth which includes these and which these
constitute. The special act of attention by which we satisfy
ourselves that the facts *are* so included,—that the general
truth *is* so constituted,—then affords little room for error,
with moderate attention and clearness of thought.

11. We may find an example of this *act of attention*
thus required, at any one of the steps of induction in our
Tables; for instance, at the step in the early progress of
astronomy at which it was inferred, that the earth is a
globe, and that the sphere of the heavens performs a diurnal
revolution round this globe of the earth. How was this
established in the belief of the Greeks, and how is it fixed
in our conviction? As to the globular form, we find that
as we travel to the north the apparent pole of the heavenly
motions, and the constellations which are near it, seem to
mount higher, and as we proceed southwards they descend.
Again, if we proceed from two different points consi-
derably to the east and west of each other, and travel
directly northwards from each, as from the south of Spain
to the north of Scotland, and from Greece to Scandinavia,
these two north and south lines will be much nearer to
each other in their northern than in their southern parts.
These and similar facts, as soon as they are clearly estimated and connected in the mind, are *seen to be consis-*

tent with a convex surface of the earth, and with no other: and this notion is further confirmed by observing that the boundary of the earth's shadow upon the moon is always circular; it being supposed to be already established that the moon receives her light from the sun, and that lunar eclipses are caused by the interposition of the earth. As for the assertion of the diurnal revolution of the starry sphere, it is merely putting the visible phenomena in an exact geometrical form: and thus we establish and verify the doctrine of the revolution of the sphere of the heavens about the globe of the earth, by contemplating it so as to *see* that it does really and exactly include the particular facts from which it is collected.

We may, in like manner, illustrate this mode of verification by any of the other steps of the same Table. Thus if we take the great Induction of Copernicus, the heliocentric scheme of the solar system, we find it in the Table exhibited as including and explaining, *first*, the diurnal revolution just spoken of; *second*, the motions of the moon among the fixed stars; *third*, the motions of the planets with reference to the fixed stars and the sun; *fourth*, the motion of the sun in the ecliptic. And the scheme being clearly conceived, we *see* that all the particular facts *are* faithfully represented by it; and this agreement, along with the simplicity of the scheme, in which respect it is so far superior to any other conception of the solar system, persuade us that it is really the plan of nature.

In exactly the same way, if we attend to any of the several remarkable discoveries of Newton, which form the principal steps in the latter part of the Table, as for instance, the proposition that the sun attracts all the planets with a force which varies inversely as the square of the distance, we find it proved by its including three other propositions previously established;—*first*, that the sun's mean force on different planets follows the specified variation (which is proved from Kepler's third law); *second*, that the force by which each planet is acted upon in different parts of its orbit tends to the sun (which is proved by the equable description of areas); *third*, that this force in different parts of the same orbit is also inversely as the square of the distance (which is proved from the elliptical form of the orbit). And the Newtonian generalization, when its consequences are mathematically traced, is *seen* to agree with each of these particular propositions, and thus is fully established.

12. But when we say that the more general proposition *includes* the several more particular ones, we must recollect what has before been said, that these particulars form the general truth, not by being merely enumerated and added together, but by being seen *in a new light*. No mere verbal recitation of the particulars can decide whether the general proposition is true; a special act of thought is requisite in order to determine how truly each is included in the supposed induction. In this respect the Inductive Table is not like a mere schedule of accounts, where the rightness of each part of the reckoning is tested by mere addition of the particulars. On the contrary, the Inductive truth is never the mere *sum* of the facts. It is made into something more by the introduction of a new mental element; and the mind, in order to be able to supply this element, must have peculiar endowments and discipline. Thus looking back at the instances noticed in the last article, how are we to see that a convex surface of the earth is necessarily implied by the convergence of meridians towards the north, or by the visible descent of the north pole of the heavens as we travel south? Manifestly the student, in order to see this, must have clear conceptions of the relations of space,

either naturally inherent in his mind, or established there by geometrical cultivation,—by studying the properties of circles and spheres. When he is so prepared, he will feel the force of the expressions we have used, that the facts just mentioned are *seen to be consistent* with a globular form of the earth; but without such aptitude he will not see this consistency: and if this be so, the mere assertion of it in words will not avail him in satisfying himself of the truth of the proposition.

In like manner, in order to perceive the force of the Copernican induction, the student must have his mind so disciplined by geometrical studies, or otherwise, that he sees clearly how absolute motion and relative motion would alike produce apparent motion. He must have learnt to cast away all prejudices arising from the seeming fixity of the earth; and then he will see that there is nothing which stands in the way of the induction, while there is much which is on its side. And in the same manner the Newtonian induction of the law of the sun's force from the elliptical form of the orbit, will be evidently satisfactory to him only who has such an insight into Mechanics as to see that a curvilinear path must arise from a constantly deflecting force; and who is able to follow the steps of geometrical reasoning by which, from the properties of the ellipse, Newton proves this deflection to be in the proportion in which he asserts the force to be. And thus in all cases the inductive truth must indeed be verified by comparing it with the particular facts; but then this comparison is possible for him only whose mind is properly disciplined and prepared in the use of those conceptions, which, in addition to the facts, the act of induction requires.

13. In the Tables some indication is given, at several of the steps, of the act which the mind must thus perform, besides the mere conjunction of facts, in order to

attain to the inductive truth. Thus in the cases of the
Newtonian inductions just spoken of, the inferences are
stated to be made " By Mechanics;" and in the case of
the Copernican induction, it is said that, " By the nature
of motion, the apparent motion is the same, whether the
heavens or the earth have a diurnal motion; and the
latter is more simple." But these verbal statements are
to be understood as mere hints*: they cannot supersede
the necessity of the student's contemplating for himself
the mechanical principles and the nature of motion thus
referred to.

14. In the Common or Syllogistic Logic, a certain
Formula of language is used in stating the reasoning, and
is useful in enabling us more readily to apply the Criterion of Form to alleged demonstrations. This formula
is the usual Syllogism ; with its members, Major Premiss,
Minor Premiss, and Conclusion. It may naturally be
asked whether in Inductive Logic there is any such Formula? Whether there is any standard form of words in
which we may most properly express the inference of a
general truth from particular facts?

At first it might be supposed that the formula of
Inductive Logic need only be of this kind : "These particulars, and all known particulars of the same kind, are
exactly included in the following general proposition."
But a moment's reflection on what has just been said will
show us that this is not sufficient: for the particulars are
not merely *included* in the general proposition. It is not
enough that they appertain to it by enumeration. It is,
for instance, no adequate example of Induction to say,
" Mercury describes an elliptical path, so does Venus, so
do the Earth, Mars, Jupiter, Saturn, Uranus; therefore
all the Planets describe elliptical paths." This is, as we
have seen, the mode of stating the *evidence* when the

* In the Inductive Tables they are marked by an asterisk.

proposition is once suggested; but the Inductive step consists in the *suggestion* of a conception not before apparent. When Kepler, after trying to connect the observed places of the planet Mars in many other ways, found at last that the conception of an ellipse would include them all, he obtained a truth by induction: for this conclusion was not obviously included in the phenomena, and had not been applied to these facts previously. Thus in our Formula, besides stating that the particulars are included in the general proposition, we must also imply that the generality is constituted by a new Conception,—new at least in its application.

Hence our Inductive Formula might be something like the following: "These particulars, and all known particulars of the same kind, are exactly expressed by adopting the Conceptions and Statement of the following Proposition." It is of course requisite that the Conceptions should be perfectly clear, and should precisely embrace the facts, according to the explanation we have already given of those conditions.

15. It may happen, as we have already stated, that the Explication of a Conception, by which it acquires its due distinctness, leads to a Definition, which Definition may be taken as the summary and total result of the intellectual efforts to which this distinctness is due. In such cases, the Formula of Induction may be modified according to this condition; and we may state the inference by saying, after an enumeration and analysis of the appropriate facts, "These facts are completely and distinctly expressed by adopting the following Definition and Proposition."

This Formula has been adopted in stating the Inductive Propositions which constitute the basis of the science of Mechanics, in a work intitled *The Mechanical Euclid*. The fundamental truths of the subject are expressed in

Inductive Pairs of Assertions, consisting each of a Definition and a Proposition, such as the following :

Def.—A *Uniform Force* is that which acting in the
direction of the body's motion, adds or subtracts equal
velocities in equal times.

Prop.—Gravity is a Uniform Force.

Again,

Def.—Two *Motions* are *compounded* when each produces its separate effect in a direction parallel to itself.

Prop.—When any Force acts upon a body in motion,
the motion which the Force would produce in the body
at rest is compounded with the previous motion of the
body.

And in like manner in other cases.

In these cases the proposition is, of course, established,
and the definition realized, by an enumeration of the
facts. And in the case of inferences made in such a form,
the Definition of the Conception and the Assertion of
the Truth are both requisite and are correlative to one
another. Each of the two steps contains the verification
and justification of the other. The Proposition derives its
meaning from the Definition; the Definition derives its
reality from the Proposition. If they are separated, the
Definition is arbitrary or empty, the Proposition vague or
ambiguous.

16. But it must be observed that neither of the preceding Formulæ expresses the full cogency of the inductive proof. They declare only that the results *can* be
clearly explained and rigorously deduced by the employment of a certain Definition and a certain Proposition.
But in order to make the conclusion demonstrative, which
in perfect examples of Induction it is, we ought to be
able to declare that the results can be clearly explained
and rigorously declared *only* by the Definition and Proposition which we adopt. And in reality, the conviction

of the sound inductive reasoner does reach to this point. The Mathematician asserts the Laws of Motion, seeing clearly that they (or laws equivalent to them) afford the only means of clearly expressing and deducing the actual facts. But this conviction, that the inductive inference is not only consistent with the facts, but necessary, finds its place in the mind gradually, as the contemplation of the consequences of the proposition, and the various relations of the facts, becomes steady and familiar. It is scarcely possible for the student at once to satisfy himself that the inference is thus inevitable. And when he arrives at this conviction, he sees also, in many cases at least, that there may be other ways of expressing the substance of the truth established, besides that special Proposition which he has under his notice.

We may, therefore, without impropriety, renounce the undertaking of conveying in our formula this final conviction of the necessary truth of our inference. We may leave it to be thought, without insisting upon saying it, that in such cases what *can* be true, *is* true. But if we wish to express the ultimate significance of the Inductive Act of thought, we may take as our Formula for the Colligation of Facts by Induction, this:—"The several Facts are exactly expressed as one Fact if, *and only if*, we adopt the Conception and the Assertion" of the inductive inference.

17. I have said that the mind must be properly disciplined in order that it may see the necessary connexion between the facts and the general proposition in which they are included. And the perception of this connexion, though treated as *one step* in our inductive inference, may imply *many steps* of demonstrative proof. The connexion is this, that the particular case is included in the general one, that is, may be *deduced* from it: but this deduction may often require many links of reasoning.

Thus in the case of the inference of the law of the force from the elliptical form of the orbit by Newton, the proof that in the ellipse the deflection from the tangent is inversely as the square of the distance from the focus of the ellipse, is a ratiocination consisting of several steps, and involving several properties of Conic Sections; these properties being supposed to be previously established by a geometrical system of demonstration on the special subject of the Conic Sections. In this and similar cases the Induction involves many steps of Deduction. And in such cases, although the Inductive step, the Invention of the Conception, is really the most important, yet since, when once made, it occupies a familiar place in men's minds; and since the Deductive Demonstration is of considerable length and requires intellectual effort to follow it at every step; men often admire the deductive part of the proposition, the geometrical or algebraical demonstration, far more than that part in which the philosophical merit really resides.

18. Deductive reasoning is virtually a collection of syllogisms, as has already been stated; and in such reasoning, the general principles, the Definitions and Axioms, necessarily stand at the *beginning* of the demonstration. In an inductive inference, the Definitions and Principles are the *final result* of the reasoning, the ultimate effect of the proof. Hence when an Inductive Proposition is to be established by a proof involving several steps of demonstrative reasoning, the enunciation of the Proposition will contain, explicitly or implicitly, principles which the demonstration proceeds upon as axioms, but which are really inductive inferences. Thus in order to prove that the force which retains a planet in an ellipse varies inversely as the square of the distance, it is taken for granted that the Laws of Motion are true, and that they apply to the planets. Yet the doctrine that this

is so, as well as the law of the force, were established only by this and the like demonstrations. The doctrine which is the hypothesis of the deductive reasoning, is the inference of the inductive process. The special facts which are the basis of the inductive inference, are the conclusion of the train of deduction. And in this manner the deduction establishes the induction. The principle which we gather from the facts is true, because the facts can be derived from it by rigorous demonstration. Induction moves upwards and deduction downwards on the same stair.

But still there is a great difference in the character of their movements. Deduction descends steadily and methodically, step by step: Induction mounts by a leap which is out of the reach of method. She bounds to the top of the stair at once; and then it is the business of Deduction, by trying each step in order, to establish the solidity of her companion's footing. Yet these must be processes of the same mind. The Inductive Intellect makes an assertion which is subsequently justified by demonstration; and it shows its sagacity, its peculiar character, by enunciating the proposition when as yet the demonstration does not exist: but then it shows that it *is* sagacity, by also producing the demonstration.

It has been said that inductive and deductive reasoning are contrary in their scheme; that in Deduction we infer particular from general truths; while in Induction we infer general from particular: that Deduction consists of many steps, in each of which we apply known general propositions in particular cases; while in Induction we have a single step, in which we pass from many particular truths to one general proposition. And this is truly said; but though contrary in their motions, the two are the operation of the same mind travelling over the same ground. Deduction is a necessary part of Induction. Deduction justifies by calculation what Induction had happily guessed. Induction recognizes the ore of truth by its weight; Deduction confirms the recognition by chemical analysis. Every step of Induction must be confirmed by rigorous deductive reasoning, followed into such detail as the nature and complexity of the relations (whether of quantity or any other) render requisite. If not so justified by the supposed discoverer, it is *not* Induction.

19. Such Tabular arrangements of propositions as we have constructed may be considered as the *Criterion of Truth* for the doctrines which they include. They are the Criterion of Inductive Truth, in the same sense in which Syllogistic Demonstration is the Criterion of Necessary Truth,—of the certainty of conclusions, depending upon evident First Principles. And that such Tables are really a Criterion of the truth of the propositions which they contain, will be plain by examining their structure. For if the connexion which the inductive process assumes be ascertained to be in each case real and true, the assertion of the general proposition merely collects together ascertained truths; and in like manner each of those more particular propositions is true, because it merely expresses collectively more special facts: so that the most general theory is only the assertion of a great body of facts, duly classified and subordinated. When we assert the truth of the Copernican theory of the motions of the solar system, or of the Newtonian theory of the forces by which they are caused, we merely assert the groups of propositions which, in the Table of Astronomical Induction, are included in these doctrines; and ultimately, we may consider ourselves as merely asserting at once so many Facts, and therefore, of course, expressing an indisputable truth.

20. At any one of these steps of Induction in the Table, the inductive proposition is a *Theory* with regard

to the Facts which it includes, while it is to be looked upon as a *Fact* with respect to the higher generalizations in which it is included. In any other sense, as was formerly shown, the opposition of Fact and Theory is untenable, and leads to endless perplexity and debate. Is it a Fact or a Theory that the planet Mars revolves in an Ellipse about the Sun? To Kepler, employed in endeavouring to combine the separate observations by the conception of an Ellipse, it is a Theory; to Newton, engaged in inferring the law of force from a knowledge of the elliptical motion, it is a Fact. There are, as we have already seen, no special attributes of Theory and Fact which distinguish them from one another. Facts are phenomena apprehended by the aid of conceptions and mental acts, as Theories also are. We commonly call our observations *Facts*, when we apply, without effort or consciousness, conceptions perfectly familiar to us: while we speak of Theories, when we have previously contemplated the Facts and the connecting Conception separately, and have made the connexion by a conscious mental act. The real difference is a difference of relation; as the same proposition in a demonstration is the *premiss* of one syllogism and the *conclusion* in another;—as the same person is a father and a son. Propositions are Facts and Theories, according as they stand above or below the Inductive Brackets of our Tables.

21. To obviate mistakes I may remark that the terms *higher* and *lower*, when used of generalizations, are unavoidably represented by their opposites in our Inductive Tables. The highest generalization is that which includes all others; and this stands the lowest on our page, because, reading downwards, that is the place which we last reach.

There is a distinction of the knowledge acquired by Scientific Induction into two kinds, which is so important that we shall consider it in the succeeding chapter.

Chapter VII.

OF LAWS OF PHENOMENA AND OF CAUSES.

1. In the first attempts at acquiring an exact and connected knowledge of the appearances and operations which nature presents, men went no further than to learn *what* takes place, not *why* it occurs. They discovered an Order which the phenomena follow, Rules which they obey; but they did not come in sight of the Powers by which these rules are determined, the Causes of which this order is the effect. Thus, for example, they found that many of the celestial motions took place as if the sun and stars were carried round by the revolutions of certain celestial spheres; but what causes kept these spheres in constant motion, they were never able to explain. In like manner in modern times, Kepler discovered that the planets describe ellipses, before Newton explained why they select this particular curve, and describe it in a particular manner. The laws of reflection, refraction, dispersion, and other properties of light have long been known; the causes of these laws are at present under discussion. And the same might be said of many other sciences. The discovery of *the Laws of Phenomena* is, in all cases, the first step in exact knowledge; these Laws may often for a long period constitute the whole of our science; and it is always a matter requiring great talents and great efforts, to advance to a knowledge of the *Causes* of the phenomena.

Hence the larger part of our knowledge of nature, at least of the certain portion of it, consists of the knowledge of the Laws of Phenomena. In Astronomy indeed, besides knowing the rules which guide the appearances, and resolving them into the real motions from which they arise, we can refer these motions to the forces which pro-

duce them.   In Optics, we have become acquainted with
a vast number of laws by which varied and beautiful
phenomena are governed ; and perhaps we may assume,
since the evidence of the undulatory theory has been so
fully developed, that we know also the Causes of the
Phenomena.   But in a large class of sciences, while we
have learnt many Laws of Phenomena, the causes by
which these are produced are still unknown or disputed.
Are we to ascribe to the operation of a fluid or fluids,
and if so, in what manner, the facts of heat, magnetism,
electricity, galvanism?   What are the forces by which
the elements of chemical compounds are held together?
What are the forces, of a higher order, as we cannot help
believing, by which the course of vital action in organized
bodies is kept up?   In these and other cases, we have
extensive departments of science; but we are as yet
unable to trace the effects to their causes; and our
science, so far as it is positive and certain, consists
entirely of the laws of phenomena.

2. In those cases in which we have a division of the
science which teaches us the doctrine of the causes, as
well as one which states the rules which the effects follow, I have distinguished the two portions of the science
by certain terms.   I have thus spoken of *Formal* Astronomy and *Physical* Astronomy.   The latter phrase has long
been commonly employed to describe that department
of Astronomy which deals with those forces by which the
heavenly bodies are guided in their motions ; the former
adjective appears well suited to describe a collection of
rules depending on those ideas of space, time, position,
number, which are, as we have already said, the *forms* of
our apprehension of phenomena.   The laws of phenomena
may be considered as *formulæ*, expressing results in
terms of those ideas.   In like manner, I have spoken of
Formal Optics and Physical Optics ; the latter division

including all speculations concerning the machinery by which the effects are produced. Formal Acoustics and Physical Acoustics may be distinguished in like manner, although these two portions of science have been a good deal mixed together by most of those who have treated of them. Formal Thermotics, the knowledge of the laws of the phenomena of heat, ought in like manner to lead to Physical Thermotics, or the Theory of Heat with reference to the mode in which its effects are produced; —a branch of science which as yet can hardly be said to exist.

3. What *kinds of cause* are we to admit in science? This is an important, and by no means an easy question. In order to answer it, we must consider in what manner our progress in the knowledge of causes has hitherto been made. By far the most conspicuous instance of success in such researches, is the discovery of the causes of the motions of the heavenly bodies. In this case, after the formal laws of the motions,—their conditions as to space and time,—had become known, men were enabled to go a step further; to reduce them to the familiar and general cause of motion—mechanical force; and to determine the laws which this force follows. That this was a step in addition to the knowledge previously possessed, and that it was a real and peculiar truth, will not be contested. And a step in any other subject which should be analogous to this in astronomy;—a discovery of causes and forces as certain and clear as the discovery of universal gravitation;—would undoubtedly be a vast advance upon a body of science consisting only of the laws of phenomena.

4. But although physical astronomy may well be taken as a standard in estimating the value and magnitude of the advance from the knowledge of phenomena to the knowledge of causes; the peculiar features of the

transition from formal to physical science in that subject must not be allowed to limit too narrowly our views of the nature of this transition in other cases. We are not, for example, to consider that the step which leads us to the knowledge of causes in any province of nature must necessarily consist in the discovery of centres of forces, and collections of such centres, by which the effects are produced. The discovery of the causes of phenomena may imply the detection of a fluid by whose undulations, or other operations, the results are occasioned. The phenomena of acoustics are, we know, produced in this manner by the air; and in the cases of light, heat, magnetism, and others, even if we reject all the theories of such fluids which have hitherto been proposed, we still cannot deny that such theories are intelligible and possible, as the discussions concerning them have shown. Nor can it be doubted that if the assumption of such a fluid, in any case, were as well evidenced as the doctrine of universal gravitation is, it must be considered as a highly valuable theory.

5. But again; not only must we, in aiming at, the formation of a Causal Section in each Science of Phenomena, consider fluids and their various modes of operation admissible, as well as centres of mechanical force; but we must be prepared, if it be necessary, to consider the forces, or powers to which we refer the phenomena, under still more general aspects, and invested with characters different from mere mechanical force.   For example; the forces by which the chemical elements of · bodies are bound together, and from which arise, both their sensible texture, their crystalline form, and their chemical composition, are certainly forces of a very different nature from the mere attraction of matter according to its mass. The powers of assimilation and reproduction in plants and animals are obviously still more removed from mere

mechanism; yet these powers are not on that account less real, nor a less fit and worthy subject of scientific inquiry.

6. In fact, these forces—mechanical, chemical and vital,—as we advance from one to the other, each bring into our consideration new characters; and what these characters are, has appeared in the survey which we have made of the Fundamental Ideas of the various sciences. It was then shown that the forces by which chemical effects are produced necessarily involve the Idea of Polarity,—they are polar forces; the particles tend together in virtue of opposite properties which in the combination neutralize each other. Hence, in attemping to advance to a theory of Causes in chemistry, our task is by no means to invent laws of *mechanical* force, and collections of forces, by which the effects may be produced. We know beforehand that no such attempt can succeed. Our aim must be to conceive such new kinds of force, including polarity among their characters, as may best render the results intelligible.

7. Thus in advancing to a Science of Cause in any subject, the labour and the struggle is, not to analyse the phenomena according to any preconceived and already familiar ideas, but to form distinctly new conceptions, such as do really carry us to a more intimate view of the processes of nature. Thus in the case of astronomy, the obstacle which deferred the discovery of the true causes from the time of Kepler to that of Newton, was the difficulty of taking hold of mechanical conceptions and axioms with sufficient clearness and steadiness; which, during the whole of that interval, mathematicians were learning to do. In the question of causation which now lies most immediately in the path of science, that of the causes of electrical and chemical phenomena, the business of rightly fixing and limiting the conception of

polarity, is the proper object of the efforts of discoverers. Accordingly a large portion of Mr. Faraday's recent labours* is directed, not to the attempt at discovering new laws of phenomena, but to the task of throwing light upon the conception of polarity, and of showing how it must be understood, so that it shall include electrical induction and other phenomena, which have commonly been ascribed to forces acting mechanically at a distance. He is by no means content, nor would it answer the ends of science that he should be, with stating the results of his experiments; he is constantly, in every page, pointing out the interpretation of his experiments, and showing how the conception of polar forces enters into this interpretation. "I shall," he says†, "use every opportunity which presents itself of returning to that strong test of truth, experiment; but," he adds, "I shall necessarily have occasion to speak theoretically, and even hypothetically." His hypothesis that electrical inductive action always takes place by means of a continuous line of polarized particles, and not by attraction and repulsion at a distance, if established, cannot fail to be a great step on our way towards a knowledge of causes, as well as phenomena, in the subjects under his consideration.

8. The process of obtaining new conceptions is, to most minds, far more unwelcome than any labour in employing old ideas. The effort is indeed painful and oppressive; it is feeling in the dark for an object which we cannot find. Hence it is not surprising that we should far more willingly proceed to seek for new causes by applying conceptions borrowed from old ones. Men were familiar with solid frames, and with whirlpools of fluid, when they had not learnt to form any clear con-

---

* Eleventh, Twelfth, and Thirteenth Series of Researches, *Phil. Trans.*, 1837 and 8.

† Art. 1318.

ception of attraction at a distance. Hence they at first imagined the heavenly motions to be caused by crystalline spheres, and vortices. At length they were taught to conceive central forces, and then they reduced the solar system to these. But having done this, they fancied that all the rest of the machinery of nature must be central forces. We find Newton expressing this conviction[*], and the mathematicians of the last century acted upon it very extensively. We may especially remark Laplace's labours in this field. Having explained, by such forces, the phenomena of capillary attraction, he attempted to apply the same kind of explanation to the reflection, refraction, and double refraction of light;—to the constitution of gases;—the operation of heat. It was soon seen that the explanation of refraction was arbitrary, and that of double refraction illusory; while polarization entirely eluded the grasp of this machinery. Centres of force would no longer represent the modes of causation which belonged to the phenomena. Polarization required some other contrivance, such as the undulatory theory supplied. No theory of light can be of any avail in which the fundamental idea of polarity is not clearly exhibited.

9. The sciences of magnetism and electricity have given rise to theories in which this relation of polarity is exhibited by means of two opposite fluids[†];—a positive and a negative fluid, or a vitreous and a resinous, for electricity, and a boreal and an austral fluid for magnetism. The hypothesis of such fluids gives results agreeing in a remarkable manner with the facts and their measures, as Coulomb and others have shown. It may be asked how far we may, in such a case, suppose that we have discovered the true cause of the phenomena, and whether it is sufficiently proved that these fluids really exist. The

---

[*] Multa me movent, &c.. Pref. to the *Principia*, already quoted.
[†] *Hist. Ind. Sci.*, iii., 23,

right answer seems to be, that the hypothesis certainly represents the truth so far as regards the polar relation of the two energies, and the laws of the attractive and repulsive forces of the particles in which these energies reside; but that we are not entitled to assume that the vehicles of these energies possess other attributes of material fluids, or that the forces thus ascribed to the particles are the primary elementary forces from which the action originates. We are the more bound to place this cautious limit to our acceptance of the Coulombian theory, since in electricity Faraday has in vain endeavoured to bring into view one of the polar fluids without the other: whereas such a result ought to be possible if there were two separable fluids. The impossibility of this separate exhibition of one fluid appears to show that the fluids are real only so far as they are polar. And Faraday's view above mentioned, according to which the attractions at a distance are resolved into the action of lines of polarized particles of air, appears still further to show that the conceptions hitherto entertained of electrical forces, according to the Coulombian theory, do not penetrate to the real and intimate nature of the causation belonging to this case.

10. Since it is thus difficult to know when we have seized the true cause of the phenomena in any department of science, it may appear to some persons that physical inquirers are imprudent and unphilosophical in undertaking this research of causes; and that it would be safer and wiser to confine ourselves to the investigation of the laws of phenomena, in which field the knowledge which we obtain is definite· and certain. Hence there have not been wanting those who have laid it down as a maxim that " science must study only the laws of phenomena, and never the mode of production*." But it is

* COMTE, *Philosophic Positive.*

easy to see that such a maxim would confine the breadth
and depth of scientific inquiries to a most scanty and
miserable limit. Indeed, such a rule would defeat its
own object; for the laws of phenomena, in many cases,
cannot be even expressed or understood without some
hypothesis respecting their mode of production. How
could the phenomena of polarization have been conceived
or reasoned upon, except by imagining a polar arrangement of particles, or transverse vibrations, or some equivalent hypothesis? The doctrines of fits of easy transmission, the doctrine of moveable polarization, and the like,
even when erroneous as representing the whole of the
phenomena, were still useful in combining some of them
into laws; and without some such hypotheses the facts could
not have been followed out. The doctrine of a fluid caloric
may be false; but without imagining such a fluid, how
could the movement of heat from one part of a body to
another be conceived? It may be replied that Fourier,
Laplace, Poisson, who have principally cultivated the
Theory of Heat, have not conceived it as a fluid, but
have referred conduction to the radiation of the molecules of bodies, which they suppose to be separate points.
But this molecular constitution of bodies is itself an
assumption of the mode in which the phenomena are
produced; and the radiation of heat suggests inquiries
concerning a fluid emanation, no less than its conduction
does. In like manner, the attempts to connect the laws
of phenomena of heat and of gases, have led to hypotheses respecting the constitution of gases, and the combination of their particles with those of caloric, which
hypotheses may be false, but are probably the best means
of discovering the truth.

To debar science from inquiries like these, on the
ground that it is her business to inquire into facts, and
not to speculate about causes, is a curious example of

that barren caution which hopes for truth without daring to venture upon the quest of it. This temper would have stopped with Kepler's discoveries, and would have refused to go on with Newton to inquire into the mode in which the phenomena are produced. It would have stopped with Newton's optical facts, and would have refused to go on with him and his successors to inquire into the mode in which these phenomena are produced. And, as we have abundantly shown, it would, on that very account, have failed in seeing what the phenomena really are.

In many subjects the attempt to study the laws of phenomena, independently of any speculations respecting the causes which have produced them, is neither possible for human intelligence nor for human temper. Men cannot contemplate the phenomena without clothing them in terms of some hypothesis, and will not be schooled to suppress the questionings which at every moment rise up within them concerning the causes of the phenomena. Who can attend to the appearances which come under the notice of the geologist;—strata regularly bedded, full of the remains of animals such as now live in the depths of the ocean, raised to the tops of mountains, broken, contorted, mixed with rocks such as still flow from the mouths of volcanos;—who can see phenomena like these, and imagine that he best promotes the progress of our knowledge of the earth's history, by noting down the facts, and abstaining from all inquiry whether these are really proofs of past states of the earth and of subterraneous forces, or merely an accidental imitation of the effects of such causes? In this and similar cases, to proscribe the inquiry into causes would be to annihilate the science.

Finally, this caution does not even gain its own single end, the escape from hypotheses. For, as we have said, those who will not seek for new and appropriate causes of newly-studied phenomena, are almost inevitably led to ascribe the facts to modifications of causes already familiar. They may declare that they will not hear of such causes as vital powers, elective affinities, electric, or calorific, or luminiferous ethers or fluids ; but they will not the less on that account assume hypotheses equally un-authorized; for instance—universal mechanical forces; a molecular constitution of bodies; solid, hard, inert mat-ter;—and will apply these hypotheses in a manner which is arbitrary in itself as well as quite insufficient for its purpose.

11. It appears, then, to be required, both by the analogy of the most successful efforts of science in past times and by the irrepressible speculative powers of the human mind, that we should attempt to discover both the *laws of phenomena,* and their *causes.* In every department of science, when prosecuted far enough, these two great steps of investigation must succeed each other. The laws of phenomena must be known before we can speculate concerning causes ; the causes must be inquired into when the phenomena have been reduced to rule. In both these speculations the suppositions and concep-tions which occur must be constantly tested by reference to observation and experiment. In both we must, as far as possible, devise hypotheses which, when we thus test them, display those characters of truth of which we have already spoken ;—an agreement with facts such as will stand the most patient and rigid inquiry; a provision for predicting truly the results of untried cases; a consilience of inductions from various classes of facts; and a pro-gressive tendency of the scheme to simplicity and unity.

12. We shall attempt hereafter to give several rules of a more precise and detailed kind for the discovery of the causes, and still more, of the laws of phe-nomena. But it will be useful in the first place to point out the Classification of the Sciences which results from

the principles already established in this work. And for this purpose we must previously decide the question, whether the practical Arts, as Medicine and Engineering, must be included in our list of Sciences.

CHAPTER VIII.

OF ART AND SCIENCE.

1. THE distinction of Arts and Sciences very materially affects all classifications of the departments of Human Knowledge. It is often maintained, expressly or tacitly, that the Arts are a part of our knowledge, in the same sense in which the Sciences are so; and that Art is the application of Science to the purposes of practical life. It will be found that these views require some correction, when we understand *Science* in the exact sense in which we have throughout endeavoured to contemplate it, and in which alone our examination of its nature can instruct us in the true foundations of our knowledge.

When we cast our eyes upon the early stages of the histories of nations, we cannot fail to be struck with the consideration, that in many countries the Arts of life already appear, at least in some rude form or other, when, as yet, nothing of science exists. A practical knowledge of astronomy such as enables them to reckon months and years, is found among all nations except the mere savages. A practical knowledge of mechanics must have existed in those nations which have left us the gigantic monuments of early architecture. The pyramids and temples of Egypt and Nubia, the Cyclopean walls of Italy and Greece, the temples of Magna Græcia and Sicily, the obelisks and edifices of India, the cromlechs and Druidical circles of countries formerly Celtic,—must have demanded no small practical mechanical skill and power. Yet those modes of reckoning time must have preceded the rise of speculative astronomy; these structures must have been erected before the theory of mechanics was known. To suppose, as some have done, a great body of science, now lost, to have existed in the remote ages to which these remains belong, is not only quite gratuitous and contrary to all analogy, but is a supposition which cannot be extended so far as to explain all such cases. For it is impossible to imagine that *every* art has been preceded by the science which renders a reason for its processes. Certainly men formed wine from the grape, before they possessed a science of fermentation; the first instructor of every artificer in brass and iron can hardly be supposed to have taught the chemistry of metals as a science; the inventor of the square and the compasses had probably no more knowledge of demonstrated geometry than have the artisans who now use those implements; and finally the use of speech, the employment of the inflections and combinations of words, must needs be assumed as having been prior to any general view of the nature and analogy of language. Even at this moment, the greater part of the arts which exist in the world are not accompanied by the sciences on which they theoretically depend. Who shall state to us the general chemical truths to which the manufactures of glass, and porcelain, and iron, and brass, owe their existence? Do not almost all artisans practise many successful artifices long before science explains the ground of the process? Do not arts at this day exist, in a high state of perfection, in countries in which there is no science, as China and India? These countries and many others have no theories of mechanics, of optics, of chemistry, of physiology; yet they construct and use mechanical and optical instruments, make chemical combinations, take advantage of physiological laws. It is too evident to need further illustration that art may exist without science;—that it has usually been anterior to it, and even now commonly advances independently, leaving science to follow as it can.

2. We here mean by *Science*, that exact, general, speculative knowledge, of which we have, throughout this work, been endeavouring to exhibit the nature and rules. Between such science and the *practical Arts* of life, the points of difference are sufficiently manifest. The object of Science is *Knowledge;* the object of Art are *Works.* The latter is satisfied with producing its material results; to the former, the operations of matter, whether natural or artificial, are interesting only so far as they can be embraced by intelligible principles. The end of art is the beginning of science; for when it is seen *what* is done, then comes the question *why* it is done. Art may have fixed general rules, stated in words; but she has these merely as means to an end: to Science, the propositions which she obtains are each, in itself, a sufficient end of the effort by which it is acquired. When Art has brought forth her product, her task is finished; Science is constantly led by one step of her path to another. Each proposition which she obtains impels her to go onwards to other propositions more general, more profound, more simple. Art puts elements together, without caring to know what they are, or why they coalesce. Science analyses the compound, and at every such step strives not only to perform, but to understand the analysis. Art advances in proportion as she becomes able to bring forth products more multiplied, more complex, more various; but Science, straining her eyes to penetrate more and more deeply into the nature of things, reckons her success in proportion as she sees, in all the phenomena, however multiplied, complex, and varied, the results of one or two simple and general laws.

3. There are many acts which man, as well as animals, performs by the guidance of nature, without seeing or seeking the reason why he does so; as the acts by which he balances himself in standing or moving, and those by which he judges of the form and position of the objects around him. These actions have their reason in the principles of geometry and mechanics; but of such reasons he who thus acts is unaware: he works blindly, under the impulse of an unknown principle which we call *Instinct*. When man's speculative nature seeks and finds the reasons why he should act thus or thus;—why he should stretch out his arm to prevent his falling, or assign a certain position to an object in consequence of the angles under which it is seen;—he may perform the same actions as before, but they are then done by the aid of a different faculty, which, for the sake of distinction, we may call *Insight*. Instinct is a purely active principle; it is seen in deeds alone; it has no power of looking inwards; it asks no questions; it has no tendency to discover reasons or rules; it is the opposite of Insight.

4. Art is not identical with Instinct: on the contrary, there are broad differences. Instinct is stationary; Art is progressive. Instinct is mute; it acts, but gives no rules for acting: Art can speak; she can lay down rules. But though Art is thus separate from Instinct, she is not essentially combined with Insight. She can see what to do, but she needs not to see why it is done. She may lay down rules, but it is not her business to give reasons. When man makes *that* his employment, he enters upon the domain of science. Art takes the phenomena and laws of nature as she finds them: that they are multiplied, complex, capricious, incoherent, disturbs her not. She is content that the rules of nature's operations should be perfectly arbitrary and unintelligible, provided they are constant, so that she can depend upon their effects. But

Science is impatient of all appearance of caprice, inconsistency, irregularity, in nature. She will not believe in the existence of such characters. She resolves one apparent anomaly after another; her task is not ended till every thing is so plain and simple, that she is tempted to believe she sees that it could by no possibility have been otherwise than it is.

5. It may be said that, after all, Art does really involve the knowledge which Science delivers;—that the artisan who raises large weights, practically *knows* the properties of the mechanical powers;—that he who manufactures chemical compounds is virtually acquainted with the laws of chemical combination. To this we reply, that it might on the same grounds be asserted, that he who acts upon the principle that two sides of a triangle are greater than the third is really acquainted with geometry; and that he who balances himself on one foot knows the properties of the centre of gravity. But this is an acquaintance with geometry and mechanics which even brute animals possess. It is evident that it is not of such knowledge as this that we have here to treat. It is plain that this mode of possessing principles is altogether different from that contemplation of them on which science is founded We neglect the most essential and manifest differences, if we confound our unconscious assumptions with our demonstrative reasonings.

6. The real state of the case is, that the principles which Art *involves*, Science alone *evolves*. The truths on which the success of Art depends, lurk in the artist's mind in an undeveloped state; guiding his hand, stimulating his invention, balancing his judgment, but not appearing in the form of enunciated propositions. Principles are not to him direct objects of meditation: they are secret Powers of Nature, to which the forms which tenant the world owe their constancy, their movements,

their changes, their luxuriant and varied growth, but which he can nowhere directly contemplate. That the creative and directive principles which have their lodgment in the artist's mind, when *unfolded* by our speculative powers into systematic shape, become science, is true; but it is precisely this process of development which gives to them their character of science. In practical Art, principles are unseen guides, leading us by invisible strings through paths where the end alone is looked at: it is for Science to direct and purge our vision so that these airy ties, these principles and laws, generalizations and theories, become distinct objects of vision. Many may feel the intellectual monitor, but it is only to her favourite heroes that the Goddess of Wisdom visibly reveals herself.

7. Thus Art, in its earlier stages at least, is widely different from Science, independent of it, and anterior to it. At a later period, no doubt, Art may borrow aid from Science; and the discoveries of the philosopher may be of great value to the manufacturer and the artist. But even then, this application forms no essential part of the science: the interest which belongs to it is not an intellectual interest. The augmentation of human power and convenience may impel or reward the physical philosopher; but the processes by which man's repasts are rendered more delicious, his journeys more rapid, his weapons more terrible, are not, therefore, Science. They may involve principles which are of the highest interest to science; but as the advantage is not practically more precious because it results from a beautiful theory, so the theoretical principle has no more conspicuous place in science because it leads to convenient practical consequences. The nature of science is purely intellectual; knowledge alone,—exact general truth,—is her object; and we cannot mix with such materials, as matters of the

same kind, the merely empirical maxims of art, without introducing endless confusion into the subject, and making it impossible to attain any solid footing in our philosophy.

8. I shall therefore not place, in our Classification of the Sciences, the Arts, as has generally been done; nor shall I notice the applications of sciences to art, as forming any separate portion of each science. The sciences, considered as bodies of general speculative truths, are what we are here concerned with; and applications of such truths, whether useful or useless, are important to us only as illustrations and examples. Whatever place in human knowledge the Practical Arts may hold, they are not Sciences. And it is only by this rigorous separation of the Practical from the Theoretical, that we can arrive at any solid conclusions respecting the nature of truth, and the mode of arriving at it, such as it is our object to attain.

Chapter IX.

OF THE CLASSIFICATION OF SCIENCES.

1. The Classification of Sciences has its chief use in pointing out to us the extent of our powers of arriving at truth, and the analogies which may obtain between those certain and lucid portions of knowledge with which we are here concerned, and those other portions, of a very different interest and evidence, which we here purposely abstain to touch upon. The classification of human knowledge will, therefore, have a more peculiar importance when we can include in it the moral, political, and metaphysical, as well as the physical portions of our knowledge. But such a survey does not belong to our

present undertaking: and a general view of the connexion and order of the branches of sciences which our
review has hitherto included, will even now possess some
interest; and may serve hereafter as an introduction to a
more complete scheme of the general body of human
knowledge.

2. In this, as in any other case, a sound classification
must be the result, not of any assumed principles impe-
ratively applied to the subject, but of an examination of
the objects to be classified;—of an analysis of them
into the principles in which they agree and differ. The
Classification of Sciences must result from the consideration of their nature and contents. Accordingly, that
review of the sciences in which the History of them
engaged us, led to a Classification, of which the main
features are indicated in that work. The Classification
thus obtained, depends neither upon the faculties of the
mind to which the separate parts of our knowledge owe
their origin, nor upon the objects which each science
contemplates; but upon a more natural and fundamental
element;—namely, the *Ideas* which each science involves.
The Ideas regulate and connect the facts, and are the
foundations of the reasoning, in each science: and having
in the present work more fully examined these Ideas, we
are now prepared to state here the classification to which
they lead. If we have rightly traced each science to the
Conceptions which are really fundamental *with regard to
it*, and which give rise to the first principles on which it
depends, it is not necessary for our purpose that we
should decide whether these Conceptions are absolutely
ultimate principles of thought, or whether, on the contrary, they can be further resolved into other Fundamental Ideas. We need not now suppose it determined
whether or not Number is a mere modification of the
Idea of Time, and Force a mere modification of the Idea

of Cause: for however this may be, our Conception of
Number is the foundation of Arithmetic, and our Conception of Force is the foundation of Mechanics.   It is to
be observed also that in our classification, each Science
may involve, not only the Ideas or Conceptions which are
placed opposite to it in the list, but also all which *precede*
it.   Thus Formal Astronomy involves not only the Conception of Motion, but also those which are the foundation of Arithmetic and Geometry.   In like manner,
Physical Astronomy employs the Sciences of Statics and
Dynamics, and thus rests on their foundations; and they,
in turn, depend upon the Ideas of Space and of Time as
well as of Cause.

3. We may further observe, that this arrangement of
Sciences according to the Fundamental Ideas which they
involve, points out the transition from those parts of
human knowledge which have been included in our
History and Philosophy, to other regions of speculation
into which we have not entered.   We have repeatedly
found ourselves upon the borders of inquiries of a psychological, or moral, or theological nature.   Thus the History
of Physiology* led us to the consideration of Life, Sensation, and Volition; and at these Ideas we stopped, that
we might not transgress the boundaries of our subject as
then predetermined.   It is plain that the pursuit of
such conceptions and their consequences, would lead us
to the sciences (if we are allowed to call them sciences)
which contemplate not only animal, but human principles of action, to Anthropology and Psychology.   In
other ways, too, the Ideas which we have examined,
although manifestly the foundations of sciences such as
we have here treated of, also plainly pointed to speculations of a different order;   thus the Idea of a Final Cause
is an indispensable guide in Biology, as we have seen;

---

* *Hist. Ind. Sci.*, iii. 431.

but the conception of Design as directing the order of
nature, once admitted, soon carries us to higher contemplations.   Again, the Class of Palætiological Sciences
which we were in the History led to construct, although
we there admitted only one example of the Class, namely
Geology, does in reality include many vast lines of
research ; as the history and causes of the diffusion of
plants and animals, the history of languages, arts, and
consequently of civilization.   Along with these researches,
comes the question how far these histories point backwards to a natural or a supernatural origin ; and the Idea
of a First Cause is thus brought under our consideration.
Finally, it is not difficult to see that as the Physical
Sciences have their peculiar governing Ideas, which support and shape them, so the Moral and Political Sciences
also must similarly have their fundamental and formative
Ideas, the source of universal and certain truths, each of
their proper kind.   But to follow out the traces of this
analogy, and to verify the existence of those Fundamental
Ideas in Morals and Politics, is a task quite out of the
sphere of the work in which we are here engaged.

4. We may now place before the reader our Classification of the Sciences, adding in the list a few not belonging to our present subject, that the nature of the transition by which we are to extend our philosophy into
a wider and higher region may be in some measure
perceived.

We may observe that the term *Physics*, when confined
to a peculiar class of Sciences, is usually understood to
exclude the Mechanical Sciences on the one side, and
Chemistry on the other; and thus embraces the Secondary
Mechanical and Analytico-Mechanical Sciences.   But
the adjective *Physical* applied to any science and opposed
to *Formal*, as in Astronomy and Optics, implies those
speculations in which we consider not only the Laws of

Phenomena but their Causes; and generally, as in those cases, their Mechanical Causes.

Fundamental Ideas or Conceptions.	Sciences.	Classification.
Space	Geometry	
Time		
Number	Arithmetic	Pure Mathematical Sciences.
Sign	Algebra	
Limit	Differentials	
Motion	Pure Mechanism	
	Formal Astronomy	Pure Motional Sciences.
Cause		
Force		
Matter	Statics	
Inertia	Dynamics	
Fluid Pressure	Hydrostatics	Mechanical Sciences.
	Hydrodynamics	
	Physical Astronomy	
Outness		
Medium *of Sensation*	Acoustics	
Intensity *of Qualities*	Formal Optics	Secondary Mechanical Sciences.
Scales of Qualities	Physical Optics	(Physics.)
	Thermotics	
	Atmology	
Polarity	Electricity	Analytico-Mechanical Sciences.
	Magnetism	
	Galvanism	(Physics.)
Element (*Composition*)		
Chemical Affinity		
Substance (*Atoms*)	Chemistry	Analytical Science.
Symmetry	Crystallography	Analytico-Classificatory
Likeness	Systematic Mineralogy	Sciences.
Degrees of Likeness	Systematic Botany	
	Systematic Zoology	Classificatory Sciences.
Natural Affinity	Comparative Anatomy	
(*Vital Powers*)		
Assimilation		
Irritability		
(*Organization*)	Biology	Organical Sciences.
Final Cause		
Instinct		
Emotion	Psychology	
Thought		
Historical Causation	Geology	
	Distribution of Plants and Animals	Palætiological Sciences.
	Glossology	
	Ethnography	
First Cause	Natural Theology.	

In the next Book, we shall trace the opinions of some of the most eminent writers, respecting the sources of our knowledge of nature and the rules which may aid us in seeking it. For the knowledge of a true Scientific Method is a science resembling other sciences; and the ideas and views which it involves have been in some measure gradually developed into clearness and certainty by successive attempts. We may, therefore, acquire a more confident persuasion of the right direction of our path, by seeing how far it coincides with that which has been pointed out, with more or less distinctness, by many of the most sagacious and vigorous intellects, who have bestowed their attention upon this inquiry.

INDUCTIVE TABLE OF ASTRONOMY.

THE TRUTH OF UNIVERSAL GRAVITATION.
(All bodies attract each other with a Force of Gravity which is inversely as the squares of the distances.)

INDUCTIVE TABLE OF OPTICS.

A PDF of the colour image originally positioned here can be downloaded
from the web address given on page iv of this book,
by clicking on 'Resources Available'.

BOOK XII.

REVIEW OF OPINIONS ON THE NATURE OF KNOWLEDGE AND THE METHODS OF SEEKING IT

CHAPTER I.

INTRODUCTION.

By the examination of the elements of human thought in which we have been engaged, and by a consideration of the history of the most clear and certain parts of our knowledge, we have been led to certain doctrines respecting the progress of that exact and systematic knowledge which we call Science; and these doctrines we have endeavoured to lay before the reader in the preceding Book. The questions on which we have thus ventured to pronounce have had a strong interest for man, from the earliest period of his intellectual progress, and have been the subjects of lively discussion and bold speculation in every age. We conceive that in the doctrines to which our researches have conducted us, we have a far better hope that we possess a body of permanent truths, than the earlier essays on the same subjects could furnish. For we have not taken our examples of knowledge at hazard, as earlier speculators did, and were almost compelled to do; but have drawn our materials from the vast store of unquestioned truths which modern science offers to us: and we have formed our judgment concerning the nature and progress of knowledge by considering what such science is, and how it has reached its present condition.   But though we have thus pursued our speculations concerning knowledge with advantages which earlier writers did not possess, it is still both interesting and instructive for us to regard the opinions upon this subject which have been delivered by the philosophers of past times.   It is especially interesting to see some of the truths which we have endeavoured to expound, gradually dawning in men's minds, and assuming the clear and permanent form in which we can now contemplate them. I shall therefore, in this Book, pass in review many of the opinions of the writers of various ages concerning the mode by which man best acquires the truest knowledge; and I shall endeavour, as we proceed, to appreciate the real value of such judgments, and their place in the progress of sound philosophy.

In this estimate of the opinions of others, I shall be guided by those general doctrines which I have, as I trust, established in the preceding part of this work. And without attempting here to give any summary of these doctrines, I may remark that there are two main principles by which speculations on such subjects in all ages are connected and related to each other; namely, the opposition of *Ideas* and *Sensations*, and the distinction of *practical* and *speculative* knowledge.   The opposition of Ideas and Sensations is exhibited to us in the antithesis of Theory and Fact, which are necessarily considered as distinct and of opposite natures, and yet necessarily identical, and constituting Science by their identity.   In like manner, although practical knowledge is in substance identical with speculative, (for all knowledge is speculation,) there is a distinction between the two in their history, and in the subjects by which they are exemplified, which distinction is quite essential in judging of the philosophical views of the ancients.   The alternatives of identity and diversity, in these two antitheses, the successive separation, opposition, and reunion of principles which thus arise, have produced, (as they may easily be imagined capable of doing,) a long and varied series of systems concerning the nature of knowledge; among which we shall have to guide our course by the aid of the views already presented.

I am far from undertaking, or wishing, to review the whole series of opinions which thus comes under our view; and I do not even attempt to examine all the principal authors who have written on such subjects. I merely wish to select some of the most considerable forms which such opinions have assumed, and to point out in some measure the progress of truth from age to age. In doing this, I can only endeavour to seize some of the most prominent features of each time and of each step; and I must pass rapidly from classical antiquity to those which we have called the dark ages, and from them to modern times. At each of these periods the modifications of opinion, and the speculations with which they were connected, formed a vast and tangled maze, into the byways of which our plan does not allow us to enter. We shall esteem ourselves but too fortunate, if we can discover the single track by which ancient led to modern philosophy.

I must also repeat that my survey of philosophical writers is here confined to this one point,—their opinions on the nature of knowledge and the method of science. I with some effort avoid entering upon other parts of the philosophy of those of whom I speak; I knowingly pass by those portions of their speculations which are in many cases the most interesting and celebrated;—their opinions concerning the human soul, the Divine governor of the world, the foundations or leading doctrines of politics, religion, and general philosophy. I am desirous that my reader should bear this in mind, since he must otherwise be offended with the scanty and partial view which I give in this place of the philosophers whom I enumerate.

CHAPTER II.

PLATO.

THERE would be small advantage in beginning our examination earlier than the period of the Socratic School at Athens; for although the spirit of inquiry on such subjects had awoke in Greece at an earlier period, and although the peculiar aptitude of the Grecian mind for such researches had shown itself repeatedly in subtle distinctions and acute reasonings, all the positive results of these early efforts were contained in a more definite form in the reasonings of the Platonic age. Anterior to that time, the Greeks did not possess plain and familiar examples of exact knowledge, such as the truths of Arithmetic, Geometry, Astronomy, and Optics, became in the school of Plato; nor were the antitheses of which we spoke above, so distinctly and fully unfolded as we find them in Plato's works.

The question which hinges upon one of these antitheses, occupies a prominent place in several of the Platonic dialogues;—namely, whether our knowledge be obtained by means of Sensation or of Ideas. One of the doctrines which Plato most earnestly inculcated upon his countrymen was, that we do not *know* concerning sensible objects, but concerning ideas. The first attempts of the Greeks at metaphysical analysis had given rise to a school which maintained that material objects are the only realities. In opposition to this, arose another school,

which taught that material objects have no permanent
reality, but are ever waxing and waning, constantly
changing their substance. "And hence," as Aristotle
says*, "arose the doctrine of ideas which the Platonists
held. For they assented to the opinion of Heraclitus,
that all sensible objects are in a constant state of flux.
So that if there is to be any knowledge and science, it
must be concerning some permanent natures, different
from the sensible natures of objects; for there can be no
permanent science respecting that which is perpetually
changing. It happened that Socrates turned his speculations to the moral virtues, and was the first philosopher
who endeavoured to give universal definitions of such
matters. He wished to reason systematically, and therefore
he tried to establish definitions, for definitions are the
basis of systematic reasoning. There are two things
which may justly be looked upon as steps in philosophy
due to Socrates; inductive reasonings, and universal definitions;—both of them steps which belong to the foundations of science. Socrates, however, did not make
universals, or definitions separable from the objects; but
his followers separated them, and these essences they
termed *Ideas*." And the same account is given by other
writers†. "Some existences are sensible, some intelligible: and according to Plato, they who wish to understand
the principles of things, must first separate the ideas from
the things, such as the ideas of Similarity, Unity, Number,
Magnitude, Position, Motion: second, that we must
assume an absolute Fair, Good, Just, and the like: third,
that we must consider the ideas of relation, as Knowledge,
Power: recollecting that the things which we perceive
have this or that appellation applied to them because
they partake of this or that Idea; those things being *just*
which participate in the idea of The Just, those being

* *Metaph.*, xii. 4. † Diog. Laert. *Vit. Plat.*

beautiful, which contain the idea of The Beautiful." And many of the arguments by which this doctrine was maintained are to be found in the Platonic dialogues. Thus the opinion that true knowledge consists in sensation, which had been asserted by Protagoras and others, is refuted in the *Theætetus:* and we may add, so victoriously refuted, that the arguments there put forth have ever since exercised a strong influence upon the speculative world. It may be remarked that in the minds of Plato and of those who have since pursued the same paths of speculation, the interest of such discussions as those we are now referring to, was by no means limited to their bearing upon mere theory; but was closely connected with those great questions of morals which have always a practical import. Those who asserted that the only foundation of knowledge was sensation, asserted also that the only foundation of virtue was the desire of pleasure. And in Plato, the metaphysical part of the disquisitions concerning knowledge in general, though independent in its principles, always seems to be subordinate in its purpose to the questions concerning the knowledge of our duty.

Since Plato thus looked upon the Ideas which were involved in each department of knowledge as forming its only essential part, it was natural that he should look upon the study of Ideas as the true mode of pursuing knowledge. This he himself describes in the *Philebus* *. " The best way of arriving at truth is not very difficult to point out, but most hard to pursue. All the arts which have ever been discovered, were revealed in this manner. It is a gift of the gods to man, which, as I conceive, they sent down by some Prometheus, in a blaze of light; and the ancients, more clear-sighted than we, and less removed from the gods, handed down this traditionary doctrine:

* T. ii. p. 16, c, d. ed. Bekker, t. v. p. 437.

that whatever is said to be, comes of One and of Many, and comprehends in itself the Finite and the Infinite in coalition (being one kind, and consisting of infinite individuals). And this being the state of things, we must, in each case, endeavour to seize the One Idea (the idea of the kind) as the chief point; for we shall find that it is there. And when we have seized this one thing, we may then consider how it comprehends in itself two, or three, or any other number; and, again, examine each of these ramifications separately; till at last we perceive, not only that One is at the same time One and Many, but also how many: And when we have thus filled up the interval between the Infinite and the One, we may consider that we have done with each one. The gods then, as I have said, taught us by tradition thus to contemplate, and to learn, and to teach one another. But the philosophers of the present day seize upon the One, at hazard, too soon or too late, and then immediately snatch at the Infinite; but the intermediate steps escape them, by which the subject is subdivided, so that it can be the subject of logical exposition and discussion."

It would seem that what the author here describes as the most perfect form of exposition, is that which refers each object to its place in a classification containing a complete series of subordinations, and which gives a definition of each class. We have repeatedly remarked that, in sciences of classification, each new definition which gives a tenable and distinct separation of classes is an important advance in our knowledge; but that such definitions are rather the last than the first step in each advance. In the progress of real knowledge, these definitions are always the results of a laborious study of individual cases, and are never arrived at by a pure effort of thought, which is what Plato appears to have imagined as the true mode of philosophizing. And still less do the advances of other sciences consist in seizing at once upon the highest generality, and filling in afterwards all the intermediate steps between that and the special instances. On the contrary, as we have seen, the ascents from particular to general are all successive; and each step of this ascent requires time, and labour, and a patient examination of actual facts and objects.

It would, of course, be absurd to blame Plato for having inadequate views of the nature of progressive knowledge, at the time when knowledge could hardly be said to have begun its progress. But we already find in his speculations, as appears in the passages just quoted from his writings, several points brought into view which will require our continued attention as we proceed. In overlooking the necessity of a gradual and successive advance from the less general to the more general truths, Plato shared in a dimness of vision which prevailed among philosophers to the time of Francis Bacon. In thinking too slightly of the study of actual nature, he manifested a bias from which the human intellect freed itself in the vigorous struggles which terminated the dark ages. In pointing out that all knowledge implies a unity of what we observe as manifold, which unity is given by the mind, Plato taught a lesson which has of late been too obscurely acknowledged, the recoil by which men repaired their long neglect of facts having carried them for a while so far as to think that facts were the whole of our knowledge. And in analysing this principle of Unity, by which we thus connect sensible things, into various Ideas, such as Number, Magnitude, Position, Motion, he made a highly important step, which it has been the business of philosophers in succeeding times to complete and to follow out.

But the efficacy of Plato's speculations in their bearing upon physical science, and upon theory in general, was much weakened by the confusion of practical with theoretical knowledge, which arose from the ethical propensities of the Socratic school. In the Platonic Dialogues, Art and Science are constantly spoken of indiscriminately. The skill possessed by the Painter, the Architect, the Shoemaker, is considered as a just example of human science, no less than the knowledge which the geometer or the astronomer possesses of the theoretical truths with which he is conversant. Not only so; but traditionary and mythological tales, mystical imaginations and fantastical etymologies, are mixed up, as no less choice ingredients, with the most acute logical analyses, and the most exact conduct of metaphysical controversies. There is no distinction made between the knowledge possessed by the theoretical psychologist and the physician, the philosophical teacher of morals and the legislator or the administrator of law. This, indeed, is the less to be wondered at, since even in our own time the same confusion is very commonly made by persons not otherwise ignorant or uncultured.

On the other hand, we may remark finally, that Plato's admiration of Ideas was not a barren imagination, even so far as regarded physical science. For, as we have seen *, he had a very important share in the introduction of the theory of epicycles, having been the first to propose to astronomers in a distinct form, the problem of which that theory was the solution; namely, " to explain the celestial phenomena by the combination of equable circular motions." This demand of an ideal hypothesis which should exactly express the phenomena (as well as they could then be observed), and from which, by the interposition of suitable steps, all special cases might be deduced, falls in well with those views respecting the proper mode of seeking knowledge which we have quoted from the *Philebus*. And the Idea which could thus

---

* *Hist. Ind. Sci.*, i. 104.

represent and replace all the particular Facts, being not only sought but found, we may readily suppose that the philosopher was, by this event, strongly confirmed in his persuasion that such an Idea was indeed what the inquirer ought to seek. In this conviction all his genuine followers up to modern times have participated; and thus, though they have avoided the error of those who hold that facts alone are valuable as the elements of our knowledge, they have frequently run into the opposite error of too much despising and neglecting facts, and of thinking that the business of the inquirer after truth was only a profound and constant contemplation of the conceptions of his own mind. But of this hereafter.

Chapter III.

ARISTOTLE.

The views of Aristotle with regard to the foundations of human knowledge are very different from those of his tutor Plato, and are even by himself put in opposition to them. He dissents altogether from the Platonic doctrine that Ideas are the true materials of our knowledge; and after giving, respecting the origin of this doctrine, the account which we quoted in the last chapter, he goes on to reason against it. "Thus," he says*, "they devised Ideas of all things which are spoken of as universals: much as if any one having to count a number of objects, should think that he could not do it while they were few, and should expect to count them by making them more numerous. For the kinds of things are almost more numerous than the special sensible objects, by seeking the causes of which they were led to their Ideas." He

* *Metaph.* xii. 4.

then goes on to urge several other reasons against the assumption of Ideas and the use of them in philosophical researches.

Aristotle himself establishes his doctrines by trains of reasoning. But reasoning must proceed from certain First Principles; and the question then arises, Whence are these First Principles obtained? To this he replies, that they are the result of *Experience,* and he even employs the same technical expression by which we at this day describe the process of collecting these principles from observed facts;—that they are obtained by *Induction.* I have already quoted passages in which this statement is made*. "The way of reasoning," he says†, "is the same in philosophy, and in any art or science: we must collect the *facts* (τὰ ὑπαρχόντα), and the things to which the facts happen, and must have as large a supply of these as possible, and then we must examine them according to the terms of our syllogisms." . . . . "There are peculiar principles in each science; and in each case these principles must be obtained from *experience.* Thus astronomical observation supplies the principles of astronomical science. For the phenomena being rightly taken, the demonstrations of astronomy were discovered; and the same is the case with any other Art or Science. So that if the facts in each case be taken, it is our business to construct the demonstrations. For if *in our natural history* (κατὰ τὴν ἱστορίαν) we have omitted none of the facts and properties which belong to the subject, we shall learn what we can demonstrate and what we cannot." And, again‡, "It is manifest that if any sensation be wanting, there must be some knowledge wanting, which we are thus prevented from having. For we acquire knowledge either *by Induction* (ἐπαγωγῇ) or by Demonstration: and

* *Hist. Ind. Sci.,* i. 74.   † *Analyt. Prior.,* i. 30.
‡ *Analyt. Post ,* i. 18.

Demonstration is from universals, but Induction from particulars. It is impossible to have universal theoretical propositions except by Induction: and we cannot make inductions without having sensation; for sensation has to do with particulars."

It is easy to show that Aristotle uses the term Induction, as we use it, to express the process of collecting a general proposition from particular cases in which it is exemplified. Thus in a passage which we have already quoted*, he says, "Induction, and Syllogism from Induction is when we attribute one extreme term to the middle by means of the other." The import of this technical phraseology will further appear by the example which he gives: "We find that several animals which are deficient in bile are longlived, as man, the horse, the mule; hence we infer that *all* animals which are deficient in bile are longlived."

We may observe, however, that both Aristotle's notion of induction, and many other parts of his philosophy, are obscure and imperfect, in consequence of his refusing to contemplate ideas as something distinct from sensation. It thus happens that he always assumes the ideas which enter into his propositions as *given;* and considers it as the philosopher's business to determine whether such propositions are true or not: whereas the most important feature in induction is, as we have said, the *introduction* of a new idea, and not its employment when once introduced. That the mind in this manner gives unity to that which is manifold;—that we are thus led to speculative principles which have an evidence higher than any others,—and that a peculiar sagacity in some men seizes upon the conceptions by which the facts may be bound into true propositions,—are doctrines which form no essential part of the philosophy of the Stagirite, although

---

* *Anal. Pri.*, ii. 23, περὶ τῆς ἐπαγωγῆς.

such views are sometimes recognized, more or less clearly, in his expressions. Thus he says*, "There can be no knowledge when the sensation does not continue in the mind. For this purpose, it is necessary both to perceive, and to have some *unity* in the mind; (αἰσθανομένοις ἔχειν "EN TI ἐν τῇ ψυχῇ) and many such perceptions having taken place, some difference is then perceived: and from the remembrance of these arises Reason. Thus from Sensation comes Memory, and from Memory of the same thing often repeated comes Experience: for many acts of Memory make up one Experience. And from Experience, or from any Universal Notion which takes a permanent place in the mind,—from the *unity in the manifold*, the same some one thing being found in many facts,—springs the first principle of Art and of Science; of Art if it be employed about production; of Science, if about existence."

I will add to this, Aristotle's notice of *Sagacity;* since, although little or no further reference is made to this quality in his philosophy, the passage fixes our attention upon an important step in the formation of knowledge. "Sagacity," (ἀγχίνοια) he says†, "is a hitting by guess (εὐστοχία τις) upon the middle term (the conception common to two cases) in an inappreciable time. As for example, if any one seeing that the bright side of the moon is always towards the sun, suddenly perceives why this is; namely, because the moon shines by the light of the sun:—or if he sees a person talking with a rich man, he guesses that he is borrowing money;—or conjectures that two persons are friends, because they are enemies of the same person."—To consider only the first of these examples;—the conception here introduced, that of a body shining by the light which another casts upon it, is not contained in the observed facts, but introduced

* *Anal. Post.*, ii. 19.   † *Ib.*, i. 34.

by the mind. It is, in short, that conception which, in the act of induction, the mind superadds to the phenomena as they are presented by the senses: and to invent such appropriate conceptions, such "eustochies," is, indeed, the precise office of inductive sagacity.

At the end of this work (the *Later Analytics*) Aristotle ascribes our knowledge of principles to Intellect, (νοῦς) or, as it appears necessary to translate the word, *Intuition**. "Since, of our intellectual habits by which we aim at truth, some are always true, but some admit of being false, as Opinion and Reasoning, but Science and Intuition are always true; and since there is nothing which is more certain than Science except Intuition; and since Principles are better known to us than the Deductions from them; and since all Science is connected by reasoning, we cannot have Science respecting Principles. Considering this then, and that the beginning of Demonstration cannot be Demonstration, nor the beginning of Science, Science; and since, as we have said, there is no other kind of truth, Intuition must be the beginning of Science."

What is here said, is, no doubt, in accordance with the doctrines which we have endeavoured to establish respecting the nature of Science, if by this *Intuition* we understand that contemplation of certain Fundamental Ideas, which is the basis of all rigorous knowledge. But notwithstanding this apparent approximation, Aristotle was far from having an habitual and practical possession of the principles which he thus touches upon. He did not, in reality, construct his philosophy by giving Unity to that which was manifold, or by seeking in Intuition principles which might be the basis of Demonstration; nor did he collect, in each subject, fundamental propositions by an induction of particulars. He rather endea-

* *Anal. Post.*, ii. **19**.

voured to divide than to unite; he employed himself, not in combining facts, but in analysing notions; and the criterion to which he referred his analysis was, not the facts of our experience, but our habits of language. Thus his opinions rested, not upon sound inductions, gathered in each case from the phenomena by means of appropriate Ideas; but upon the loose and vague generalizations which are implied in the common use of speech.

Yet Aristotle was so far consistent with his own doctrine of the derivation of knowledge from experience, that he made in almost every province of human knowledge, a vast collection of such special facts as the experience of his time supplied. These collections are almost unrivalled, even to the present day, especially in Natural History; in other departments, when to the facts we must add the right Inductive Idea, in order to obtain truth, we find little of value in the Aristotelic works. But in those parts which refer to Natural History, we find not only an immense and varied collection of facts and observations, but a sagacity and acuteness in classification which it is impossible not to admire. This indeed appears to have been the most eminent faculty in Aristotle's mind.

The influence of Aristotle in succeeding ages will come under our notice shortly.

Chapter IV.

THE LATER GREEKS.

THUS while Plato was disposed to seek the essence of our knowledge in Ideas alone, Aristotle, slighting this source of truth, looked to Experience as the beginning of Science; while he attempted to obtain, by division and deduction, all that Experience did not immediately supply. And thus, with these two great names, began that struggle of opposite opinions which has ever since that time agitated the speculative world, as men have urged the claims of Ideas or of Experience to our respect, and as alternately each of these elements of knowledge has been elevated above its due place, while the other has been unduly depressed. We shall see the successive turns of this balanced struggle in the remaining portions of this review.

But we may observe that practically the influence of Plato predominated rather than that of Aristotle, in the remaining part of the history of ancient philosophy. It was, indeed, an habitual subject of dispute among men of letters, whether the sources of true knowledge are to be found in the Senses or in the Mind; the Epicureans taking one side of this alternative, and the Academics another, while the Stoics in a certain manner included both elements in their view. But none of these sects showed their persuasion that the materials of knowledge were to be found in the domain of Sense, by seeking them there. No one appears to have thought of following the example of Aristotle, and gathering together a store of observed facts. We may except, perhaps, assertions belonging to some provinces of Natural History, which were collected by various writers: but in these, the mixed character of the statements, the want of discrimination in the estimate of evidence, the credulity and love of the marvellous which the authors for the most part displayed, showed that instead of improving upon the example of Aristotle, they were wandering further and further from the path of real knowledge. And while they thus collected, with so little judgment, such statements as offered themselves, it hardly appears to have occurred to any one to enlarge the stores of observation by the aid of experiment; and to learn what the laws of nature were, by trying what were their results in particular cases. They used no instruments for obtaining an insight into the constitution of the universe, except logical distinctions and discussions; and proceeded as if the phenomena familiar to their predecessors must contain all that was needed as a basis for natural philosophy. By thus contenting themselves with the facts which the earlier philosophers had contemplated, they were led also to confine themselves to the ideas which those philosophers had put forth. For all the most remarkable alternatives of hypothesis, so far as they could be constructed with a slight and common knowledge of phenomena, had been promulgated by the acute and profound thinkers who gave the first impulse to philosophy: and it was not given to man to add much to the original inventions of *their* minds till he had undergone anew a long discipline of observation, and of thought employed upon observation. Thus the later authors of the Greek Schools became little better than commentators on the earlier; and the common places with which the different schools carried on their debates, —the constantly recurring argument, with its known attendant answer,—the distinctions drawn finer and finer and leading to nothing,—render the speculations of those times a *scholastic* philosophy, in the same sense in which we employ the term when we speak of the labours of the middle ages. It will be understood that I now refer to that which is here my subject, the opinions concerning our knowledge of nature, and the methods in use for the purpose of obtaining such knowledge. Whether the moral speculations of the ancient world were of the same stationary kind, going their round in a limited circle, like their metaphysics and physics, must be considered on some other occasion.

As a specimen of the later Greek reasonings on physical philosophy, I may take a passage from Galen's Commentary on the Treatise of Hippocrates, *On the Elements.* "What, then," he asks*, "is the method of discovering these Elements? To me it seems there can be no other than that which was introduced by Hippocrates. For we must reason first, considering if an Element be a thing which is one, according to its idea; (ἕν τι τὴν ἰδέαν;) and next, if many and various and dissimilar, how many, and of what kind they are, and how related by their association. Now that the First Element is not one only, comprising both our bodies and other things, Hippocrates shows. For if man were one Element only, he could not fall sick; for there would be nothing which could derange his health, if he were of one Element only." We have seen, in the History of Science, that Galen is one of the greatest names in ancient Physiology: but when he makes the attempt to pass at one step from the most familiar facts to the ultimate constitution of the universe, it is not wonderful that his reasonings are of no real value or import.

Before we quit the ancients we may observe some peculiarities in the Roman disciples of the Greek philosophy, which may be worthy our notice.

Chapter V.

THE ROMANS.

The Romans had no philosophy but that which they borrowed from the Greeks; and what they thus received, they hardly made entirely their own. The vast and profound question of which we have been speaking, the relation between Existence and our Knowledge of what

---

* Lib. i. c. 2.

exists, they never appear to have fathomed, even so far as to discern how wide and deep it is. In the development of the ideas by which nature is to be understood, they went no further than their Greek masters had gone, nor indeed was more to be looked for. And in the practical habit of accumulating observed facts as materials for knowledge, they were much less discriminating and more credulous than their Greek predecessors. The descent from Aristotle to Pliny, in the judiciousness of the authors and the value of their collections of facts, is immense.

Since the Romans were thus servile followers of their Greek teachers, and little acquainted with any example of new truths collected from the world around them, it was not to be expected that they could have any just conception of that long and magnificent ascent from one set of truths to others of higher order and wider compass, which the history of science began to exhibit when the human mind recovered its progressive habits. Yet some dim presentiment of the splendid career thus destined for the intellect of man appears from time to time to have arisen in their minds. Perhaps the circumstance which most powerfully contributed to suggest this vision, was the vast intellectual progress which they were themselves conscious of having made, through the introduction of the Greek philosophy; and to this may be added, perhaps, some other features of national character. Their temper was too stubborn to acquiesce in the absolute authority of the Greek philosophy, although their minds were not inventive enough to establish a rival by its side. And the wonderful progress of their political power had given them a hope in the progress of man which the Greeks never possessed. The Roman, as he believed the fortune of his State to be destined for eternity, believed also in the immortal destiny and endless advance of that

Intellectual Republic of which he had been admitted a
denizen.

It is easy to find examples of such feelings as I have
endeavoured to describe. The enthusiasm with which
Lucretius and Virgil speak of physical knowledge, manifestly arises in a great measure from the delight which
they had felt in becoming acquainted with the Greek
theories.

> Me vero primum dulces ante omnia musæ
> Quarum sacra fero ingenti perculsus amore,
> Accipiant, cœlique vias et sidera monstrent,
> Defectus solis varios, Lunæque labores!
>
> Felix qui potuit rerum cognoscere causas !

Ovid* expresses a similar feeling.

> Felices animos quibus hæc cognoscere primis
>   Inque domos superas scandere cura fuit !. . .
> Admovere oculis distantia sidera nostris
>   Ætheraque ingenio supposuere suo.
> Sic petitur cœlum : non ut ferat Ossam Olympus
>   Summaque Peliacus sidera tanget apex.

And from the whole tenour of these and similar passages, it is evident that the intellectual pleasure which
arises from our first introduction to a beautiful physical
theory had a main share in producing this enthusiasm
at the contemplation of the victories of science; although
undoubtedly the moral philosophy, which was never separated from the natural philosophy, and the triumph over
superstitious fears which a knowledge of nature was
supposed to furnish, added warmth to the feeling of
exultation.

We may trace a similar impression in the ardent
expressions which Pliny† makes use of in speaking of the
early astronomers, and which we have quoted in the
History. "Great men! elevated above the common
standard of human nature, by discovering the laws which

---

* L. i., *Fast.*   † i. 75.

celestial occurrences obey, and by freeing the wretched mind of man from the fears which eclipses inspired."

This exulting contemplation of what science had done, naturally led the mind to an anticipation of further achievements still to be performed. Expressions of this feeling occur in Seneca, and are of the most remarkable kind, as the following example will show*.

" Why do we wonder that comets, so rare a pheno-mena, have not yet had their laws assigned?—that we should know so little of their beginning and their end, when their recurrence is at wide intervals? It is not yet fifteen hundred years since Greece,

> Stellis numeros et nomina fecit,

reckoned the stars, and gave them names. There are still many nations which are acquainted with the heavens by sight only; which do not yet know why the moon dis-appears, why she is eclipsed. It is but lately that among us philosophy has reduced these matters to a certainty. The day shall come when the course of time and the labour of a maturer age shall bring to light what is yet concealed. One generation, even if it devoted itself to the skies, is not enough for researches so extensive. How then can it be so, when we divide this scanty allowance of years into no equal shares between our studies and our vices? These things then must be explained by a long succession of inquiries. We have but just begun to know how arise the morning and evening appearances, the stations, the progressions, and the retrogradations of the fixed stars which put themselves in our way;—which appearing perpetually in another and another place com-pel us to be curious. Some one will hereafter demon-strate in what region the comets wander; why they move so far asunder from the rest; of what size and nature they are. Let us be content with what we have dis-

* *Quæst. Nat.*, vii. 25.

covered: let posterity contribute its share to truth."
Again he adds* in the same strain. "Let us not
wonder that what lies so deep is brought out so slowly.
How many animals have become known for the first time
in this age! And the members of future generations
shall know many of which we are ignorant. Many
things are reserved for ages to come, when our memory
shall have passed away. The world would be a small
thing indeed, if it did not contain matter of inquiry *for*
all the world. Eleusis reserves something for the second
visit of the worshipper. *So too Nature does not at once
disclose all* HER *mysteries.* We think ourselves initiated,
we are but in the vestibule. The arcana are not thrown
open without distinction and without reserve. This age
will see some things; that which comes after us, others."

While we admire the happy coincidence of these
conjectures with the soundest views which the history of
science teaches us, we must not forget that they are
merely conjectures, suggested by very vague impressions,
and associated with very scanty conceptions of the laws
of nature. Seneca's *Natural Questions,* from which the
above extract is taken, contains a series of dissertations
on various subjects of Natural Philosophy; as Meteors,
Rainbows, Lightning, Springs, Rivers, Snow, Hail, Rain,
Wind, Earthquakes and Comets. In the whole of these
dissertations, the statements are loose, and the explanations of little or no value. Perhaps it may be worth our
while to notice a case in which he refers to an observation of his own, although his conclusion from it be erroneous. He is arguing† against the opinion that Springs
arise from the water which falls in rain. "In the first
place," he says, "I, a very diligent digger in my vineyard,
affirm that no rain is so heavy as to moisten the earth to
the depth of more than ten feet. All the moisture is

* *Quæst. Nat.,* vii. 30, 31.   † *Ib.,* iii. 7.

consumed in this outer crust, and descends not to the lower part." We have here something of the nature of an experiment; and indeed, as we may readily conceive, the instinct which impels man to seek truth by experiment can never be altogether extinguished. Seneca's experiment was deprived of its value by the indistinctness of his ideas, which led him to rest in the crude conception of the water being "consumed" in the superficial crust of the earth.

It is unnecessary to pursue further the reasonings of the Romans on such subjects, and we now proceed to the ages which succeeded the fall of their empire.

CHAPTER VI.

THE SCHOOLMEN OF THE MIDDLE AGES.

IN the History of the Sciences I have devoted a Book to the state of Science in the middle ages, and have endeavoured to analyse the intellectual defects of that period. Among the characteristic features of the human mind during those times, I have noticed Indistinctness of Ideas, a Commentatorial Spirit, Mysticism, and Dogmatism. The account there given of this portion of the history of man belongs, in reality, rather to the present work than to the History of Progressive Science. For, as we have there remarked, theoretical Science was, during the period of which we speak, almost entirely stationary; and the investigation of the causes of such a state of things may be considered as a part of that review, in which we are now engaged, of the vicissitudes of man's acquaintance with the methods of discovery. But when we offered to the world a history of science, to leave so large a chasm unexplained, would have made the series of events seem defective and broken; and the survey of the Middle Ages was therefore inserted. I would beg to refer to that portion of the former work the reader who wishes for information in addition to what is here given.

The Indistinctness of Ideas and the Commentatorial Disposition of those ages have already been here brought under our notice. Viewed with reference to the opposition between Experience and Ideas, on which point, as we have said, the succession of opinions in a great measure turns, it is clear that the commentatorial method belongs to the ideal side of the question: for the commentator seeks for such knowledge as he values, by analysing and illustrating what his author has said; and, content with this material of speculation, does not desire to add to it new stores of experience and observation. And with regard to the two other features in the character which we gave to those ages, we may observe that Dogmatism demands for philosophical theories the submission of mind, due to those revealed religious doctrines which are to guide our conduct and direct our hopes: while Mysticism elevates ideas into realities, and offers them to us as the objects of our religious regard. Thus the Mysticism of the middle ages and their Dogmatism alike arose from not discriminating the offices of theoretical and practical philosophy. Mysticism claimed for ideas the dignity and reality of principles of moral action and religious hope: Dogmatism imposed theoretical opinions respecting speculative points with the imperative tone of rules of conduct and faith.

If, however, the opposite claims of theory and practice interfered with the progress of science by the confusion they thus occasioned, they did so far more by drawing men away altogether from mere physical speculations. The Christian religion, with its precepts, its hopes, and its promises, became the leading subject of men's thoughts; and the great active truths thus revealed, and the duties thus enjoined, made all inquiries of mere curiosity appear frivolous and unworthy of man. The Fathers of the Church sometimes philosophized ill; but far more commonly they were too intent upon the great lessons which they had to teach, respecting man's situation in the eyes of his Heavenly Master, to philosophize at all respecting things remote from the business of life and of no importance in man's spiritual concerns.

Yet man has his intellectual as well as his spiritual wants. He has faculties which demand systems and reasons, as well as precepts and promises. The Christian doctor, who knew so much more than the heathen philosopher respecting the Creator and Governor of the universe, was not long content to know or to teach less, respecting the universe itself. While it was still maintained that Theology was the only really important study, Theology was so extended and so fashioned as to include all other knowledge: and after no long time, the Fathers of the Church themselves became the authors of systems of universal knowledge.

But when this happened, the commentatorial spirit was still in its full vigour. The learned Christians could not, any more than the later Greeks or the Romans, devise, by the mere force of their own invention, new systems, full, comprehensive, and connected, like those of the heroic age of philosophy. The same mental tendencies which led men to look for speculative coherence and completeness in the view of the universe, led them also to admire and dwell upon the splendid and acute speculations of 'the Greeks. They were content to find, in these immortal works, the answers to the questions which their curiosity prompted; and to seek what further satisfaction they might require, in analysing and unfolding the doctrines promulgated by those great masters of knowledge. Thus the Christian doctors became, as to general philosophy, commentators upon the ancient Greek teachers.

Among these, they selected Aristotle as their peculiar object of admiration and study. The vast store, both of opinions and facts, which his works contain, his acute distinctions, his cogent reasons in some portions of his speculations, his symmetrical systems in almost all, naturally commended him to the minds of subtle and curious men. We may add that Plato, who taught men to contemplate Ideas separate from Things, was not so well fitted for general acceptance as Aristotle, who rejected this separation. For although the due apprehension of this opposition of ideas and sensations is a necessary step in the progress of true philosophy, it requires a clearer view and a more balanced mind than the common herd of students possess; and Aristotle, who evaded the necessary perplexities in which this antithesis involves us, appeared, to the temper of those times, the easier and the plainer guide of the two.

The Doctors of the middle ages having thus adopted Aristotle as their master in philosophy, we shall not be surprised to find them declaring, after him, that experience is the source of our knowledge of the visible world. But though, like the Greeks, they thus talked of experiment, like the Greeks, they showed little disposition to discover the laws of nature by observation of facts. This barren and formal recognition of experience or sensation as one source of knowledge, not being illustrated by a practical study of nature, and by real theoretical truths obtained by such a study, remained ever vague, wavering, and empty. Such a mere acknowledgement cannot, in any times, ancient or modern, be considered as indicating a just apprehension of the true basis and nature of science.

In imperfectly perceiving how, and how far, experience is the source of our knowledge of the external world, the teachers of the middle ages were in the dark; but so on this subject have been almost all the writers of all ages, with the exception of those who in recent times have had their minds enlightened by contemplating philosophically the modern progress of science. The opinions of the doctors of the middle ages on such subjects generally had those of Aristotle for their basis; but the subject was often still further analysed and systematized, with an acute and methodical skill hardly inferior to that of Aristotle himself.

The Stagirite, in the beginning of his *Physics*, had made the following remarks. " In all bodies of doctrine which involve principles, causes, or elements, Science and Knowledge arise from the knowledge of these; (for we then consider ourselves to *know* respecting any subject, when we know its first cause, its first principles, its ultimate elements.) It is evident, therefore, that in seeking a knowledge of nature, we must first know what are its principles. But the course of our knowledge is, from the things which are better known and more manifest to us, to the things which are more certain and evident in nature. For those things which are most evident in truth, are not most evident to us. [And consequently we must advance from things obscure in nature, but manifest to us, towards the things which are really in nature more clear and certain.] The things which are first obvious and apparent to us are complex; and from these we obtain, by analysis, principles and elements. We must proceed from universals to particulars. For the whole is better known to our senses than the parts, and for the same reason, the universal better known than the particular. And thus words signify things in a large and indiscriminate way, which is afterwards analysed by definition; as we see that

the children at first call all men *father*, and all women
mother, but afterwards learn to distinguish."

There are various assertions contained in this extract
which came to be considered as standard maxims, and
which occur constantly in the writers of the middle ages.
Such are, for instance, the maxim, "Vere scire est per
causas scire;" the remark, that compounds are known to us
before their parts, and the illustration from the expressions
used by children. Of the mode in which this subject was
treated by the schoolmen, we may judge by looking at
passages of Thomas Aquinas which treat of the subject
of the human understanding. In the *Summa Theologiæ*,
the eighty-fifth Question is *On the manner and order of
understanding*, which subject he considers in eight Articles; and these must, even now, be looked upon as exhibiting many of the most important and interesting points
of the subject. They are, *First*, Whether our understanding understands by abstracting ideas (*species*) from
appearances; *Second*, Whether intelligible species abstracted from appearances are related to our understanding as that *which* we understand, or that *by which* we
understand; *Third*, Whether our understanding does
naturally understand universals first; *Fourth*, Whether
our understanding can understand many things at once;
Fifth, Whether our understanding understands by com-
pounding and dividing; *Sixth*, Whether the understanding can err; *Seventh*, Whether one person can understand
the same thing better than another; *Eighth*, Whether
our understanding understands the indivisible sooner than
the divisible. And in the discussion of the last point, for
example, reference is made to the passage of Aristotle
which we have already quoted. "It may seem," he says,
"that we understand the indivisible before the divisible;
for *the Philosopher* says that we understand and know
by knowing principles and elements; but indivisibles are

the principles and elements of divisible things. But to this we may reply, that in our receiving of science, principles and elements are not always first; for sometimes from the sensible effects we go on to the knowledge of intelligible principles and causes." We see that both the objection and the answer are drawn from Aristotle.

We find the same close imitation of Aristotle in Albertus Magnus, who, like Aquinas, flourished in the thirteenth century. Albertus, indeed, wrote treatises corresponding to almost all those of the Stagirite, and was called the *Ape of Aristotle.* In the beginning of his *Physics,* he says, " Knowledge does not always begin from that which is first according to the nature of things, but from that of which the knowledge is easiest. For the human intellect, on account of its relation to the senses (*propter reflexionam quam habet ad sensum*), collects science from the senses; and thus it is easier for our knowledge to begin from that which we can apprehend by sense, imagination, and intellect, than from that which we apprehend by intellect alone." We see that he has somewhat systematized what he has borrowed.

This disposition to dwell upon and systematize the leading doctrines of metaphysics assumed a more definite and permanent shape in the opposition of the Realists and Nominalists. The opposition involved in this controversy is, in fact, that fundamental antithesis of Sense and Ideas about which philosophy has always been engaged; and of which we have marked the manifestation in Plato and Aristotle. The question, What is the object of our thoughts when we reason concerning the external world? must occur to all speculative minds: and the difficulties of the answer are manifest. We must reply either that our own Ideas, or that Sensible Things, are the elements of our knowledge of nature. And then the scruples again occur,—how we have any *general* knowledge if our thoughts are fixed on particular objects; and, on the other hand,—how we can attain to any *true* knowledge of nature by contemplating ideas which are not identical with objects in nature. The two opposite opinions maintained on this subject were, on the one side,—that our general propositions refer to objects which are *real*, though divested of the peculiarities of individuals; and, on the other side,—that in such propositions, individuals are not represented by any reality, but bound together by a *name*. These two views were held by the Realists and Nominalists respectively: and thus the Realist manifested the adherence to Ideas, and the Nominalist the adherence to the impressions of Sense, which have always existed as opposite yet correlative tendencies in man.

The Realists were the prevailing sect in the Scholastic times: for example, both Thomas Aquinas and Duns Scotus, the *Angelical* and the *Subtile* Doctor, held this opinion, although opposed to each other in many of their leading doctrines on other subjects. And as the Nominalist, fixing his attention upon sensible objects, is obliged to consider what is the *principle of generalization*, in order that the possibility of any general proposition may be conceivable; so on the other hand, the Realist, beginning with the contemplation of universal ideas, is compelled to ask what is the *principle of individuation*, in order that he may comprehend the application of general propositions in each particular instance. This inquiry concerning the principle of individuation was accordingly a problem which occupied all the leading minds among the Schoolmen*. It will be apparent from what has been said, that it is only one of the many forms of the fundamental antithesis of the Ideas and the Senses, which we have constantly before us in this review.

---

* See the opinion of Aquinas, DEGERANDO, *Hist. Com. des Syst.*, iv. 499; of Duns Scotus, *ib.*, iv. 523.

The recognition of the derivation of our knowledge, in part at least, from Experience, though always loose and incomplete, appears often to be independent of the Peripatetic traditions. Thus Richard of St. Victor, a writer of contemplative theology in the twelfth century, says *, that "there are three sources of knowledge, experience, reason, faith. Some things we prove by experiment, others we collect by reasoning, the certainty of others we hold by believing. And with regard to temporal matters, we obtain our knowledge by actual experience; the other guides belong to divine knowledge." Richard also propounds a division of human knowledge which is clearly not derived directly from the ancients, and which shows that considerable attention must have been paid to such speculations. He begins by laying down clearly and broadly the distinction, which, as we have seen, is of primary importance, between *practice* and *theory*. *Practice*, he says, includes seven mechanical arts; those of the clothier, the armourer, the navigator, the hunter, the physician, and the player. *Theory* is threefold, divine, natural, doctrinal; and is thus divided into Theology, Physics, and Mathematics. *Mathematics*, he adds, treats of the invisible *forms* of visible things. We have seen that by many profound thinkers this word *forms* has been selected as best fitted to describe those relations of things which are the subject of mathematics. Again, Physics discovers causes from their effects and effects from their causes. It would not be easy at the present day to give a better account of the object of physical science. But Richard of St. Victor makes this account still more remarkably judicious, by the examples to which he alludes; which are earthquakes, the tides, the virtues of plants, the instincts of animals, the classification of minerals, plants and reptiles.

* *Liber Excerptionum*, l. i. c. 1.

Unde tremor terris, quâ vi maria alta tumescant,
Herbarum vires, animos irasque ferarum,
Omne genus fruticum, lapidum quoque, reptiliumque.

He further adds*, "Physical science ascends from effects to causes, and descends again from causes to effects." This declaration Francis Bacon himself might have adopted. It is true, that Richard would probably have been little able to produce any clear and definite instances of knowledge, in which this ascent and descent were exemplified; but still the statement, even considered as a mere conjectural thought, contains a portion of that sagacity and comprehensive power which we admire so much in Bacon.

Richard of St. Victor, who lived in the twelfth century, thus exhibits more vigour and independence of speculative power than Thomas Aquinas, Albertus Magnus, and Duns Scotus in the thirteenth. In the interval, about the end of the twelfth century, the writings of Aristotle had become generally known in the West; and had been elevated into the standard of philosophical doctrine, by the divines mentioned above, who felt a reverent sympathy with the systematizing and subtle spirit of the Stagirite as soon as it was made manifest to them. These doctors, following the example of their great forerunner, reduced every part of human knowledge to a systematic form; the systems which they thus framed were presented to men's minds as the only true philosophy, and dissent from them was no longer considered to be blameless. It was an offence against religion as well as reason to reject the truth, and the truth could be but one. In this manner arose that claim which the Doctors of the Church put forth to control men's opinions upon all subjects, and which we have spoken of in the History of Science as the Dogmatism of the

* *Tr. Ex.*, l. i. c. 7.

Middle Ages. There is no difficulty in giving examples of this characteristic. We may take for instance a Statute of the University of Paris, occasioned by a Bull of Pope John XXI., in which it is enacted, "that no Master or Bachelor of any faculty, shall presume to read lectures upon any author in a private room, on account of the many perils which may arise therefrom; but shall read in public places, where all may resort, and may faithfully report what is there taught; excepting only books of Grammar and Logic, in which there can be no presumption." And certain errors of Brescain are condemned in a Rescript* of the papal Legate Odo, with the following expressions: "Whereas, as we have been informed, certain Logical professors treating of Theology in their disputations, and Theologians treating of Logic, contrary to the command of the law are not afraid to mix and confound the lots of the Lord's heritage; we exhort and admonish your University, all and singular, that they be content with the landmarks of the Sciences and Faculties which our Fathers have fixed; and that having due fear of the curse pronounced in the law against him who removeth his neighbour's landmark, you hold such sober wisdom according to the Apostles, that ye may by no means incur the blame of innovation or presumption."

The account which, in the History of Science, I gave of Dogmatism as a characteristic of the middle ages, has been indignantly rejected by a very pleasing modern writer, who has, with great feeling and great diligence, brought into view the merits and beauties of those times, termed by him *Ages of Faith*. He urges† that religious authority was never claimed for physical science: and he quotes from Thomas Aquinas, a passage in which the

---

* TENNEMAN, viii. 461.

† *Mores Catholici, or Ages of Faith*, viii. p. 247.

author protests against the practice of confounding opinions of philosophy with doctrines of faith. We might quote in return the Rescript* of Stephen, bishop of Paris, in which he declares that there can be but one truth, and rejects the distinction of things being true according to philosophy and not according to the Catholic faith; and it might be added, that among the errors condemned in this document are some of Thomas Aquinas himself. We might further observe, that if no physical doctrines were condemned in the times of which we now speak, this was because, on such subjects, no new opinions were promulgated, and not because opinion was free. As soon as new opinions, even on physical subjects, attracted general notice, they were prohibited by authority, as we see in the case of Galileo†.

But this disinclination to recognize philosophy as independent of religion, and this disposition to find in new theories, even in physical ones, something contrary to religion or scripture, are, it would seem, very natural

---

* TENNEMAN, viii. 460.

† If there were any doubt on this subject, we might refer to the writers who afterwards questioned the supremacy of Aristotle, and who with one voice assert that an infallible authority had been claimed for him. Thus Laurentius Valla: "Quo minus ferendi sunt recentes Peripatetici, qui nullius sectæ hominibus interdicunt libertate ab Aristotle dissentiendi, quasi sophos hic, non philosophus." *Pref. in Dial.* (TENNEMAN, ix. 29.) So Ludovicus Vives: "Sunt ex philosophis et ex theologis qui non solem quo Aristoteles pervenit extremum esse aiunt naturæ, sed quâ pervenit eam rectissimam esse omnium et certissimam in natura viam." (TENNEMAN, ix. 43.) We might urge too, the evasions practised by Reformers, through fear of the dogmatism to which they had to submit; for example, the protestation of Telesius at the end of the Proem to his work, *De Rerum Natura:* "Nec tamen, si quid eorum quæ nobis posita sunt, sacris literis, Catholicæve ecclesiæ decretis non cohæreat, tenendum id, quin penitus rejiciendum asseveramus contendimusque. Neque enim *humana* modo *ratio* quævis, sed *ipse* etiam *sensus* illis *posthabendus*, et si illis non congruat, abnegandus omnino et ipse etiam est sensus."

tendencies of theologians; and it would be unjust to assert that these propensities were confined to the periods when the authority of papal Rome was highest; or that the spirit which has in a great degree controlled and removed such habits was introduced by the Reformation of religion in the sixteenth century. We must trace to other causes, the clear and general recognition of Philosophy, as distinct from Theology, and independent of her authority. In the earlier ages of the Church, indeed, this separation had been acknowledged. St. Augustin says, " A Christian should beware how he speaks on questions of natural philosophy, as if they were doctrines of Holy Scripture; for an infidel who should hear him deliver absurdities could not avoid laughing. Thus the Christian would be confused, and the infidel but little edified; for the infidel would conclude that our authors really entertained these extravagant opinions, and therefore they would despise them, to their own eternal ruin. Therefore the opinions of philosophers should never be proposed as dogmas of faith, or rejected as contrary to faith, when it is not certain that they are so." These words are quoted with approbation by Thomas Aquinas, and it is said*, are cited in the same manner in every encyclopedical work of the middle ages. This warning of genuine wisdom was afterwards rejected, as we have seen; and it it only in modern times that its value has again been fully recognized. And this improvement we must ascribe, mainly to the progress of physical science. For a great body of undeniable truths on physical subjects being accumulated, such as had no reference to nor connexion with the truths of religion, and yet such as possessed a strong interest for most men's minds, it was impossible longer to deny that there were wide provinces of knowledge which

---

* *Ages of Faith*, viii. 247: to the author of which I am obliged for this quotation.

were not included in the dominions of Theology, and over
which she had no authority. In the fifteenth and sixteenth
centuries, the fundamental doctrines of mechanics, hydrostatics, optics, magnetics, chemistry, were established and
promulgated; and along with them, a vast train of consequences, attractive to the mind by the ideal relations
which they exhibited, and striking to the senses by the
power which they gave man over nature. Here was a
region in which philosophy felt herself entitled and impelled to assert her independence. From this region,
there is a gradation of subjects in which philosophy advances more and more towards the peculiar domain of
religion; and at some intermediate points there have
been, and probably will always be, conflicts respecting
the boundary line of the two fields of speculation. For
the limit is vague and obscure, and appears to fluctuate
and shift with the progress of time and knowledge.

Our business at present is not with the whole extent
and limits of philosophy, but with the progress of physical
science more particularly, and the methods by which it
may be attained: and we are endeavouring to trace historically the views which have prevailed respecting such
methods, at various periods of man's intellectual progress.
Among the most conspicuous of the revolutions which
opinions on this subject have undergone, is the transition
from an implicit trust in the internal powers of man's
mind to a professed dependence upon external observation;
and from an unbounded reverence for the wisdom of the
past, to a fervid expectation of change and improvement. The origin and progress of this disposition of
mind; the introduction of a state of things in which men
not only obtained a body of indestructible truths from
experience, and increased it from generation to generation, but professedly, and we may say, ostentatiously,

declared such to be the source of their knowledge, and such their hopes of its destined career;—the rise, in short, of Experimental Philosophy, not only as a habit, but as a Philosophy of Experience, is what we must now endeavour to exhibit.

CHAPTER VII.

THE INNOVATORS OF THE MIDDLE AGES.

General Remarks. — In the rise of Experimental Philosophy, understanding the term in the way just now stated, two features have already been alluded to: the disposition to cast off the prevalent reverence for the opinions and methods of preceding teachers with an eager expectation of some vast advantage to be derived from a change; and the belief that this improvement must be sought by drawing our knowledge from external observation rather than from mere intellectual efforts;—*the Insurrection against Authority,* and *the Appeal to Experience.* These two movements were closely connected; but they may easily be distinguished, and in fact, persons were very prominent in the former part of the task, who had no comprehension of the latter principle, from which alone the change derives its value. There were many Malcontents who had no temper, talent or knowledge, which fitted them to be Reformers.

The authority which was questioned, in the struggles of which we speak, was that of the Scholastic System, the combination of Philosophy with Theology; of which Aristotle, presented in the form and manner which the Doctors of the Church had imposed upon him, is to be considered the representative. When there was demanded of men a submission of the mind, such as this system claimed, the natural love of freedom in man's bosom, and the speculative tendencies of his intellect, rose in rebellion, from time to time, against the ruling oppression. We find in all periods of the scholastic ages examples of this disposition of man to resist overstrained authority; the tendency being mostly, however, combined with a want of solid thought, and showing itself in extravagant pretensions and fantastical systems put forwards by the insurgents. We have pointed out one such opponent* of the established systems, even among the Arabian schoolmen, a more servile race than ever the Europeans were. We may here notice more especially an extraordinary character who appeared in the thirteenth century, and who may be considered as belonging to the Prelude of the Reform in Philosophy, although he had no share in the Reform itself.

Raymond Lully.—Raymond Lully is perhaps traditionally best known as an Alchemist, of which art he appears to have been a cultivator. But this was only one of the many impulses of a spirit ardently thirsty of knowledge and novelty. He had†, in his youth, been a man of pleasure, but was driven by a sudden shock of feeling to resolve on a complete change of life. He plunged into solitude, endeavoured to still the remorse of his conscience by prayer and penance, and soon had his soul possessed by visions which he conceived were vouchsafed him. In the feeling of religious enthusiasm thus excited, he resolved to devote his life to the diffusion of Christian truth among Heathens and Mahomedans. For thus purpose, at the age of thirty he betook himself to the study of Grammar, and of the Arabic language. He breathed earnest supplications for an illumination from above; and these were answered by his

---

* Algazel. See *Hist. Ind. Sci.*, i., 251.
† TENNEMAN, viii. 830.

receiving from heaven, as his admirers declare, his *Ars Magna*, by which he was able without labour or effort to learn and apply all knowledge. The real state of the case is, that he put himself in opposition to the established systems, and propounded a New Art, from which he promised the most wonderful results; but that his Art really is merely a mode of combining ideal conceptions without any reference to real sources of knowledge, or any possibility of real advantage. In a Treatise addressed, in A.D. 1310, to King Philip of France, entitled *Liber Lamentationis Duodecim Principiorum Philosophiæ contra Averroistas*, Lully introduces Philosophy, accompanied by her twelve Principles, (Matter, Form, Generation, &c.) uttering loud complaints against the prevailing system of doctrine; and represents her as presenting to the king a petition that she may be upheld and restored by her favourite, the Author. His *Tabula Generalis ad omnes Scientias applicabilis* was begun the 15th September, 1292, in the Harbour of Tunis, and finished in 1293, at Naples. In order to frame an Art of thus tabulating all existing sciences, and indeed all possible knowledge, he divides into various classes the conceptions with which he has to deal. The first class contains nine *Absolute Conceptions*: Goodness, Greatness, Duration, Power, Wisdom, Will, Virtue, Truth, Majesty. The second class has nine *Relative Conceptions*: Difference, Identity, Contrariety, Beginning, Middle, End, Majority, Equality, Minority. The third class contains nine *Questions*: Whether? What? Whence? Why? How great? How circumstanced? When? Where? and How? The fourth class contains the nine *Most General Subjects*: God, Angel, Heaven, Man, *Imaginativum*, *Sensitivum*, *Vegetativum*, *Elementativum*, *Instrumentativum*. Then come nine *Prædicaments*, nine *Moral Qualities*, and so on. These conceptions are arranged in the compartments of certain concentric moveable circles, and give various combinations by means of triangles and other figures, and thus propositions are constructed.

It must be clear at once, that no real knowledge, which is the union of facts and ideas, can result from this machinery for shifting about, joining and disjoining, empty conceptions. This, and all similar schemes go upon the supposition that the logical combinations of notions do of themselves compose knowledge; and that really existing things may be arrived at by a successive system of derivation from our most general ideas. It is imagined that by distributing the nomenclature of abstract ideas according to the place which they can hold in our propositions, and by combining them according to certain conditions, we may obtain formulæ including all possible truths, and thus fabricate a science in which all sciences are contained. We thus obtain the means of talking and writing upon all subjects, without the trouble of thinking: the revolutions of the emblematical figures are substituted for the operations of the mind. Both exertion of thought, and knowledge of facts, become superfluous. And this reflection, adds an intelligent author*, explains the enormous number of books which Lully is said to have written; for he might have written those even during his sleep, by the aid of a moving power which should keep his machine in motion. Having once devised this invention for manufacturing science, Lully varied it in a thousand ways, and followed it into a variety of developments. Besides Synoptical Tables, he employs Genealogical Trees, which he dignifies with the name of the Tree of Science. The only requisite for the application of his System was a certain agreement in the numbers of the classes into which different subjects were

* Degerando, iv. 535.

distributed; and as this symmetry does not really exist in the operations of our thoughts, some violence was done to the natural distinction and subordination of conceptions, in order to fit them for the use of the System.

Thus Lully, while he professed to teach an Art which was to shed new light upon every part of science, was in fact employed in a pedantic and trifling repetition of known truths or truisms; and while he complained of the errors of existing methods, he proposed in their place one which was far more empty, barren, and worthless, than the customary processes of human thought. Yet his method is spoken of* with some praise by Leibnitz, who indeed rather delighted in the region of ideas and words, than in the world of realities. But Francis Bacon speaks far otherwise and more justly on this subject†. "It is not to be omitted that some men, swollen with emptiness rather than knowledge, have laboured to produce a certain Method, not deserving the name of a legitimate Method, since it is rather a method of imposture: which yet is doubtless highly grateful to certain would-be philosophers. This method scatters about certain little drops of science in such a manner that a smatterer may make a perverse and ostentatious use of them with a certain show of learning. Such was the Art of Lully, which consisted of nothing but a mass and heap of the words of each science; with the intention that he who can readily produce the words of any science shall be supposed to know the science itself. Such collections are like a rag shop, where you find a patch of everything, but nothing which is of any value."

Roger Bacon.—We now come to a philosopher of a very different character, who was impelled to declare his dissent from the reigning philosophy by the abundance of his knowledge, and by his clear apprehension of the

* *Opera*, v. 16.   † *Works*, vii. 296.

mode in which real knowledge had been acquired and must be increased.

Roger Bacon was born in 1214, near Ilchester, in Somersetshire, of an old family. In his youth he was a student at Oxford, and made extraordinary progress in all branches of learning. He then went to the University of Paris, as was at that time the custom of learned Englishmen, and there received the degree of Doctor of Theology. At the persuasion of Robert Grostête, bishop of Lincoln, he entered the brotherhood of Franciscans in Oxford, and gave himself up to study with extraordinary fervour. He was termed by his brother monks *Doctor Mirabilis*. We know from his own works, as well as from the traditions concerning him, that he possessed an intimate acquaintance with all the science of his time which could be acquired from books; and that he had made many remarkable advances by means of his own experimental labours. He was acquainted with Arabic, as well as with the other languages common in his time. In the titles of his works, we find the whole range of science and philosophy, Mathematics and Mechanics, Optics, Astronomy, Geography, Chronology, Chemistry, Magic, Music, Medicine, Grammar, Logic, Metaphysics, Ethics, and Theology; and judging from those which are published, these works are full of sound and exact knowledge. He is, with good reason, supposed to have discovered, or to have had some knowledge of, several of the most remarkable inventions which were made generally known soon afterwards, as gunpowder, lenses, burning specula, telescopes, clocks, the correction of the calendar, and the explanation of the rainbow.

Thus possessing, in the acquirements and habits of his own mind, abundant examples of the nature of knowledge and of the process of invention, Bacon felt also a deep interest in the growth and progress of science, a spirit of inquiry respecting the causes which produced or prevented its advance, and a fervent hope and trust in its future destinies; and these feelings impelled him to speculate worthily and wisely respecting a reform of the method of philosophizing. The manuscripts of his works have existed for nearly six hundred years in many of the libraries of Europe, and especially in those of England; and for a long period the very imperfect portions of them which were generally known, left the character and attainments of the author shrouded in a kind of mysterious obscurity. About a century ago, however, his *Opus Majus* was published* by Dr. S. Jebb, principally from a manuscript in the library of Trinity College, Dublin; and this contained most or all of the separate works which were previously known to the public, along with others still more peculiar and characteristic. We are thus able to judge of Bacon's knowledge and of his views, and they are in every way well worthy our attention.

The *Opus Majus* is addressed to Pope Clement the Fourth, whom Bacon had known when he was legate in England as Cardinal-bishop of Sabina, and who admired the talents of the monk, and pitied him for the persecutions to which he was exposed. On his elevation to the papal chair, this account of Bacon's labours and views was sent, at the earnest request of the pontiff. Besides the *Opus Majus*, he wrote two others, the *Opus Minus* and *Opus Tertium;* which were also sent to the pope, as the author says†, "on account of the danger of roads, and the possible loss of the work." These works still exist unpublished, in the Cottonian and other libraries.

* *Fratris Rogeri Bacon Ordinis Minorum* Opus Majus *ad Clementem Quartum, Pontificem Romanum, ex MS. Codice Dubliniensi cum aliis quibusdam collato nunc primum edidit* S. Jebb, M.D. Londini, 1733. † *Opus Majus*, Præf.

The *Opus Majus* is a work equally wonderful with regard to its general scheme, and to the special treatises with which the outlines of the plan are filled up. The professed object of the work is to urge the necessity of a reform in the mode of philosophizing, to set forth the reasons why knowledge had not made a greater progress, to draw back attention to the sources of knowledge which had been unwisely neglected, to discover other sources which were yet almost untouched, and to animate men in the undertaking, by a prospect of the vast advantages which it offered. In the development of this plan, all the leading portions of science are expounded in the most complete shape which they had at that time assumed; and improvements of a very wide and striking kind are proposed in some of the principal of these departments. Even if the work had had no leading purpose, it would have been highly valuable as a treasure of the most solid knowledge and soundest speculations of the time; even if it had contained no such details, it would have been a work most remarkable for its general views and scope. It may be considered as, at the same time, the *Encyclopedia* and the *Novum Organon* of the thirteenth century.

Since this work is thus so important in the history of inductive philosophy I shall give, in a note, a view* of

* Contents of Roger Bacon's *Opus Majus.*

Part   I. On the four causes of human ignorance : — Authority, Custom, Popular Opinion, and the Pride of supposed Knowledge.

Part II. On the source of perfect wisdom in the Sacred Scripture.

Part III. On the Usefulness of Grammar.

Part IV. On the Usefulness of Mathematics.

      (1.) The necessity of Mathematics in Human Things (published separately as the *Specula Mathematica*).

      (2.) The necessity of Mathematics in Divine Things.—1°. This study has occupied holy men: 2°. Geography : 3°. Chronology: 4°. Cycles; the Golden Number, &c. : 5°. Natural phenomena, as the Rainbow :. 6°. Arithmetic: 7°. Music.

its divisions and contents. But I must now endeavour to point out more especially the way in which the various principles, which the reform of scientific method involved, are here brought into view.

One of the first points to be noticed for this purpose, is the resistance to authority; and at the stage of philosophical history with which we here have to do, this means resistance to the authority of Aristotle, as adopted and interpreted by the Doctors of the Schools. Bacon's work* is divided into Six Parts; and of these Parts, the First is, Of the four universal Causes of all Human Ignorance. The causes thus enumerated† are:—the force of unworthy authority;—traditionary habit;—the imperfection of the undisciplined senses;—and the disposition to conceal our ignorance and to make an ostentatious show of our knowledge. These influences involve every man, occupy every condition. They prevent our obtaining the most useful and large and fair doctrines of wisdom, the secrets of all sciences and arts. He then proceeds to argue, from the testimony of philosophers themselves, that the authority of antiquity, and especially of Aristotle, is not infallible. "We find‡ their books full of doubts, obscurities, and perplexities. They scarce agree with each other in one

    (3.) The Necessity of Mathematics in Ecclesiastical Things. 1°. The Certification of Faith: 2°. The Correction of the Calendar.

    (4.) The Necessity of Mathematics in the State.—1°. Of Climates: 2°. Hydrography: 3°. Geography: 4°. Astrology.

Part V. On Perspective (published separately as *Perspectiva*).

    (1.) The organs of vision.

    (2.) Vision in straight lines.

    (3.) Vision reflected and refracted.

    (4.) De multiplicatione specierum (on the propagation of the impressions of light, heat, &c.)

Part VI. On Experimental Science.

  * *Op. Maj.*, p. 1.   † *Ib.*, p. 2.   ‡ *Ib.*, p. 10.

empty question or one worthless sophism, or one operation
of science, as one man agrees with another in the practical
operations of medicine, surgery, and the like arts of secu-
lar men.   Indeed," he adds, " not only the philosophers,
but the saints have fallen into errors which they have
afterwards retracted," and this he instances in Augustin,
Jerome, and others.   He gives an admirable sketch of the
progress of philosophy from the Ionic school to Aristotle ;
whom he speaks of with great applause. " Yet," he adds*,
" those who came after him corrected him in some things,
and added many things to his works, and shall go on
adding to the end of the world." Aristotle, he adds, is
now called peculiarly † the Philosopher, " yet there was a
time when his philosophy was silent and unregarded, either
on account of the rarity of copies of his works, or their
difficulty, or from envy ; till after the time of Mahomet,
when Avicenna and Averroes, and others, recalled this
philosophy into the full light of exposition.   And although
the Logic, and some other works were translated by
Boethius from the Greek, yet the philosophy of Aristotle
first received a quick increase among the Latins at the
time of Michael Scot ; who, in the year of our Lord
1230, appeared, bringing with him portions of the
books of Aristotle on Natural Philosophy and Mathematics.   And yet a small part only of the works of this
author is translated, and a still smaller part is in the
hands of common students." He adds further‡ (in the
Third Part of the *Opus Majus*, which is a Dissertation on
Language) that the translations which are current of these
writings, are very bad and imperfect. With these views,
he is moved to express himself somewhat impatiently §

---

* *Op. Maj.*, p. 36.   † Autonomaticé.   ‡ *Op. Maj.*, p. 46.

§ See *Pref.* to Jebb's edition. The passages there quoted, however,
are not extracts from the *Opus Majus*, but (apparently) from the *Opus
Minus (MS. Cott.* Tib. c. 5.) "Si haberem potestatem supra libros

respecting these works: "If I had," he says, "power over the works of Aristotle, I would have them all burnt; for it is only a loss of time to study in them, and a course of error, and a multiplication of ignorance beyond expression." "The common herd of students," he says, "with their heads, have no principle by which they can be excited to any worthy employment; and hence they mope and make asses of themselves over their bad translations, and lose their time, and trouble, and money."

The remedies which he recommends for these evils, are, in the first place, the study of that only perfect wisdom which is to be found in the sacred Scripture*, in the next place, the study of mathematics and the use of experiment†. By the aid of these methods, Bacon anticipates the most splendid progress for human knowledge. He takes up the strain of hope and confidence which we have noticed as so peculiar in the Roman writers; and quotes some of the passages of Seneca which we adduced in illustration of this:—that the attempts in science were at first rude and imperfect, and were afterwards improved;—that the day will come, when what is still unknown shall be brought to light by the progress of time and the labours of a longer period;—that one age does not suffice for inquiries so wide and various;—that the people of future times shall know many things unknown to us;—and the time shall arrive when posterity will wonder that we overlooked what was so obvious. Bacon himself adds anticipations more peculiarly in the spirit of his own time. "We have seen," he says, at the end of

Aristotelis, ego facerem omnes cremari; quia non est nisi temporis amissio studere in illis, et cause erroris, et multiplicatio ignorantiæ, ultra id quod valeat explicari. .... Vulgus studentum cum capitibus suis non habet unde excitetur ad aliquid dignum, et ideo languet et asininat circa male translata, et tempus et studium amittit in omnibus et expensis."

* Part ii.   † Parts iv., v. and vi.

the work, "how Aristotle, by the ways which wisdom teaches, could give to Alexander the empire of the world. And this the Church ought to take into consideration against the infidels and rebels, that there may be a sparing of Christian blood, and especially on account of the troubles that shall come to pass in the days of Antichrist; which by the grace of God, it would be easy to obviate, if prelates and princes would encourage study, and join in searching out the secrets of nature and art."

It may not be improper to observe here that this belief in the appointed progress of knowledge, is not combined with any overweening belief in the unbounded and independent power of the human intellect. On the contrary, one of the lessons which Bacon draws from the state and prospects of knowledge, is the duty of faith and humility. "To him," he says*, "who denies the truth of the faith because he is unable to understand it, I will propose in reply the course of nature, and as we have seen it in examples." And after giving some instances, he adds, "These, and the like, ought to move men and to excite them to the reception of divine truths. For if, in the vilest objects of creation, truths are found, before which the inward pride of man must bow, and believe though it cannot understand, how much more should man humble his mind before the glorious truths of God!" He had before said†: "Man is incapable of perfect wisdom in this life; it is hard for him to ascend towards perfection, easy to glide downwards to falsehoods and vanities: let him then not boast of his wisdom, or extol his knowledge. What he knows is little and worthless, in respect of that which he believes without knowing; and still less, in respect of that which he is ignorant of. He is mad who thinks highly of his wisdom; he most mad, who exhibits

* *Op. Maj.*, p. 476.   † *Ib.*, p. 15.

it as something to be wondered at." He adds, as another reason for humility, that he has proved by trial, he could teach in one year, to a poor boy, the marrow of all that the most diligent person could acquire in forty years laborious and expensive study.

To proceed somewhat more in detail with regard to Roger Bacon's views of a Reform in Scientific inquiry, we may observe that by making Mathematics and Experiment the two great points of his recommendation, he directed his improvement to the two essential parts of all knowledge, Ideas and Facts, and thus took the course which the most enlightened philosophy would have suggested. He did not urge the prosecution of experiment, to the comparative neglect of the existing mathematical sciences and conceptions; a fault which there is some ground for ascribing to his great namesake and successor Francis Bacon: still less did he content himself with a mere protest against the authority of the schools, and a vague demand for change, which was almost all that was done by those who put themselves forward as reformers in the intermediate time. Roger Bacon holds his way steadily between the two poles of human knowledge; which, as we have seen, it is far from easy to do. "There are two modes of knowing," says he*; "by argument, and by experiment. Argument concludes a question; but it does not make us feel certain, or acquiesce in the contemplation of truth, except the truth be also found to be so by experience." It is not easy to express more decidedly the clearly seen union of exact conceptions with certain facts, which, as we have explained, constitutes real knowledge.

* *Op. Maj.*, p. 445, see also p. 448. "Scientiæ aliæ sciunt sua principia invenire per experimenta, sed conclusiones per argumenta facta ex principiis inventis. Si vero debeant habere experientiam conclusionum suarum particularem et completam, tunc oportet quod habeant per adjutorium istius scientiæ nobilis, (experimentalis.)"

One large division of the *Opus Majus* is "On the Usefulness of Mathematics," which is shown by a copious enumeration of existing branches of knowledge, as Chronology, Geography, the Calendar, and (in a separate Part) Optics. There is a chapter * "in which it is proved by reason, that all science requires mathematics." And the arguments which are used to establish this doctrine, show a most just appreciation of the office of mathematics in science. They are such as follows:—That other sciences use examples taken from mathematics as the most evident:—That mathematical knowledge is, as it were, innate in us, on which point he refers to the well known dialogue of Plato, as quoted by Cicero :—That this science, being the easiest, offers the best introduction to the more difficult:—That in mathematics, things as known to us are identical with things as known to nature:—That we can here entirely avoid doubt and error, and obtain certainty and truth :—That mathematics is prior to other sciences in nature, because it takes cognizance of quantity, which is apprehended by intuition, (*intuitu intellectus.*) "Moreover," he adds,† "there have been found famous men, as Robert, bishop of Lincoln, and Brother Adam Marshman, (de Marisco) and many others, who by the power of mathematics have been able to explain the causes of things; as may be seen in the writings of these men, for instance, concerning the Rainbow and Comets, and the generation of heat, and climates, and the celestial bodies."

But undoubtedly the most remarkable portion of the *Opus Majus* is the Sixth and last Part, which is entitled " De Scientia experimentali." It is indeed an extraordinary circumstance to find a writer of the thirteenth century, not only recognizing experiment as one source of knowledge, but urging its claims as far more important than men had yet been aware, exemplifying its value by

---

* *Op. Maj.*, p. 60.   † *Ib.*, p. 64.

striking and just examples, and speaking of its authority with a dignity of diction which sounds like a foremurmur of the Baconian sentences uttered nearly four hundred years later. Yet this is the character of what we here find*. " Experimental science, the sole mistress of speculative sciences, has three great Prerogatives among other parts of knowledge: First she tests by experiment the noblest conclusions of all other sciences : Next she discovers respecting the notions which other sciences deal with, magnificent truths to which these sciences of themselves can by no means attain : her Third dignity is, that she by her own power and without respect of other sciences, investigates the secrets of nature."

The examples which Bacon gives of these "Prerogatives" are very curious, exhibiting, among some error and credulity, sound and clear views. His leading example of the First Prerogative, is the Rainbow, of which the cause, as given by Aristotle, is tested by reference to experiment with a skill which is, even to us now, truly admirable. The examples of the Second Prerogative are three :—*first*, the art of making an artificial sphere which shall move with the heavens by natural influences, which Bacon trusts may be done, though astronomy herself cannot do it—" et tunc," he says, " thesaurum unius regis valeret hoc instrumentum ;" —*secondly*, the art of prolonging life, which experiment may teach, though medicine has no means of securing it except by regimen†;—*thirdly*, the art of making gold finer

* " Veritatis magnificas in terminis aliarum scientiarum in quas per nullam viam possunt illæ scientia, hæc sola scientiarum domina speculativarum, potest dare." *Op. Maj.*, p. 465.

† One of the ingredients of a preparation here mentioned, is the flesh of a dragon, which, it appears, is used as food by the Ethiopians. The mode of preparing this food cannot fail to amuse the reader. " Where there are good flying dragons, by the art which they possess, they draw them out of their dens, and have bridles and saddles in readiness, and they ride upon them, and make them bound about in

than fine gold, which goes beyond the power of alchemy. The Third Prerogative of experimental science, arts independent of the received sciences, is exemplified in many curious examples, many of them whimsical traditions. Thus it is said that the character of a people may be altered by altering the air*. Alexander, it seems, applied to Aristotle to know whether he should exterminate certain nations which he had discovered, as being irreclaimably barbarous; to which the philosopher replied, "If you can alter their air, permit them to live, if not, put them to death." In this part, we find the suggestion that the fire-works made by children, of saltpetre, might lead to the invention of a formidable military weapon.

It could not be expected that Roger Bacon, at a time when experimental science hardly existed, could give any *precepts* for the discovery of truth by experiment, But nothing can be a better *example* of the method of such investigation, than his inquiry concerning the cause of the Rainbow. Neither Aristotle, nor Avicenna, nor Seneca, he says, have given us any clear knowledge of this matter, but experimental science can do so. Let the experimenter (experimentator) consider the cases in which he finds the same colours, as the hexagonal crystals from Ireland and India; by looking into these he will see colours like these of the rainbow. Many think that this arises from some special virtue of these stones and their hexagonal figure; let therefore the experimenter go on, and he will find the same in other transparent stones, in dark ones as well as in light coloured. He will find the same effect also in other forms than the hexagon, if they be furrowed in the surface, as the Irish crystals are. Let him consider too,

the-air in a violent manner, that the hardness and toughness of the flesh may be reduced, as boars are hunted and bulls are baited before they are killed for eating." *Op. Maj.*, p. 470.

* *Ib.*, p. 473.

that he sees the same colours in the drops which are
dashed from oars in the sunshine ;—and in the spray thrown
by a mill wheel ;—and in the dew drops which lie on the
grass in a meadow on a summer morning ;—and if a man
takes water in his mouth and projects it on one side into
a sunbeam ;—and if in an oil lamp hanging in the air, the
rays fall in certain positions upon the surface of the oil ;
—and in many other ways, are colours produced. We
have here a collection of instances, which are almost all
examples of the same kind as the phenomenon under
consideration ; and by the help of a principle collected
by induction from these facts, the colours of the rainbow
were afterwards really explained.

With regard to the form and other circumstances of
the bow he is still more precise. He bids us measure the
height of the bow and of the sun, to show that the centre
of the bow is exactly opposite to the sun. He explains
the circular form of the bow,—its being independent of
the form of the cloud, its moving when we move, its flying
when we follow,—by its consisting of the reflections from
a vast number of minute drops. He does not, indeed,
trace the course of the rays through the drop, or account
for the precise magnitude which the bow assumes ; but
he approaches to the verge of this part of the explanation ;
and must be considered as having given a most happy
example of experimental inquiry into nature, at a time
when such examples were exceedingly scanty. In this
respect, he was more fortunate than Francis Bacon, as we
shall hereafter see.

We know but little of the biography of Roger Bacon,
but we have every reason to believe that his influence
upon his age was not great. He was suspected of magic,
and is said to have been put into close confinement in
consequence of this charge. In his work he speaks of

Astrology, as a science well worth cultivating. "But," says he, "Theologians and Decretists, not being learned in such matters, and seeing that evil as well as good may be done, neglect and abhor such things, and reckon them among Magic Arts." We have already seen, that at the very time when Bacon was thus raising his voice against the habit of blindly following authority, and seeking for all science in Aristotle, Thomas Aquinas was employed in fashioning Aristotle's tenets into that fixed form in which they became the great impediment to the progress of knowledge. It would seem, indeed, that something of a struggle between the progressive and stationary powers of the human mind was going on at this time Bacon himself says[*], "Never was there so great an appearance of wisdom, nor so much exercise of study in so many Faculties, in so many regions, as for this last forty years. Doctors are dispersed everywhere, in every castle, in every burgh, and especially by the students of two Orders, (he means the Franciscans and Dominicans, who were almost the only religious orders that distinguished themselves by an application to study[†],) which has not happened except for about forty years. And yet there was never so much ignorance, so much error." And in the part of his work which refers to Mathematics, he says of that study[‡], that it is the door and the key of the sciences; and that the neglect of it for thirty or forty years has entirely ruined the studies of the Latins. According to these statements, some change, disastrous to the fortunes of science, must have taken place about 1230, soon after the foundation of the Dominican and Franciscan Orders[§]. Nor can we doubt that the adoption

---

[*] Quoted by Jebb, *Pref.* to *Op. Maj.*
[†] MOSHEIM, *Hist.* iii. 161.   [‡] *Op. Maj.*, p. 57.
[§] MOSHEIM, iii. 161.

of the Aristotelian philosophy by these two Orders, in the form in which the Angelical Doctor had systematized it, was one of the events which most tended to defer, for three centuries, the reform which Roger Bacon urged as a matter of crying necessity in his own time.

CHAPTER VIII.

THE REVIVAL OF PLATONISM.

Causes of Delay in the Advance of Knowledge.—IN the insight possessed by learned men into the method by which truth was to be discovered, the fourteenth and fifteenth centuries went backwards, rather than forwards, from the point which had been reached in the thirteenth. Roger Bacon had urged them to have recourse to experiment; but they returned with additional and exclusive zeal to the more favourite employment of reasoning upon their own conceptions. He had called upon them to look at the world without; but their eyes forthwith turned back upon the world within. In the constant oscillation of the human mind between Ideas and Facts, after having for a moment touched the latter, it seemed to swing back more impetuously to the former. Not only was the philosophy of Aristotle firmly established for a considerable period, but when men began to question its authority, they attempted to set up in its place a philosophy still more purely ideal, that of Plato. It was not till the actual progress of experimental knowledge for some centuries had given it a vast accumulation of force, that it was able to break its way fully into the circle of speculative science. The new Platonist schoolmen had to run their course, the practical discoverers had to prove their merit by their works, the Italian innovators had to utter their aspirations for a change, before the second Bacon could truly declare that the time for a fundamental reform was at length arrived.

It cannot but seem strange, to any one who attempts to trace the general outline of the intellectual progress of man, and who considers him as under the guidance of a Providential sway, that he should thus be permitted to wander so long in a wilderness of intellectual darkness; and even to turn back, by a perverse caprice as it might seem, when on the very border of the brighter and better land which was his destined inheritance. We do not attempt to solve this difficulty: but such a course of things naturally suggests the thought, that a progress in physical science is not the main object of man's career, in the eyes of the Power who directs the fortunes of our race. We can easily conceive that it may have been necessary to man's general welfare that he should continue to turn his eyes inwards upon his own heart and faculties, till Law and Duty, Religion and Government, Faith and Hope, had been fully incorporated with all the past acquisitions of human intellect; rather than that he should have rushed on into a train of discoveries tending to chain him to the objects and operations of the material world. The systematic Law* and philosophical Theology which acquired their ascendancy in men's minds at the time of which we speak, kept them engaged in a region of speculations which perhaps prepared the way for a profounder and wider civilization, for a more elevated and spiritual character, than might have been possible without such a preparation. The great Italian poet of the fourteenth century speaks with strong admiration of the founders of the system which prevailed in his time.

---

* Gratian published the *Decretals* in the twelfth century; and the Canon and Civil Law became a regular study in the universities soon afterwards.

Thomas, Albert, Gratian, Peter Lombard, occupy distinguished places in the Paradise. The first, who is the poet's instructor, says,—

> Io fui degli agni della santa greggia
> Che Domenico mena per cammino
> U' ben s'impingua se non si vaneggia.
>   Questo che m'è a destra piu vicino
> Frate e maestro fummi; ed esso Alberto
> E di Cologna, ed io Tomas d'Aquino. . . . .
>   Quell' altro fiammeggiar esce del riso
> De Grazian, che l'uno et l'altro foro
> Ajutò si che piace in Paradiso.

> I, then, was of the lambs that Dominic
> Leads, for his saintly flock, along the way
> Where well they thrive not swoln with vanity.
> He nearest on my right hand brother was
> And master to me; Albert of Cologne
> Is this; and of Aquinum Thomas, I. . . . .
> That next resplendence issues from the smile
> Of Gratian who to either forum lent
> Such help as favour wins in Paradise.

It appears probable that neither poetry, nor painting, nor the other arts which require for their perfection a lofty and spiritualized imagination, would have appeared in the noble and beautiful forms which they assumed in the fourteenth and fifteenth century, if men of genius had, at the beginning of that period, made it their main business to discover the laws of nature, and to reduce them to a rigorous scientific form. Yet who can doubt that the absence of these touching and impressive works would have left one of the best and purest parts of man's nature without its due nutriment and development? It may perhaps be a necessary condition in the progress of man, that the Arts which aim at beauty reach their excellence before the Sciences which seek speculative truth; and if this be so, we inherit, from the middle ages, treasures which may well reconcile us to the delay which took place in their cultivation of experimental science.

However this may be, it is our business at present to trace the circumstances of this very lingering advance. We have already noticed the contest of the Nominalists and Realists, which was one form, though, with regard to scientific methods, an unprofitable one, of the antithesis of Ideas and Things. Though, therefore, this struggle continued, we need not dwell upon it. The Nominalists denied the real existence of Ideas, which doctrine was to a great extent implied in the prevailing systems; but the controversy in which they thus engaged, did not lead them to seek for knowledge in a new field and by new methods. The arguments which Occam the Nominalist opposes to those of Duns Scotus the Realist, are marked with the stamp of the same system, and consist only in permutations and combinations of the same elementary conceptions. It was not till the impulse of external circumstances was added to the discontent, which the more stirring intellects felt towards the barren dogmatism of their age, that the activity of the human mind was again called into full play, and a new career of progression entered upon, till then undreamt of, except by a few prophetic spirits.

Causes of Progress.—These circumstances were principally the revival of Greek and Roman literature, the invention of printing, the Protestant Reformation, and a great number of curious discoveries and inventions in the arts, which were soon succeeded by important steps in speculative physical science. Connected with the first of these events, was the rise of a party of learned men who expressed their dissatisfaction with the Aristotelian philosophy, as it was then taught, and manifested a strong preference for the views of Plato. It is by no means suitable to our plan to give a detailed account of this new Platonic school; but we may notice a few of the writers who belong to it, so far at least as to indicate its influence upon the Methods of pursuing science.

In the fourteenth century*, ·the frequent intercourse of the most cultivated persons of the Eastern and Western Empire, the increased study of the Greek language in Italy, the intellectual activity of the Italian States, the discovery of manuscripts of the classical authors, were circumstances which excited or nourished a new and zealous study of the works of Greek and Roman genius. The genuine writings of the ancients, when presented in their native life and beauty, instead of being seen only in those lifeless fragments and dull transformations which the scholastic system had exhibited, excited an intense enthusiasm. Europe, at that period, might be represented by Plato's beautiful allegory, of a man who, after being long kept in a dark cavern, in which his knowledge of the external world is gathered from the images which stream through the chinks of his prison, is at last led forth into the full blaze of day. It was inevitable that such a change should animate men's efforts and enlarge their faculties. Greek literature became more and more known, especially by the influence of learned men who came from Constantinople into Italy: these teachers, though they honoured Aristotle, reverenced Plato no less, and had never been accustomed to follow with servile submission of thought either these or any other leaders. The effect of such influences soon reveals itself in the works of that period. Dante has woven into his *Divina Comedia* some of the ideas of Platonism. Petrarch, who had formed his mind by the study of Cicero, and had thus been inspired with a profound admiration for the literature of Greece, learnt Greek from Barlaam, a monk who came as ambassador from the Emperor of the

---

* TENNEMAN, ix. 14.

East to the Pope, in 1339. With this instructor, the
poet read the works of Plato; struck by their beauty,
he contributed, by his writings and his conversation, to
awake in others an admiration and love for that philosopher, which soon became strongly and extensively prevalent among the learned in Italy.

Hermolaus Barbarus, &c.—Along with this feeling
there prevailed also, among those who had learnt to relish
the genuine beauties of the Greek and Latin writers, a
strong disgust for the barbarisms in which the scholastic
philosophy was clothed. Hermolaus Barbarus*, who
was born in 1454, at Venice, and had formed his taste
by the study of classical literature, translated, among
other learned works, Themistius's paraphrastic exposition
of the Physics of Aristotle; with the view of trying
whether the Aristotelian Natural Philosophy could not
be presented in good Latin, which the scholastic teachers
denied. In his Preface he expresses great indignation
against those philosophers who have written and disputed
on philosophical subjects in barbarous Latin, and in
an uncultured style, so that all refined minds are
repelled from these studies by weariness and disgust.
They have, he says, by this barbarism, endeavoured to
secure to themselves, in their own province, a supremacy
without rivals or opponents. Hence they maintain that
mathematics, philosophy, jurisprudence, cannot be expounded in correct Latin;—that between these sciences
and the genuine Latin language there is a great gulf, as
between things that cannot be brought together: and
on this ground they blame those who combine the study
of philology and eloquence with that of science. This
opinion, adds Hermolaus, perverts and ruins our studies;
and is highly prejudicial and unworthy in respect to
the state. Hermolaus awoke in others, as for instance

* TENNEMAN, ix. 25.

in John Picus of Mirandula, the same dislike to the reigning school philosophy. As an opponent of the same kind, we may add Marius Nizolius of Bersallo, a scholar who carried his admiration of Cicero to an exaggerated extent, and who was led, by a controversy with the defenders of the scholastic philosophy, to publish (1553) a work *On the True Principles and True Method of Philosophizing.* In the title of this work, he professes to give "the true principles of almost all arts and sciences, refuting and rejecting almost all the false principles of the Logicians and Metaphysicians." But although, in the work, he attacks the scholastic philosophy, he does little or nothing to justify the large pretensions of his title; and he excited, it is said, little notice. It is therefore curious that Leibnitz should have thought it worth his while to re-edit this work, which he did in 1670, adding remarks of his own.

Nicolaus Cusanus.—Without dwelling upon this opposition to the scholastic system on the ground of taste, I shall notice somewhat further those writers who put forwards Platonic views, as fitted to complete or to replace the doctrines of Aristotle. Among these, I may place Nicolaus Cusanus, so called from Cus, a village on the Moselle, where he was born in 1401; who was afterwards raised to the dignity of cardinal. We might, indeed, at first be tempted to include Cusanus among those persons who were led to reject the old philosophy by being themselves agents in the progressive movement of physical science. For he published, before Copernicus, and independently of him, the doctrine that the earth is in motion*. But it should be recollected that in order to see the possibility of this doctrine, and its claims to acceptance, no new reference to observation was requisite.

* " Jam nobis manifestum est terram istam in veritate moveri," &c. —*De Doctâ Ignorantiâ,* lib. ii. cap. 12.

The Heliocentric System was merely a new mode of representing to the mind facts with which all astronomers had long been familiar. The system might very easily have been embraced and inculcated by Plato himself; as indeed it is said to have been actually taught by Pythagoras. The mere adoption of the Heliocentric view, therefore, without attempting to realize the system in detail, as Copernicus did, cannot entitle a writer of the fifteenth century to be looked upon as one of the authors of the discoveries of that period; and we must consider Cusanus as a speculative anti-Aristotelian, rather than as a practical reformer.

The title of Cusanus's book, *De Doctâ Ignorantiâ*, shows how far he was from agreeing with those who conceived that, in the works of Aristotle, they had a full and complete system of all human knowledge. At the outset of this book*, he says, after pointing out some difficulties in the received philosophy, " If, therefore, the case be so, (as even the most profound Aristotle, in his *First Philosophy*, affirms,) that in things most manifest by nature, there is a difficulty, no less than for an owl to look at the sun; since the appetite of knowledge is not implanted in us in vain, we ought to desire to know that we are ignorant. If we can fully attain to this, we shall arrive at *Instructed Ignorance*." How far he was from placing the source of knowledge in experience, as opposed to ideas, we may see in the following passage† from another work of his, *On Conjectures*. " Conjectures must proceed from our mind, as the real world proceeds from the infinite Divine Reason. For since the human mind, the lofty likeness of God, participates, as it may, in the fruitfulness of the creative nature, it doth from itself, as the image of the Omnipotent Form, bring forth reasonable thoughts which have a similitude to real existences.

---

* *De Doct. Ignor.*, lib. i. c. 1.   † *De Conjecturis*, l. i. c. 3, 4.

Thus the Human Mind exists as a conjectural form of the world, as the Divine Mind is its real form." We have here the Platonic or ideal side of knowledge put prominently and exclusively forwards.

Marsilius Ficinus, &c.—A person who had much more influence on the diffusion of Platonism was Marsilius Ficinus, a physician of Florence. In that city there prevailed, at the time of which we speak, the greatest enthusiasm for Plato. George Gemistius Pletho, when in attendance upon the Council of Florence, had imparted to many persons the doctrines of the Greek philosopher; and, among others, had infused a lively interest on this subject into the elder Cosmo, the head of the family of the Medici. Cosmo formed the plan of founding a Platonic academy. Ficinus*, well instructed in the works of Plato, Plotinus, Proclus, and other Platonists, was selected to further this object, and was employed in translating the works of these authors into Latin. It is not to our present purpose to consider the doctrines of this school, except so far as they bear upon the nature and methods of knowledge; and therefore I must pass by, as I have in other instances done, the greater part of their speculations, which related to the nature of God, the immortality of the soul, the principles of Goodness and Beauty, and other points of the same order. The object of these and other Platonists of this school, however, was not to expel the authority of Aristotle by that of Plato. Many of them had come to the conviction that the highest ends of philosophy were to be reached only by bringing into accordance the doctrines of Plato and of Aristotle. Of this opinion was John Picus, Count of Mirandula and Concordia; and under this persuasion he employed the whole of his life in labouring upon a work, *De Concordiá Platonis et Aristotelis,* which was not

* Born in 1433.

completed at the time of his death, in 1494; and has never been published. But about a century later, another writer of the same school, Francis Patricius*, pointing out the discrepancies between the two Greek teachers, urged the propriety of deposing Aristotle from the supremacy he had so long enjoyed. " Now all these doctrines, and others not a few," he says†, "since they are Platonic doctrines, philosophically most true, and consonant with the Catholic faith, whilst the Aristotelian tenets are contrary to the faith, and philosophically false, who will not, both as a Christian and a philosopher, prefer Plato to Aristotle? And why should not hereafter, in all the colleges and monasteries of Europe, the reading and study of Plato be introduced? Why should not the philosophy of Aristotle be forthwith exiled from such places? Why must men continue to drink the mortal poison of impiety from that source?" with much more in the same strain.

The Platonic school, of which we have spoken, had, however, reached its highest point of prosperity before this time, and was already declining. About 1500, the Platonists appeared to triumph over the Peripatetics‡; but the death of their great patron, Cardinal Bessarion, about this time, and we may add, the hollowness of their system in many points, and its want of fitness for the wants and expectations of the age, turned men's thoughts partly back to the established Aristotelian doctrines, and partly forwards to schemes of bolder and fresher promise.

Francis Patricius.—Patricius, of whom we have just spoken, was one of those who had arrived at the conviction that the formation of a new philosophy, and not merely the restoration of an old one, was needed. In 1593, appeared his *Nova de Universis Philosophia;* and

---

* Born 1529, died 1597.   † *Aristoteles Exotericus*, p. 50.
‡ TIRABOSCHI, t. vii. part ii. p. 411.

the mode in which it begins* can hardly fail to remind us of the expressions which Francis Bacon soon afterwards used in the opening of a work of the same nature. " Francis Patricius, being about to found anew the true philosophy of the universe, dared to begin by announcing the following indisputable principles." Here, however, the resemblance between Patricius and true inductive philosophers ends. His principles are barren *à priori* axioms; and his system has one main element, *Light,* (*Lux,* or *Lumen,*) to which all operations of nature are referred. In general cultivation, and practical knowledge of nature, he was distinguished among his contemporaries. In various passages of his works he relates† observations which he had made in the course of his travels, in Cyprus, Corfu, Spain, the mountains of the Modenese, and Dalmatia, which was his own country; his observations relate to light, the saltness of the sea, its flux and reflux, and other points of astronomy, meteorology, and natural history. He speaks of the sex of plants‡; rejects judicial astrology; and notices the astronomical systems of Copernicus, Tycho, Fracastoro, and Torre. But the mode in which he speaks of experiments proves, what indeed is evident from the general scheme of his system, that he had no due appreciation of the place which observation must hold in real natural philosophy.

Picus, Agrippa, &c.—It had been seen in the later

---

* " Franciscus Patricius, novam veram integram de universis conditurus philosophiam, sequentia uti verissima prænuntiare est ausus. Prænunciata ordine persecutus, divinis oraculis, geometricis rationibus, clarissimisque experimentis comprobavit.

Ante primum nihil,
Post primum omnia,
A principio omnia," &c.
His other works are *Panaugia, Pancosmia, Dissertationes Peripateticæ.*

† TIRABOSCHI, t. vii. part ii. p. 411.

‡ *Dissert. Peripatet.,* t. ii. lib. v. sub fin.

philosophical history of Greece, how readily the ideas of the Platonic school lead on to a system of unfathomable and unbounded mysticism. John Picus, of Mirandula*, added to the study of Plato and the Neoplatonists, a mass of allegorical interpretations of the Scriptures, and the dreams of the Cabbala, a Jewish system†, which pretends to explain how all things are an emanation of the Deity. To this his nephew, Francis Picus, added a reference to inward illumination‡, by which knowledge is obtained, independently of the progress of reasoning. John Reuchlin, or Capnio, born 1455; John Baptist Helmont, born 1577; Francis Mercurius Helmont, born 1618, and others, succeeded John Picus in his admiration of the Cabbala: while others, as Jacob Bœhmen, rested upon internal revelations like Francis Picus. And thus we have a series of mystical writers, continued into modern times, who may be considered as the successors of the Platonic school; and who all exhibit views altogether erroneous with regard to the nature and origin of knowledge. Among the various dreams of this school are certain wide and loose analogies of terrestrial and spiritual things. Thus in the writings of Cornelius Agrippa (who was born 1487, at Cologne) we have such systems as the following§:—" Since there is a threefold world, elemental, celestial, and intellectual, and each lower one is governed by that above it, and receives the influence of its powers: so that the very Archetype and Supreme Author transfuses the virtues of his omnipotence into us through angels, heavens, stars, elements, animals, plants, stones,—into us, I say, for whose service he has framed and created all these things;—the Magi do not think it irrational that we should be able to ascend by the same degrees, the same worlds, to this Archetype of the world, the Author

---

* TENNEMAN, ix. 148.   † Ib., 167.   ‡ Ib., 158.
§ AGRIPPA, De Occult. Phil., lib. i. c. 1.

and First Cause of all, of whom all things are, and from whom they proceed; and should not only avail ourselves of those powers which exist in the nobler works of creation, but also should be able to attract other powers, and add them to these."

Agrippa's work, *De Vanitate Scientiarum,* may be said rather to have a sceptical and cynical, than a platonic, character. It is a declamation*, in a melancholy mood, against the condition of the sciences in his time. His indignation at the worldly success of men whom he considered inferior to himself, had, he says, metamorphosed him into a dog, as the poets relate of Hecuba of Troy, so that his impulse was to snarl and bark. His professed purpose, however, was to expose the dogmatism, the servility, the self-conceit, and the neglect of religious truth which prevailed in the reigning Schools of philosophy. His views of the nature of science, and the modes of improving its cultivation, are too imperfect and vague to allow us to rank him among the reformers of science.

Paracelsus, Fludd, &c.—The celebrated Paracelsus† put himself forwards as a reformer in philosophy, and obtained no small number of adherents. He was, in most respects, a shallow and impudent pretender; and had small knowledge of the literature or science of his time: but by the tone of his speaking and writing he manifestly belongs to the mystical school of which we are now speaking. Perhaps by the boldness with which he proposed new systems, and by connecting these with the practical doctrines of medicine, he contributed something to the introduction of a new philosophy. We have seen in the History of Chemistry that he was the author of

---

* Written in 1526.

† Philip Aurelius Theophrastus Bombastus von Hohenheim, also called Paracelsus Eremita, born at Einsiedlen in Switzerland, in 1493.

the system of Three Principles, (salt, sulphur, and mercury,) which replaced the ancient doctrine of Four Elements, and prepared the way for a true science of chemistry. But the salt, sulphur, and mercury of Paracelsus were not, he tells his disciples, the visible bodies which we call by those names, but certain invisible, astral, or sidereal elements. The astral salt is the basis of the solidity and incombustible parts in bodies; the astral sulphur is the source of combustion and vegetation; the astral mercury is the origin of fluidity and volatility. And again, these three elements are analogous to the three elements of man,—Body, Spirit, and Soul.

A writer of our own country, belonging to this mystical school, is Robert Fludd, or De Fluctibus, who was born in 1571, in Kent, and after pursuing his studies at Oxford, travelled for several years. Of all the Theosophists and Mystics, he is by much the most learned; and was engaged in various controversies with Mersenne, Gassendi, Kepler, and others. He thus brings us in contact with the next class of philosophers whom we have to consider, the practical reformers of philosophy;—those who furthered the cause of science by making, promulgating or defending the great discoveries which now began to occupy men. He adopted the principle, which we have noticed elsewhere*, of the analogy of the Macrocosm and Microcosm, the world of nature and the world of man. His system contains such a mixture and confusion of physical and metaphysical doctrines as might be expected from his ground plan, and from his school. Indeed his object, the general object of mystical speculators, is to identify physical with spiritual truths. Yet the influence of the practical experimental philosophy which was now gaining ground in the world may be traced in him. Thus he refers to experiments on distillation to

* B. ix. c. 2. s. 1. The Mystical School of Biology.

prove the existence and relation of the regions of water, air, and fire, and of the spirits which correspond to them; and is conceived, by some persons[*], to have anticipated Torricelli in the invention of the Barometer.

We need no further follow the speculations of this school. We see already abundant reason why the reform of the methods of pursuing science could not proceed from the Platonists. Instead of seeking knowledge by experiment, they immersed themselves deeper than even the Aristotelians had done in traditionary lore, or turned their eyes inwards in search of an internal illumination. Some attempts were made to remedy the defects of philosophy by a recourse to the doctrines of other sects of antiquity, when men began to feel more distinctly the need of a more connected and solid knowledge of nature than the established system gave them. Among these attempts were those of Berigard[†], Magernus, and especially Gassendi, to bring into repute the philosophy of the Ionian school, of Democritus and of Epicurus. But these endeavours were posterior in time to the new impulse given to knowledge by Copernicus, Kepler, and Galileo, and were influenced by views arising out of the success of these discoveries, and they must, therefore, be considered hereafter. In the mean time, some independent efforts (arising from speculative rather than practical reformers) were made to cast off the yoke of the Aristotelian dogmatism, and to apprehend the true form of that new philosophy which the most active and hopeful minds saw to be needed; and we must give some account of these attempts, before we can commit ourselves to the full stream of progressive philosophy.

[*] TENNEMAN, ix. 221.   [†] Ib., 265.

CHAPTER IX.

THE THEORETICAL REFORMERS OF SCIENCE.

WE have already seen that Patricius, about the middle of the sixteenth century, announced his purpose of founding anew the whole fabric of philosophy; but that, in executing this plan, he ran into wide and baseless hypotheses, suggested by *à priori* conceptions rather than by external observation; and that he was further misled by fanciful analogies resembling those which the Platonic mystics loved to contemplate. The same time, and the period which followed it, produced several other essays which were of the same nature, with the exception of their being free from the peculiar tendencies of the Platonic school: and these insurrections against the authority of the established dogmas, although they did not directly substitute a better positive system in the place of that which they assailed, shook the authority of the Aristotelian system, and led to its overthrow; which took place as soon as these theoretical were aided by other practical reformers.

Bernardinus Telesius.—Italy, always, in modern times, fertile in the beginnings of new systems, was the soil on which these innovators arose. The earliest and most conspicuous of them is Bernardinus Telesius, who was born in 1508, at Cosenza, in the kingdom of Naples. His studies, carried on with great zeal and ability, first at Milan and then at Rome, made him well acquainted with the knowledge of his times; but his own reflections convinced him that the basis of science, as then received, was altogether erroneous; and led him to attempt a reform, with which view, in 1565, he published, at Rome, his work*, " *Bernardinus Telesius, of Cosenza, on the Nature*

---

* BERNARDINI TELESII CONSENTINI *De Rerum Natura juxta propria Principia.*

of Things, according to principles of his own." In the
preface of this work he gives a short account* of the
train of reflection by which he was led to put himself in
opposition to the Aristotelian philosophy.   This kind of
autobiography occurs not unfrequently in the writings of
theoretical reformers; and shows how livelily they felt the
novelty of their undertaking.   After the storm and sack
of Rome in 1527, Telesius retired to Padua, as a peaceful
seat of the muses; and there studied philosophy and mathematics, with great zeal, under the direction of Jerom
Amalthæus and Frederic Delphinus.   In these studies he
made great progress; and the knowledge which he thus
acquired threw a new light upon his view of the Aristotelian philosophy.   He undertook a closer examination of the
Physical Doctrines of Aristotle; and as the result of this,
he was astonished how it could have been possible that so
many excellent men, so many nations, and even almost the
whole human race, should, for so long a time, have allowed
themselves to be carried away by a blind reverence for a
teacher, who had committed errors so numerous and grave
as he perceived to exist in "the philosopher."   Along with
this view of the insufficiency of the Aristotelian philosophy, arose, at an early period, the thought of erecting a
better system in its place.   With this purpose he left
Padua, when he had received the degree of Doctor, and
went to Rome, where he was encouraged in his design by
the approval and friendly exhortations of distinguished
men of letters, amongst whom were Ubaldino Bandinelli
and Giovanni della Casa.   From Rome he went to his
native place, when the incidents and occupations of a
married life for a while interrupted his philosophical project.   But after his wife was dead, and his eldest son

---

* I take this account from Tenneman : this Proem was omitted in
subsequent editions of Telesius, and is not in the one which I have
consulted. TENNEMAN, *Gesch. d. Phil.*, ix. 280.

grown to manhood, he resumed with ardour the scheme of his youth; again studied the works of Aristotle and other philosophers, and composed and published the two first books of his treatise. The opening to this work sufficiently exhibits the spirit in which it was conceived. Its object is stated in the title to be to show, that "the construction of the world, the magnitude and nature of the bodies contained in it, are not to be investigated by reasoning, which was done by the ancients, but are to be apprehended by the senses, and collected from the things themselves." And the Proem is in the same strain. "They who before us have inquired concerning the construction of this world and of the things which it contains, seem indeed to have prosecuted their examination with protracted vigils and great labour, but *never to have looked at it.*" And thus, he observes, they found nothing but error. This he ascribes to their presumption. "For, as it were, attempting to rival God in wisdom, and venturing to seek for the principles and causes of the world by the light of their own reason, and thinking they had found what they had only invented, they made an arbitrary world of their own." "*We* then," he adds, "not relying on ourselves, and of a duller intellect than they, propose to ourselves to turn our regards to the world itself and its parts."

The execution of the work, however, by no means corresponds to the announcement. The doctrines of Aristotle are indeed attacked; and the objections to these, and to other received opinions, form a large part of the work. But these objections are supported by *à priori* reasoning, and not by experiments. And thus, rejecting the Aristotelian physics, he proposes a system at least equally baseless; although, no doubt, grateful to the author from its sweeping and apparently simple character. He assumes three principles, Heat, Cold, and Matter:

Heat is the principle of motion, Cold of immobility, and Matter is the corporeal substratum, in which these incorporeal and active principles produce their effects. It is easy to imagine that, by combining and separating these abstractions in various ways, a sort of account of many natural phenomena may be given; but it is impossible to ascribe any real value to such a system. The merit of Telesius must be considered to consist in his rejection of the Aristotelian errors, in his perception of the necessity of a reform in the method of philosophizing, and in his persuasion that this reform must be founded on experiments rather than on reasoning. When he said*, "We propose to ourselves to turn our eyes to the world itself, and its parts, their passions, actions, operations and species," his view of the course to be followed was righ ; but his purpose remained but ill fulfilled, by the arbitrary edifice of abstract conceptions which his system exhibits.

Bacon, who, about half a century later, treated the subject of a reform of philosophy in a far more penetrating and masterly manner, has given us his judgment of Telesius. In his view, he considers Telesius as the restorer of the Atomic philosophy, which Democritus and Parmenides taught among the ancients; and according to his custom, he presents an image of this philosophy in an adaptation of a portion of ancient mythology†. The Celestial Cupid, who, with Cœlus, was the parent of the Gods and of the Universe, is exhibited as a representation of matter and its properties, according to the Democritean philosophy. "Concerning Telesius," says Bacon, "we think well, and acknowledge him as a lover of truth, a useful contributor to science, an amender of some tenets,

---

* Proem.

† " De Principiis atque Originibus secundum fabulas Cupidinis et Cœli : sive Parmenidis et Telesii et præcipuè Democriti Philosophia tractata in Fabula de Cupidine."

the first of recent men. But we have to do with him as the restorer of the philosophy of Parmenides, to whom much reverence is due." With regard to this philosophy, he pronounces a judgment which very truly expresses the cause of its rashness and emptiness. "It is," he says, "such a system* as naturally proceeds from the intellect, abandoned to its own impulse, and not rising from experience to theory continuously and successively." Accordingly, he says that, "Telesius, although learned in the Peripatetic philosophy (if that were anything), which, indeed, he has turned against the teachers of it, is hindered by his affirmations, and is more successful in destroying than in building."

The work of Telesius excited no small notice, and was placed in the *Index Expurgatorius*. It made many disciples, a consequence probably due to its spirit of system-making, no less than to its promise of reform, or its acuteness of argument; for till trial and reflection have taught man modesty and moderation, he can never be content to receive knowledge in the small successive instalments in which nature gives it forth to him. It is the makers of large systems, arranged with an appearance of completeness and symmetry, who, principally, give rise to Schools of philosophy.

(*Thomas Campanella.*)—Accordingly, Telesius may be looked upon as the founder of a School. His most distinguished successor was Thomas Campanella, who was born in 1568, at Stilo, in Calabria. He showed great talents at an early age, prosecuting his studies at Cosenza, the birth-place of the great opponent of Aristotle and reformer of philosophy. He, too, has given us an

---

* "Talia sunt qualia possunt esse ea quæ ab intellectu sibi permisso, nec ab experimentis continenter et gradatim sublevato, profecta videntur."

account* of the course of thought by which he was led to become an innovator. " Being afraid that not genuine truth, but falsehood in the place of truth, was the tenant of the Peripatetic School, I examined all the Greek, Latin, and Arabic commentators of Aristotle, and hesitated more and more, as I sought to learn whether what they have said were also to be read in the world itself, which I had been taught by learned men was the living book of God. And as my doctors could not satisfy my scruples, I resolved to read all the books of Plato, Pliny, Galen, the Stoics, and the Democriteans, and especially those of Telesius; and to compare them with that *first and original writing, the world;* that thus from the primary autograph, I might learn if the copies contained anything false." Campanella probably refers here to an expression of Plato, who says, " the world is God's epistle to mankind." And this image, of the natural world as an original manuscript, while human systems of philosophy are but copies, and may be false ones, became a favourite thought of the reformers, and appears repeatedly in their writings from this time. " When I held my public disputation at Cosenza," Campanella proceeds, " and still more, when I conversed privately with the brethren of the monastery, I found little satisfaction in their answers; but Telesius delighted me, on account of his freedom in philosophizing, and because he rested upon the nature of things, and not upon the assertions of men."

With these views and feelings, it is not wonderful that Campanella, at the early age of twenty-two (1590,) published a work remarkable for the bold promise of its title: " *Thomas Campanella's Philosophy demonstrated to the senses, against those who have philosophized in a arbitrary and dogmatical manner, not taking nature for their guide; in which the errors of Aristotle and his followers*

* Thom. Campanella *de Libris propriis,* as quoted in Tenneman, ix. 291.

are refuted from their own assertions and the laws of nature; and all the imaginations feigned in the place of nature by the Peripatetics are altogether rejected; with a true defence of Bernardin Telesius of Cosenza, the greatest of philosophers; confirmed by the opinions of the ancients, here elucidated and defended, especially those of the Platonists."

This work was written in answer to a book against Telesius by a Neapolitan professor named Marta; and it was the boast of the young author that he had only employed eleven months in the composition of his defence, while his adversary had been engaged eleven years in preparing his attack. Campanella found a favourable reception in the house of the Marchese Lavelli, and there employed himself in the composition of an additional work, entitled *On the Sense of Things and Magic*, and in other literary labours. These, however, are full of the indications of an enthusiastic temper, inclined to mystical devotion, and of opinions bearing the cast of pantheism. For instance, the title of the book last quoted sets forth as demonstrated in the course of the work, that " the world is the living and intelligent statue of God; and that all its parts, and particles of parts, are endowed some with a clearer, some with a more obscure sense, such as suffices for the preservation of each and of the whole." Besides these opinions, which could not fail to make him obnoxious to the religious authorities, Campanella* engaged in schemes of political revolution, which involved him in danger and calamity. He took part in a conspiracy, of which the object was to cast off the tyranny of Spain, and to make Calabria a republic. This design was discovered; and Campanella, along with others, was thrown into prison and subjected to torture. He was kept in confinement twenty-seven years; and at. last obtained his liberation

* ECONOMISTI *Italiani*, tom. i. p. xxxiii.

by the interposition of Pope Urban VIII. He was, however, still in danger from the Neapolitan Inquisition; and escaped in disguise to Paris, where he received a pension from the king, and lived in intercourse with the most eminent men of letters. He died there in 1639.

Campanella was a contemporary of Francis Bacon, whom we must consider as belonging to an epoch to which the Calabrian school of innovators was only a prelude. I shall not therefore further follow the connexion of writers of this order. Tobias Adami, a Saxon writer, an admirer of Campanella's works, employed himself, about 1620, in adapting them to the German public, and in recommending them strongly to German philosophers. Descartes, and even Bacon, may be considered as successors of Campanella; for they too were theoretical reformers; but they enjoyed the advantage of the light which had, in the mean time, been thrown upon the philosophy of science, by the great practical advances of Kepler, Galileo, and others. To these practical reformers we must soon turn our attention; but we may first notice one or two additional circumstances belonging to our present subject.

Campanella remarks that both the Peripatetics and the Platonists conducted the learner to knowledge by a long and circuitous path, which he wished to shorten by setting out from the sense. Without speaking of the methods which he proposed, we may notice one maxim* of considerable value which he propounds, and to which we have already been led. " We begin to reason from sensible objects, and definition is the end and epilogue of science. It is not the begining of our knowing, but only of our teaching."

(*Andrew Cæsalpinus.*)—The same maxim had already been announced by Cæsalpinus, a contemporary of Telesius; (he was born at Arezzo in 1520, and died at Rome

* TENNEMAN, ix. 305.

in 1603.) Cæsalpinus is a great name in science, though
professedly an Aristotelian. It has been seen in the
History of Science*, that he formed the first great
epoch of the science of botany by his systematic arrangement of plants, and that in this task he had no successor
for nearly a century. He also approached near to the
great discovery of the circulation of the blood†. He
takes a view of science which includes the remark that
we have just quoted from Campanella : " We reach perfect knowledge by three steps : Induction, Division, Definition. By Induction, we collect likeness and agreement
from observation; by Division, we collect unlikeness and
disagreement; by Definition, we learn the proper substance of each object. Induction makes universals from
particulars, and offers to the mind all intelligible matter;
Division discovers the difference of universals, and leads
to species; Definition resolves species into their principles
and elements‡." Without asserting this to be rigorously
correct, it is incomparably more true and philosophical
than the opposite view, which represents definition as the
beginning of our knowledge; and the establishment of
such a doctrine is a material step in inductive philosophy§.

(*Giordano Bruno.*)—Among the Italian innovators of
this time we must notice the unfortunate Giordano Bruno,
who was born at Nola about 1550, and burnt at Rome
in 1600. He is, however, a reformer of a different school
from Campanella ; for he derives his philosophy from
Ideas and not from Observation. He represents himself
as the author of a new doctrine, which he terms the *Nolan
Philosophy*. He was a zealous promulgator and defender
of the Copernican system of the universe, as we have
noticed in the History of Science‖. Campanella also
wrote in defence of that system.

---

* *Hist. Ind. Sci.*, iii. 280.   † *Ib.*, iii. 396.
‡ *Quæst. Peripateticæ*, i. 1.   § TENNEMAN, ix. 108.
‖ *Hist. Ind. Sci.*, i. 384.

It is worthy of remark that a thought which is often quoted from Francis Bacon, occurs in Bruno's *Cena di Cenere*, published in 1584; I mean, the notion that the later times are more aged than the earlier. In the course of the dialogue, the Pedant, who is one of the interlocutors, says, " In antiquity is wisdom;" to which the Philosophical Character replies, " If you knew what you were talking about, you would see that your principle leads to the opposite result of that which you wish to infer;—I mean, that *we* are older, and have lived longer, than our predecessors." He then proceeds to apply this, by tracing the course of astronomy through the earlier astronomers up to Copernicus.

(*Peter Ramus.*)—I will notice one other reformer of this period, who attacked the Aristotelian system on another side, on which it was considered to be most impregnable. This was Peter Ramus, (born in Picardy in 1515,) who ventured to denounce the *Logic* of Aristotle as unphilosophical and useless. After showing an extraordinary aptitude for the acquirement of knowledge in his youth, when he proceeded to the degree of Master of Arts, he astonished his examiners by choosing for the subject of the requisite disputation the thesis*, " that all which Aristotle has said is not true." This position, so startling in 1535, he defended for the whole day, without being defeated. This was, however, only a formal academical exercise, which did not necessarily imply any permanent conviction of the opinion thus expressed. But his mind was really labouring to detect and remedy the errors which he thus proclaimed. From him, as from the other reformers of this time, we have an account of this mental struggle†. He says, in a work on this subject, " I will candidly and simply explain how I was

* TENNEMAN, ix. 420.
† RAMI, *Animadversiones Aristotelicæ*, i. iv.

delivered from the darkness of Aristotle. When, according to the laws of our university, I had spent three years and a half in the Aristotelian philosophy, and was now invested with the philosophical laurel as a Master of Arts, I took an account of the time which I had consumed in this study; and considered on what subjects I should employ this logical art of Aristotle, which I had learnt with so much labour and noise. I found it made me not more versed in history or antiquities, more eloquent in discourse, more ready in verse, more wise in any subject. Alas for me! how was I overpowered, how deeply did I groan, how did I deplore my lot and my nature, how did I deem myself to be by some unhappy and dismal fate and frame of mind abhorrent from the Muses, when I found that I was one who, after all my pains, could reap no benefit from that wisdom of which I heard so much, as being contained in the Logic of Aristotle." He then relates, that he was led to the study of the Dialogues of Plato, and was delighted with the kind of analysis of the subjects discussed which Socrates is there represented as executing. "Well," he adds, "I began thus to reflect within myself—(I should have thought it impious to say it to another)—What, I pray you, prevents me from *socratizing;* and from asking, without regard to Aristotle's authority, whether Aristotle's Logic be true and correct? It may be that that philosopher leads us wrong; and if so, no wonder that I cannot find in his books the treasure which is not there. What if his dogmas be mere figments? Do I not tease and torment myself in vain, trying to get a harvest from a barren soil?" He convinced himself that the Aristotelian logic was worthless : and constructed a new system of Logic, founded mainly on the Platonic process of exhausting a subject by analytical classification of its parts. Both works, his *Animadversions on Aristotle,* and his *Logic,* appeared in 1543.

The learned world was startled and shocked to find a young man, on his first entrance into life, condemning as faulty, fallacious, and useless, that part of Aristotle's works which had always hitherto been held as a masterpiece of philosophical acuteness, and as the Organon of scientific reasoning. And in truth, it must be granted that Ramus does not appear to have understood the real nature and object of Aristotle's Logic; while his own system could not supply the place of the old one, and was not of much real value. This dissent from the established doctrines was, however, not only condemned but punished. The printing and selling of his books was forbidden through France; and Ramus was stigmatized by a sentence* which declared him rash, arrogant, impudent, and ignorant, and prohibited from teaching logic and philosophy. He was, however, afterwards restored to the office of professor: and though much attacked, persisted in his plan of reforming, not only Logic but Physics and Metaphysics. He made his position still more dangerous by adopting the reformed religion; and during the unhappy civil wars of France, he was deprived of his professorship, driven from Paris, and had his library plundered. He endeavoured, but in vain, to engage a German professor, Schegk, to undertake the reform of the Aristotelian Physics; a portion of knowledge in which he felt himself not to be strong. Unhappily for himself, he afterwards returned to Paris, where he perished in the massacre of St. Bartholomew in 1572.

Ramus's main objection to the Aristotelian Logic is, that it is not the image of the natural process of thought; an objection which shows little philosophical insight; for the course by which we obtain knowledge may well differ from the order in which our knowledge, when obtained, is exhibited. We have already seen that Ramus's contemporaries, Cæsalpinus and Campanella, had a wiser

* See *Hist. Ind. Sci.*, i. 327.

view; placing definition as the last step in knowing, but the first in teaching. But the effect which Ramus produced was by no means slight. He aided powerfully in turning the minds of men to question the authority of Aristotle on all points; and had many followers, especially among the Protestants. Among the rest, Milton, our great poet, published "Artis Logicæ plenior Institutio *ad Petri Rami methodum concinnata ;*" but this work, appearing in 1672, belongs to a succeeding period.

(*The Reformers in general.*)—It is impossible not to be struck with the series of misfortunes which assailed the reformers of philosophy of the period we have had to review. Roger Bacon was repeatedly condemned and imprisoned; and, not to speak of others who suffered under the imputation of magical arts, Telesius is said* to have been driven from Naples to his native city by calumny and envy; Cæsalpinus was accused of atheism†; Campanella was imprisoned for twenty-seven years and tortured; Giordano Bruno was burnt at Rome as a heretic; Ramus was persecuted during his life, and finally murdered by his personal enemy Jacques Charpentier, in a massacre of which the plea was religion. It is true, that for the most part these misfortunes were not principally due to the attempts at philosophical reform, but were connected rather with politics or religion. But we cannot doubt that the spirit which led men to assail the received philosophy, might readily incline them to reject some tenets of the established religion; since the boundary line of these subjects is difficult to draw. And as we have seen, there was in most of the persons of whom we have spoken, not only a well-founded persuasion of the defects of existing systems, but an eager spirit of change, and a sanguine anticipation of some wide and lofty philosophy, which was soon to elevate the minds and conditions of men. The most unfortunate were, for the most

---

* Tenneman, ix. 200.   † *Ib.* ix. 108.

part, the least temperate and judicious reformers.   Patricius, who, as we have seen, declared himself against the Aristotelian philosophy, lived and died at Rome in peace and honour*.

(*Melanchthon.*)—It is not easy to point out with precision the connexion between the efforts at a Reform in Philosophy, and the great Reformation of Religion in the sixteenth century.   The disposition to assert (practically at least) a freedom of thinking, and to reject the corruptions which tradition had introduced and authority maintained, naturally extended its influence from one subject to another; and especially in subjects so nearly connected as theology and philosophy.   The Protestants, however, did not reject the Aristotelian system; they only reformed it, by going back to the original works of the author, and by reducing it to a conformity with Scripture.   In this reform, Melanchthon was the chief author, and wrote works on Logic, Physics, Morals, and Metaphysics, which were used among Protestants.   On the subject of the origin of our knowledge, his views contained a very philosophical improvement of the Aristotelian doctrines. He recognized the importance of Ideas, as well as of Experience.   " We could not," he says†, " proceed to reason at all, except there were by nature innate in man certain fixed points, that is, principles of science;—as Number, the recognition of Order and Proportion, logical, geometrical, physical and moral Principles.   Physical principles are such as these,—everything which exists proceeds from a cause,—a body cannot be in two places at once,—time is a continued series of things or of motions,—and the like."   It is not difficult to see that such Principles partake of the nature of the Fundamental Ideas which we have attempted to arrange and enumerate in a previous part of this work.

* TENNEMAN, ix. 246.
† MELANCHTHON, *De Anima*, p. 207, quoted in TENNEMAN, ix. 121.

Before we proceed to the next chapter, which treats of the Practical Reformers of Scientific Method, let us for an instant look at the strong persuasion that the time of a philosophical revolution was at hand, implied in the titles of the works of this period. Telesius published *De Rerum Natura juxta propria principia;* Francis Helmont, *Philosophia vulgaris refutata;* Patricius, *Nova de Universis Philosophia;* Campanella, *Philosophia sensibus demonstrata, adversus errores Aristotelis:* Bruno professed himself the author of a *Nolan Philosophy;* and Ramus of a *New Logic.* The age announced itself pregnant; and the eyes of all who took an interest in the intellectual fortunes of the race, were looking eagerly for the expected offspring.

Chapter X.

THE PRACTICAL REFORMERS OF SCIENCE.

Character of the Practical Reformers.—WE now come to a class of speculators who had perhaps a greater share in bringing about the change from stationary to progressive knowledge, than those writers who so loudly announced the revolution. The mode in which the philosophers of whom we now speak produced their impressions on men's minds, was very different from the procedure of the theoretical reformers. What these talked of, they did; what these promised, they performed. While the theorists concerning knowledge proclaimed that great advances were to be made, the practical discoverers went steadily forwards. While one class spoke of a complete Reform of scientific Methods, the other, boasting little, and often thinking little of Method, proved the novelty of their instrument by obtaining new results. While the metaphysicians were exhorting men to consult experience and the senses, the physicists were examining nature by such means with unparalleled success. And while the former, even when they did for a moment refer to facts, soon rushed back into their own region of ideas, and tried at once to seize the widest generalizations, the latter, fastening their attention upon the phenomena, and trying to reduce them to laws, were carried forwards by steps measured and gradual, such as no conjectural view of scientific method had suggested; but leading to truths as profound and comprehensive as any which conjecture had dared to anticipate. The theoretical reformers were bold, self-confident, hasty, contemptuous of antiquity, ambitious of ruling all future speculations, as they whom they sought to depose had ruled the past. The practical reformers were cautious, modest, slow, despising no knowledge, whether borrowed from tradition or observation, confident in the ultimate triumph of science, but impressed with the conviction that each single person could contribute a little only to its progress. Yet though thus working rather than speculating,—dealing with particulars more than with generals,—employed mainly in adding to knowledge, and not in defining what knowledge is, or how additions are to be made to it,—these men, thoughtful, curious, and of comprehensive minds, were constantly led to important views on the nature and methods of science. And these views, thus suggested by reflections on their own mental activity, were gradually incorporated with the more abstract doctrines of the metaphysicians, and had a most important influence in establishing an improved philosophy of science. The indications of such views we must now endeavour to collect from the writings of the discoverers of the times preceding the seventeenth century.

Some of the earliest of these indications are to be found in those who dealt with Art rather than with Science.

I have already endeavoured to show that the advance of the arts which give us a command over the powers of nature, is generally prior to the formation of exact and speculative knowledge concerning those powers. But Art, which is thus the predecessor of Science, is, among nations of acute and active intellects, usually its parent, There operates, in such a cases, a speculative spirit, leading men to seek for the reasons of that which they find themselves able to do. How slowly, and with what repeated deviations men follow this leading, when under the influence of a partial and dogmatical philosophy, the late birth and slow growth of sound physical theory shows. But at the period of which we now speak, we find men, at length, proceeding in obedience to the impulse which thus drives them from practice to theory;—from an acquaintance with phenomena to a free and intelligent inquiry concerning their causes.

Leonardo da Vinci.—I have already noticed, in the History of Science, that the Indistinctness of Ideas, which was long one main impediment to the progress of science in the middle ages, was first remedied among architects and engineers. These men, so far at least as mechanical ideas were concerned, were compelled by their employments to judge rightly of the relations and properties of the materials with which they had to deal; and would have been chastised by the failure of their works, if they had violated the laws of mechanical truth. It was not wonderful, therefore, that these laws became known to them first. We have seen, in the History, that Leonardo da Vinci, the celebrated painter, who was also an engineer, is the first writer in whom we find the true view of the laws of equilibrium of the lever in the most general case. This artist, a man of a lively and discursive mind, is led to make some remarks* on the

* His works have never been published, and exist in manuscript in

formation of our knowledge, which may show the opinions on that subject that already offered themselves at the beginning of the sixteenth century*. He expresses himself as follows :—"Theory is the general, Experiments are the soldiers. The interpreter of the artifices of nature is Experience: she is never deceived. Our judgment sometimes is deceived, because it expects effects which Experience refuses to allow." And again, "We must consult Experience, and vary the circumstances till we have drawn from them general rules; for it is she who furnishes true rules. But of what use, you ask, are these rules? I reply, that they direct us in the researches of nature and the operations of art. They prevent our imposing upon ourselves and others, by promising ourselves results which we cannot obtain."

"In the study of the sciences which depend on mathematics, those who do not consult nature but authors, are not the children of nature, they are only her grandchildren. She is the true teacher of men of genius. But see the absurdity of men! They turn up their noses at a man who prefers to learn from nature herself rather than from authors who are only her clerks."

In another place, in reference to a particular case, he says, "Nature begins from the Reason and ends in Experience; but for all that, we must take the opposite course; begin from the Experiment and try to discover the Reason."

Leonardo was born forty-six years before Telesius; yet we have here an estimate of the value of experience far more just and substantial than the Calabrian school ever reached. The expressions contained in the above extracts, are well worthy our notice;—that experience is

the library of the Institute at Paris. Some extracts were published by Venturi, *Essai sur les Ouvrages de Leonard da Vinci.* Paris, 1797.

* Leonardo died in 1520, at the age of 78.

never deceived;—that we must vary our experiments, and draw from them general rules;—that nature is the original source of knowledge, and books only a derivative substitute;—with the lively image of the sons and grandsons of nature. Some of these assertions have been deemed, and not without reason, very similar to those made by Bacon a century later. Yet it is probable that the import of such expressions, in Leonardo's mind, was less clear and definite than that which they acquired by the progress of sound philosophy. When he says that theory is the general and experiments the soldiers, he probably meant that theory directs men what experiments to make; and had not in his mind the notion of a theoretical Idea ordering and brigading the Facts. When he says that Experience is the interpreter of Nature, we may recollect, that in a more correct use of this image, Experience and Nature are the writing, and the Intellect of man the interpreter. We may add, that the clear apprehension of the importance of Experience led, in this as in other cases, to an unjust depreciation of the value of what science owed to books. Leonardo would have made little progress, if he had attempted to master a complex science, astronomy for instance, by means of observation alone, without the aid of books.

But in spite of such criticism, Leonardo's maxims show extraordinary sagacity and insight; and they appear to us the more remarkable, when we find how rare such views are for a century after his time.

Copernicus.—For we by no means find, even in those practical discoverers to whom, in reality, the revolution in science, and consequently in the philosophy of science, was due, this prompt and vigorous recognition of the supreme authority of observation as a ground of belief; this bold estimate of the probable worthlessness of traditional knowledge; and this plain assertion of the reality

of theory founded upon experience. Among such discoverers, Copernicus must ever hold a most distinguished
place. The heliocentric theory of the universe, established
by him with vast labour and deep knowledge, was, for the
succeeding century, the field of discipline and exertion
of all the most active speculative minds. Men, during
that time, proved their freedom of thought, their hopeful
spirit, and their comprehensive view, by adopting, incul-
cating, and following out the philosophy which this theory
suggested. But in the first promulgation of the theory,
in the works of Copernicus himself, we find a far more
cautious and reserved temper. He does not, indeed, give
up the reality of his theory, but he expresses himself so
as to avoid shocking those who might (as some afterwards
did) think it safe to speak of it as an *hypothesis* rather
than a truth. In his preface addressed to the Pope*,
after speaking of the difficulties in the old and received
doctrines, by which he was led to his own theory, he
says, "Hence I began to think of the mobility of the
earth; and although the opinion seemed absurd, yet because I knew that to others before me this liberty had
been conceded, of imagining any kinds of circles in order
to explain the phenomena of the stars, I thought it would
also be readily granted me, that I might try whether, by
supposing the earth to be in motion, I might not arrive
at a better explanation than theirs, of the revolutions of
the celestial orbs." Nor does he anywhere assert that
the seeming absurdity had become a certain truth, or
betray any feeling of triumph over the mistaken belief
of his predecessors. And, as I have elsewhere shown,
his disciples† indignantly and justly defended him from
the charge of disrespect towards Ptolemy and other ancient astronomers. Yet Copernicus is far from compro-
mising the value or evidence of the great truths which

* Paul III., in 1543.   † *Hist. Ind. Sci.*, i. 375.

he introduced to general acceptance; and from sinking in his exposition of his discoveries below the temper which had led to them. His quotation from Ptolemy, that " He who is to follow philosophy must be a freeman in mind," is a grand and noble maxim, which it well became him to utter.

Fabricius.—In another of the great discoverers of this period, though employed on a very different subject, we discern much of the same temper. Fabricius of Acquapendente*, the tutor and forerunner of our Harvey, and one of that illustrious series of Paduan professors who were the fathers of anatomy†, exhibits something of the same respect for antiquity, in the midst of his original speculations. Thus in a dissertation‡ *On the Action of the Joints,* he quotes Aristotle's Mechanical Problems to prove that in all animal motion there must be some quiescent fulcrum; and finds merit even in Aristotle's ignorance. "Aristotle," he says§, "did not know that motion was produced by the muscle; and after staggering about from one supposition to another, at last is compelled by the facts themselves to recur to an innate spirit, which, he conceives, is contracted, and which pulls and pushes. And here we cannot help admiring the genius of Aristotle, who, though ignorant of the muscle, invents something which produces nearly the same effect as the muscle, namely, contraction and pulling." He then, with great acuteness, points out the distinction between Aristotle's opinions, thus favourably interpreted, and those of Galen. In all this, we see something of the wish to find all truths in the writings of the ancients, but nothing which materially interferes with freedom of inquiry. The anatomists have in all ages and countries been practically employed in seeking knowledge from

* Born 1537, died 1619.   † *Hist. Ind. Sci.*, iii. 396.
‡ FABRICIUS, *De Motu Locali*, p. 182.   § P. 199.

observation.   Facts have ever been to them a subject of careful and profitable study; while the ideas which enter into the wider truths of the science, are, as we have seen, even still involved in obscurity, doubt, and contest.

Maurolycus.—Francis Maurolycus of Messana, whose mathematical works were published in 1575, was one of the great improvers of the science of optics in his time. In his Preface to his Treatise on the Spheres, he speaks of previous writers on the same subject; and observes that as they have not superseded one another, they have not rendered it unfit for any one to treat the subject afresh.   " Yet," he says, " it is impossible to amend the errors of all who have preceded us.   This would be a task too hard for Atlas, although he supports the heavens. Even Copernicus is tolerated, who makes the sun to be fixed, and the earth to move round it in a circle; and who is more worthy of a whip or a scourge than of a refutation."   The mathematicians and astronomers of that time were not the persons most sensible of the progress of physical knowledge, for the bases of their science, and a great part of its substance, were contained in the writings of the ancients; and till the time of Kepler, Ptolemy's work was, very justly, looked upon as including all that was essential in the science.

Benedetti. — But the writers on Mechanics were naturally led to present themselves as innovators and experimenters; for all that the ancients had taught concerning the doctrine of motion was erroneous; while those who sought their knowledge from experiment, were constantly led to new truths.   John Baptist Benedetti, a Venetian nobleman, in 1599, published his *Speculationum Liber*, containing, among other matter, a treatise on Mechanics, in which several of the Aristotelian errors were refuted.   In the Preface to this Treatise, he says, " Many authors have written much, and with great ability, on Mechanics; but since nature is constantly bringing to light something either new, or before unnoticed, I too wished to put forth a few things hitherto unattempted, or not sufficiently explained." In the doctrine of motion he distinctly and at some length condemns and argues against all the Aristotelian doctrines concerning motion, weight, and many other fundamental principles of physics. Benedetti is also an adherent of the Copernican doctrine. He states* the enormous velocity which the heavenly bodies must have, if the earth be the centre of their motions; and adds, " which difficulty does not occur according to the beautiful theory of the Samian Aristarchus, expounded in a divine manner by Nicolas Copernicus; against which the reasons alleged by Aristotle are of no weight." Benedetti throughout shows no want of the courage or ability which were needed in order to rise in opposition against the dogmas of the Peripatetics. He does not, however, refer to experiment in a very direct manner; indeed most of the facts on which the elementary truths of mechanics rest, were known and admitted by the Aristotelians; and therefore could not be adduced as novelties. On the contrary, he begins with *à priori* maxims, which experience would not have confirmed. " Since," he says†, " we have undertaken the task of proving that Aristotle is wrong in his opinions concerning motion, there are certain absolute truths, the objects of the intellect known of themselves, which we must lay down in the first place." And then, as an example of these truths, he states this : " Any two bodies of equal size and figure, but of different materials, will have their natural velocities in the same proportion as their weights;" where by their natural velocities, he means the velocities with which they naturally fall downwards.

Gilbert.—The greatest of these practical reformers of science is our countryman, William Gilbert; if,

* *Speculationum Liber*, p. 195. † P. 169.

indeed, in virtue of the clear views of the prospects which were then opening to science, and of the methods by which her future progress was to be secured, while he exemplified those views by physical discoveries, he do not rather deserve the still higher praise of being at the same time a theoretical and a practical reformer. Gilbert's physical researches and speculations were employed principally upon subjects on which the ancients had known little or nothing; and on which therefore it could not be doubtful whether tradition or observation was the source of knowledge. Such was magnetism; for the ancients were barely acquainted with the attractive property of the magnet. Its polarity, including repulsion as well as attraction, its direction towards the north, its limited variation from this direction, its declination from the horizontal position, were all modern discoveries. Gilbert's work* on the magnet and on the magnetism of the earth, appeared in 1600; and in this, he repeatedly maintains the superiority of experimental knowledge over the physical philosophy of the ancients. His preface opens thus: "Since in making discoveries and searching out the hidden causes of things, stronger reasons are obtained from trustworthy experiments and demonstrable arguments, than from probable conjectures and the dogmas of those who philosophize in the usual manner," he has, he says, "endeavoured to proceed from common magnetical experiments to the inward constitution of the earth." As I have stated in the History of Magnetism†, Gilbert's work contains all the fundamental facts of that science, so fully stated, that we have, at this day, little to add to them. He is not, however, by the advance which

---

* GULIELMI GILBERTI, *Colcestriensis, Medici Londinensis, De Magnete, Magneticisque Corporibus, et de Magno Magnete Tellure, Physiologia Nova, plurimis et Argumentis et Experimentis demonstrata.*

† *Hist. Ind. Sci.*, iii. 45.

he thus made, led to depreciate the ancients, but only to claim for himself the same liberty of philosophizing which they had enjoyed*. "To those ancient and first parents of philosophy, Aristotle, Theophrastus, Ptolemy, Hippocrates, Galen, be all due honour; from them it was that the stream of wisdom has been derived down to posterity. But our age has discovered and brought to light many things which they, if they were yet alive, would gladly embrace. Wherefore we also shall not hesitate to expound, by probable hypotheses, those things which by long experience we have ascertained."

In this work the author not only adopts the Copernican doctrine of the earth's motion, but speaks† of the contrary supposition as utterly absurd, founding his argument mainly on the vast velocities which such a supposition requires us to ascribe to the celestial bodies. Dr. Gilbert was physician to Queen Elizabeth and to James the First, and died in 1603. Sometime after his death the executors of his brother published another work of his, *De Mundo nostro Sublunari Philosophia Nova*, in which similar views are still more comprehensively presented. In this he says, "The two lords of philosophy, Aristotle and Galen, are held in worship like gods, and rule the schools;—the former by some destiny obtained a sway and influence among philosophers, like that of his pupil Alexander among the kings of the earth;—Galen, with like success, holds his triumph among the physicians of Europe." This comparison of Aristotle to Alexander was also taken hold of by Bacon. Nor is Gilbert an unworthy precursor of Bacon in the view he gives of the History of Science, which occupies the first three chapters of his Philosophy. He traces this history from "the simplicity and ignorance of the ancients," through "the fabrication of the fable of the four elements," to Aristotle

---

* Pref. † *De Magnete*, lib. vi. c. 3, 4.

and Galen. He mentions with due disapproval the host of commentators which succeeded, the alchemists, the " shipwreck of science in the deluge of the Goths," and the revival of letters and genius in the time of " our grandfathers." " This later age," he says, " has exploded the Barbarians, and restored the Greeks and Latins to their pristine grace and honour. It remains, that if they have written aught in error, this should be remedied by better and more productive processes (*frugiferis* institutis,) not to be contemned for their novelty; (for nothing which is true is really new, but is perfect from eternity, though to weak man it may be unknown;) and that thus Philosophy may bear her fruit." The reader of Bacon will not fail to recognize, in these references to "fruit-bearing" knowledge, a similarity of expression with the *Novum Organon*.

Bacon does not appear to me to have done justice to his contemporary. He nowhere recognises in the labours of Gilbert a community of purpose and spirit with his own. On the other hand, he casts upon him a reflection which he by no means deserves. In the *Advancement of Learning*[*], he says, "Another error is, that men have used to infect their meditations, opinions, and doctrines, with some conceits which they have most admired, or some sciences to which they have most applied; and given all things else a tincture according to them, utterly untrue and unproper.......So have the alchemists made a philosophy out of a few experiments of the furnace; and Gilbertus, our countryman, hath made a philosophy out of the observations of a loadstone," (in the Latin, philosophiam etiam e magnete elicuit.) And in the same manner he mentions him in the *Novum Organon*[†], as affording an example of an empirical kind of philosophy, which appears to those daily conversant with the experi-

ments, probable, but to other persons incredible and
empty.   But instead of blaming Gilbert for disturbing
and narrowing science by a too constant reference to
magnetical rules, we might rather censure Bacon, for not
seeing how important in all natural philosophy are those
laws of attraction and repulsion of which magnetical phenomena are the most obvious illustration.   We may find
ground for such a judgment in another passage in which
Bacon speaks of Gilbert.   In the Second Book * of the
Novum Organon, having classified motions, he gives, as
one kind, what he calls, in his figurative language, *motion
for gain*, or *motion of need*, by which a body shuns heterogeneous, and seeks cognate bodies.   And he adds,
" The Electrical operation, concerning which Gilbert and
others since him have made up such a wonderful story,
is nothing else than the appetite of a body, which, excited
by friction, does not well tolerate the air, and prefers
another tangible body if it be found near."   Bacon's
notion of an appetite in the body is certainly much less
philosophical than Gilbert's, who speaks of light bodies
as drawn towards amber by certain material radii†; and
we might perhaps venture to say that Bacon here mani-
fests a want of clear mechanical ideas.   Bacon, too,
showed his inferior aptitude for physical research in
rejecting the Copernican doctrine which Gilbert adopted.
In the *Advancement of Learning*‡, suggesting a history of
the opinions of philosophers, he says that he would have
inserted in it even recent theories, as those of Paracelsus ;
of Telesius, who restored the philosophy of Parmenides ;
or Patricius, who resublimed the fumes of Platonism ;
or Gilbert, who brought back the dogmas of Philolaus.
But Bacon quotes § with pleasure Gilbert's ridicule of the
Peripatetics' definition of heat.   They had said, that heat

---

* Vol. ix. 185.   † *De Magnete*, p. 60.
‡ Book iii. c. 4.   § *Nov. Org.*, book ii. Aph. 48.

is that which separates heterogeneous and unites homogeneous matter; which, said Gilbert, is if any one were to define *man* as that which sows wheat and plants vines.

Galileo, another of Gilbert's distinguished contemporaries, had a higher opinion of him. He says*, " I extremely admire and envy this author. I think him worthy of the greatest praise for the many new and true observations which he has made, to the disgrace of so many vain and fabling authors; who write, not from their own knowledge only, but repeat everything they hear from the foolish and vulgar, without attempting to satisfy themselves of the same by experience; perhaps that they may not diminish the size of their books."

Galileo.—Galileo was content with the active and successful practice of experimental inquiry; and did not demand that such researches should be made expressly subservient to that wider and more ambitious philosophy, on which the author of the *Novum Organon* employed his powers. But still it now becomes our business to trace those portions of Galileo's views which have reference to the theory, as well as the practice of scientific investigation. On this subject, Galileo did not think more profoundly, perhaps, than several of his contemporaries; but in the liveliness of expression and illustration with which he recommended his opinions on such topics, he was unrivalled. Writing in the language of the people, in the attractive form of dialogue, with clearness, grace, and wit, he did far more than any of his predecessors had done to render the new methods, results, and prospects of science familiar to a wide circle of readers, first in Italy, and soon, all over Europe. The principal points inculcated by him were already becoming familiar to men of active and inquiring minds; such as,—that knowledge was to be sought from

* DRINKWATER's *Life of Galileo*, p. 18.

observation, and not from books;—that it was absurd to adhere to, and debate about, the physical tenets of Aristotle and the rest of the ancients. On persons who followed this latter course, Galileo fixed the epithet of Paper Philosophers*; because, as he wrote in a letter to Kepler, this sort of men fancied that philosophy was to be studied like the *Æneid* or *Odyssee*, and that the true reading of nature was to be detected by the collation of texts. Nothing so much shook the authority of the received system of Physics as the experimental discoveries, directly contradicting it, which Galileo made. By experiment, as I have elsewhere stated†, he disproved the Aristotelian doctrine that bodies fall quickly or slowly in proportion to their weight. And when he had invented the telescope, a number of new discoveries of the most striking kind (the inequalities of the moon's surface, the spots in the sun, the moon-like phases of Venus, the satellites of Jupiter, the ring of Saturn) showed, by the evidence of the eyes, how inadequate were the conceptions, and how erroneous the doctrines, of the ancients respecting the constitution of the universe. How severe the blow was to the disciples of the ancient schools, we may judge by the extraordinary forms of defence in which they tried to intrench themselves. They would not look through Galileo's glasses; they maintained that what was seen was an illusion of witchcraft; and they tried, as Galileo says‡, with logical arguments, as if with magical incantations, to charm the new planets out of the sky. No one could be better fitted than Galileo for such a warfare. His great knowledge, clear intellect, gaiety, and light irony, (with the advantage of being in the right,) enabled him to play with his adversaries as he pleased. Thus when an Aristotelian§

---

* *Life of Galileo*, p. 9.   † *Hist. Ind. Sci.*, ii. 46.
‡ *Life of Galileo*, p. 29.   § *Ib.*, p. 33.

rejected the discovery of the irregularities in the moon's surface, because, according to the ancient doctrine, her form was a perfect sphere, and held that the apparent cavities were filled with an invisible crystal substance; Galileo replied, that he had no objection to assent to this, but that then he should require his adversary in return to believe that there were on the same surface invisible crystal mountains ten times as high as those visible ones which he had actually observed and measured.

We find in Galileo many thoughts which have since become established maxims of modern philosophy. "Philosophy," he says*, "is written in that great book, I mean the Universe, which is constantly open before our eyes; but it cannot be understood, except we first know the language and learn the characters in which it is written." With this thought he combines some other lively images. One of his interlocutors says concerning another, "Sarsi perhaps thinks that philosophy is a book made up of the fancies of men, like the *Iliad* or *Orlando Furioso*, in which the matter of least importance is, that what is written be true." And again, with regard to the system of authority, he says, "I think I discover in him a firm belief that, in philosophizing, it is necessary to lean upon the opinion of some celebrated author; as if our mind must necessarily remain unfruitful and barren till it be married to another man's reason."—"No," he says, "the case is not so.—When we have the decrees of Nature, authority goes for nothing; reason is absolute†."

In the course of Galileo's controversies, questions of the logic of science came under discussion. Vincenzio di Grázia objected to a proof from induction which Galileo adduced, because *all* the particulars were not enumerated; to which the latter justly replies‡, that if induction were required to pass through all the cases, it

* *Il Saggiatore*, ii 247.   † *Ib.*, ii. 200.   ‡ *Ib.*, i. 501.

would be either useless or impossible;—impossible when the cases are innumerable; useless when they have each already been verified, since then the general proposition adds nothing to our knowledge.

One of the most novel of the characters which Science assumes in Galileo's hands is, that she becomes cautious. She not only proceeds leaning upon Experience, but she is content to proceed a little way at a time. She already begins to perceive that she must rise to the heights of knowledge by many small and separate steps. The philosopher is desirous to know much, but resigned to be ignorant for a time of that which cannot yet be known. Thus when Galileo discovered the true law of the motion of a falling body*, that the velocity increases proportionally to the time from the beginning of the fall, he did not insist upon immediately assigning the cause of this law. "The cause of the acceleration of the motions of falling bodies is not," he says, " a necessary part of the investigation." Yet the conception of this acceleration, as the result of the continued action of the force of gravity upon the falling body, could hardly fail to suggest itself to one who had formed the idea of force. In like manner, the truth that the velocities, acquired by bodies falling down planes of equal heights, are all equal, was known to Galileo and his disciples, long before he accounted for it†, by the principle, apparently so obvious, that the momentum generated is as the moving force which generates it. He was not tempted to rush at once, from an experimental truth to a universal system. Science had learnt that she must move step by step; and the gravity of her pace already indicated her approaching maturity and her consciousness of the long path which lay before her.

But besides the genuine philosophical prudence which

---

* *Hist. Ind. Sci.*, ii. 30.　　　　　† *Ib.*, ii. 49.

thus witheld Galileo from leaping hastily from one inference to another, he had perhaps a preponderating inclination towards facts; and did not feel, so much as some
other persons of his time, the need of reducing them to
ideas.   He could bear to contemplate laws of motion
without being urged by an uncontrollable desire to refer
them to conceptions of force.

Kepler.—In this respect his friend Kepler differed
from him; for Kepler was restless and unsatisfied till he
had reduced facts to laws, and laws to causes; and never
acquiesced in ignorance, though he tested with the most
rigorous scrutiny that which presented itself in the shape
of knowledge to fill the void.   It may be seen in the
History of Astronomy* with what perseverance, energy,
and fertility of invention, Kepler pursued his labours,
(enlivened and relieved by the most curious freaks of
fancy,) with a view of discovering the rules which regulate
the motions of the planet Mars.   He represents this
employment under the image of a warfare; and describes†
his object to be " to triumph over Mars, and to prepare
for him, as for one altogether vanquished, tabular prisons
and equated eccentric fetters;" and when " the enemy,
left at home a despised captive, had burst all the chains
of the equations, and broken forth of the prisons of the
tables;"—when " it was buzzed here and there that the
victory is vain, and that the war is raging anew as vio-
lently as before;"—that is, when the rules which he had
proposed did not coincide with the facts:—he by no means
desisted from his attempts, but " suddenly sent into the
field a reserve of new physical reasonings on the rout
and dispersion of the veterans," that is, tried new suppositions suggested by such views as he then entertained of
the celestial motions.   His efforts to obtain the formal

* *Hist. Ind. Sci.*, i., 426.
† *De Stell. Mart.*, p. iv. c. 51. (1609.)   DRINKWATER'S *Kepler*, p. 33.

laws of the planetary motions resulted in some of the most important discoveries ever made in astronomy; and if his physical reasonings were for the time fruitless, this arose only from the want of that discipline in mechanical ideas which the minds of mathematicians had still to undergo; for the great discoveries of Newton in the next generation showed that, in reality, the next step of the advance was in this direction. Among all Kepler's fantastical expressions, the fundamental thoughts were sound and true; namely, that it was his business, as a physical investigator, to discover a mathematical rule which governed and included all the special facts; and that the rules of the motions of the planets must conform to some conception of causation.

The same characteristics,—the conviction of rule and cause, perseverance in seeking these, inventiveness in devising hypotheses, love of truth in trying and rejecting them, and a lively Fancy playing with the Reason without interrupting her,—appear also in his work on Optics; in which he tried to discover the exact law of optical refraction*. In this undertaking he did not succeed entirely; nor does he profess to have done so. He ends his numerous attempts by saying, " Now, reader, you and I have been detained sufficiently long while I have been attempting to *collect into one fagot* the measures of different refractions."

In this and in other expressions, we see how clearly he apprehended that *colligation of facts* which is the main business of the practical discoverer. And by his peculiar endowments and habits, Kepler exhibits an essential portion of this process, which hardly appears at all in Galileo. In order to bind together facts, theory is requisite as well as observation,—the cord as well as the fagots. And the true theory is often, if not always, obtained by trying

* Published 1604. *Hist. Ind. Sci.*, ii. 346.

several and selecting the right. Now of this portion of the discoverer's exertions Kepler is a most conspicuous example. His fertility in devising suppositions, his undaunted industry in calculating the results of them, his entire honesty and candour in resigning them if these results disagreed with the facts, are a very instructive spectacle; and are fortunately exhibited to us in the most lively manner in his own garrulous narratives. Galileo urged men by precept as well as example to begin their philosophy from observation; Kepler taught them by his practice that they must proceed from observation by means of hypotheses. The one insisted upon facts; the other dealt no less copiously with ideas. In the practical, as in the speculative portion of our history, this antithesis shows itself; although in the practical part we cannot have the two elements separated, as in the speculative we sometimes have.

In the History of Science*, I have devoted several pages to the intellectual character of Kepler, inasmuch as his habit of devising so great a multitude of hypotheses, so fancifully expressed, had led some writers to look upon him as an inquirer who transgressed the most fixed rules of philosophical inquiry. This opinion has arisen, I conceive, among those who have forgotten the necessity of Ideas as well as Facts for all theory; or who have overlooked the impossibility of selecting and explicating our ideas without a good deal of spontaneous play of the mind. It must, however, always be recollected that Kepler's genius and fancy derived all their scientific value from his genuine and unmingled love of truth. These qualities appeared, not only in the judgment he passed upon hypotheses, but also in matters which more immediately concerned his reputation. Thus when Galileo's discovery of the telescope disproved several opinions

* _Hist. Ind. Sci._, i. 410.

which Kepler had published and strenuously maintained, he did not hesitate a moment to retract his assertions and range himself by the side of Galileo, whom he vigorously supported in his warfare against those who were incapable of thus cheerfully acknowledging the triumph of new facts over their old theories.

Tycho. — There remains one eminent astronomer, the friend and fellow-labourer of Kepler, whom we must not separate from him as one of the practical reformers of science. I speak of Tycho Brahe, who is, I think, not justly appreciated by the literary world in general, in consequence of his having made a retrograde step in that portion of astronomical theory which is most familiar to the popular mind. Though he adopted the Copernican view of the motion of the planets about the sun, he refused to acknowledge the annual and diurnal motion of the earth. But notwithstanding this mistake, into which he was led by his interpretation of Scripture rather than of nature, Tycho must ever be one of the greatest names in astronomy. In the philosophy of science also, the influence of what he did is far from inconsiderable; and especially its value in bringing into notice these two points :—that not only are observations the beginning of science, but that the progress of science may often depend upon the observer's pursuing his task regularly and carefully for a long time, and with well devised instruments; and again, that observed facts offer a *succession* of laws which we discover as our observations become better, and as our theories are better adapted to the observations. With regard to the former point, Tycho's observatory was far superior to all that had preceded it*, not only in the optical, but in the mechanical arrangements; a matter of almost equal consequence. And hence it was that his observations inspired in Kepler that confidence which led

* *Hist. Ind. Sci.*, ii. 266.

him to all his labours and all his discoveries. "Since," he says*, "the divine goodness has given us in Tycho Brahe an exact observer, from whose observations this error of eight minutes in the calculations of the Ptolemaic hypothesis is detected, let us acknowledge and make use of this gift of God: and since this error cannot be neglected, these eight minutes alone have prepared the way for an entire reform of Astronomy, and are to be the main subject of this work."

With regard to Tycho's discoveries respecting the moon, it is to be recollected that besides the first inequality of the moon's motion, (the *equation of the centre,* arising from the elliptical form of her orbit,) Ptolemy had discovered a second inequality, the *evection,* which, as we have observed in the History of this subject†, might have naturally suggested the suspicion that there were still other inequalities. In the middle ages, however, such suggestions, implying a constant progress in science, were little attended to; and, we have seen, that when an Arabian astronomer‡ had really discovered another inequality of the moon, it was soon forgotten, because it had no place in the established systems. Tycho not only rediscovered the lunar inequality, (the *variation,*) thus once before won and lost, but also two other inequalities; namely §, the *change of inclination* of the moon's orbit as the line of nodes moves round, and an inequality in the motion of the line of nodes. Thus, as I have elsewhere said, it appeared that the discovery of a rule is a step to the discovery of deviations from that rule, which require to be expressed in other rules. It became manifest to astronomers, and through them to all philosophers, that in the application of theory to observation, we find, not only the stated phenomena, for which the theory does

* *De Stell. Mart.*, p. 11, c. 19.   † *Hist. Ind. Sci.*, i. 217.
‡ *Ib.*, i. 228.   § MONTUCLA, i. 566.

account, but also *residual phenomena*, which are unaccounted for, and remain over and above the calculation. And it was seen further, that these residual phenomena might be, altogether or in part, exhausted by new theories.

These were valuable lessons; and the more valuable inasmuch as men were now trying to lay down maxims and methods for the conduct of science. A revolution was not only at hand, but had really taken place, in the great body of real cultivators of science. The occasion now required that this revolution should be formally recognized;—that the new intellectual power should be clothed with the forms of government;—that the new philosophical republic should be acknowledged as a sister state by the ancient dynasties of Aristotle and Plato. There was needed some great Theoretical Reformer, to speak in the name of the Experimental Philosophy; to lay before the world a declaration of its rights and a scheme of its laws. And thus our eyes are turned to Francis Bacon, and others who like him attempted this great office. We quit those august and venerable names of discoverers, whose appearance was the prelude and announcement of the new state of things then opening; and in doing so, we may apply to them the language which Bacon applies to himself* :—

Χάιρετε Κήρυκες Διὸς ἄγγελοι ἠδὲ καὶ ἀνδρῶν.
Hail Heralds, Messengers of Gods and Men!

Chapter XI.

FRANCIS BACON.

1. It is a matter of some difficulty to speak of the character and merits of this illustrious man, as regards his place in that philosophical history with which we are

* *De Augm.*, lib. iv. c. 1.

here engaged. If we were to content ourselves with estimating him according to the office which, as we have just seen, he claims for himself*, as merely the harbinger and announcer of a sounder method of scientific inquiry than that which was recognized before him, the task would be comparatively easy. For we might select from his writings those passages in which he has delivered opinions and pointed out processes, then novel and strange, but since confirmed by the experience of actual discoverers, and by the judgments of the wisest of succeeding philosophers; and we might pass by, without disrespect, but without notice, maxims and proposals which have not been found available for use ;—views so indistinct and vague, that we are even yet unable to pronounce upon their justice ;—and boundless anticipations, dictated by the sanguine hopes of a noble and comprehensive intellect. But if we thus reduce the philosophy of Bacon to that portion which the subsequent progress of science has rigorously verified, we shall have to pass over many of those declarations which have excited most notice in his writings, and shall lose sight of many of those striking thoughts which his admirers most love to dwell upon. For he is usually spoken of, at least in this country, as a teacher who not only commenced, but in a great measure completed, the Philosophy of Induction. He is considered, not only as having asserted some general principles, but laid down the special rules of scientific investigation ; as not only one of the Founders, but the supreme Legislator of the modern Republic of Science ; not only the Hercules who slew the monsters that obstructed the earlier traveller, but the Solon who established a constitution fitted for all future time.

2. Nor is it our purpose to deny that of such praise

* And in other passages : thus, "Ego enim buccinator tantum pugnam non ineo." *Nov. Org.* lib. iv. c. 1.

he deserves a share which, considering the period at which he lived, is truly astonishing. But it is necessary for us in this place to discriminate and select that portion of his system which, bearing upon *physical* science, has since been confirmed by the actual history of science. Many of Bacon's most impressive and captivating passages contemplate the extension of the new methods of discovering truth to intellectual, to moral, to political, as well as to physical science. And how far, and how, the advantages of the inductive method may be secured for those important branches of speculation, it will at some future time be a highly interesting task to examine. But our plan requires us at present to omit the consideration of these; for our purpose is to learn what the genuine course of the formation of science is, by tracing it in those portions of human knowledge, which, by the confession of all, are most exact, most certain, most complete. Hence we must here deny ourselves the dignity and interest which float about all speculations in which the great moral and political concerns of men are involved. It cannot be doubted that the commanding position which Bacon occupies in men's estimation arises from his proclaiming a reform in philosophy of so comprehensive a nature;—a reform which was to infuse a new spirit into every part of knowledge. Physical Science has tranquilly and noiselessly adopted many of his suggestions; which were, indeed, her own natural impulses, not borrowed from him; and she is too deeply and satisfactorily absorbed in contemplating her results, to talk much about the methods of obtaining them which she has thus instinctively pursued. But the philosophy which deals with mind, with manners, with morals, with polity, is conscious still of much obscurity and perplexity; and would gladly borrow aid from a system in which aid is so confidently promised. The aphorisms and phrases of the *Novum Organon* are far more frequently quoted by metaphysical, ethical, and even theological writers, than they are by the authors of works on physics.

3. Again, even as regards physics, Bacon's fame rests upon something besides the novelty of the maxims which he promulgated. That a revolution in the method of scientific research was going on, all the greatest physical investigators of the sixteenth century were fully aware, as we have shown in the last chapter. But their writings conveyed this conviction to the public at large somewhat slowly. Men of letters, men of the world, men of rank, did not become familiar with the abstruse works in which these views were published; and above all, they did not, by such occasional glimpses as they took of the state of physical science, become aware of the magnitude and consequences of this change. But Bacon's lofty eloquence, wide learning, comprehensive views, bold pictures of the coming state of things, were fitted to make men turn a far more general and earnest gaze upon the passing change. When a man of his acquirements, of his talents, of his rank and position, of his gravity and caution, poured forth the strongest and loftiest expressions and images which his mind could supply, in order to depict the "Great Instauration" which he announced; —in order to contrast the weakness, the blindness, the ignorance, the wretchedness, under which men had laboured while they followed the long beaten track, with the light, the power, the privileges, which they were to find in the paths to which he pointed;—it was impossible that readers of all classes should not have their attention arrested, their minds stirred, their hopes warmed; and should not listen with wonder and with pleasure to the strains of prophetic eloquence in which so great a subject was presented. And when it was found that the prophecy was verified; when it appeared that an immense change

in the methods of scientific research really *had* occurred;
—that vast additions to man's knowledge and power had
been acquired, in modes like those which had been spoken
of;—that further advances might be constantly looked
for;—and that a progress, seemingly boundless, was going
on in the direction in which the seer had thus pointed;
—it was natural that men should hail him as the leader
of the revolution; that they should identify him with the
event which he was the first to announce; that they should
look upon him as the author of that which he had, as they
perceived, so soon and so thoroughly comprehended.

4. For we must remark, that although (as we have
seen) he was not the only, nor the earliest writer, who
declared that the time was come for such a change, he
not only proclaimed it more emphatically, but understood
it, in its general character, much more exactly, than any
of his contemporaries. Among the maxims, suggestions
and anticipations which he threw out, there were many
of which the wisdom and the novelty were alike striking
to his immediate successors;—there are many which
even now, from time to time, we find fresh reason to
admire, for their acuteness and justice. Bacon stands
far above the herd of loose and visionary speculators who,
before and about his time, spoke of the establishment of
new philosophies. If we must select some one philosopher as the Hero of the revolution in scientific method,
beyond all doubt Francis Bacon must occupy the place
of honour.

We shall, however, no longer dwell upon these general
considerations, but shall proceed to notice some of the
more peculiar and characteristic features of Bacon's philosophy; and especially those views, which, occurring for
the first time in his writings, have been fully illustrated
and confirmed by the subsequent progress of science, and
have become a portion of the permanent philosophy of
our times.

5. (I.) The first great feature which strikes us in Bacon's philosophical views is that which we have already noticed;—his confident and emphatic announcement of a *New Era* in the progress of science, compared with which the advances of former times were poor and trifling. This was with Bacon no loose and shallow opinion, taken up on light grounds and involving only vague general notions. He had satisfied himself of the justice of such a view by a laborious course of research and reflection. In 1605, at the age of twenty-eight, he published his Treatise of the *Advancement of Learning*, in which he takes a comprehensive and spirited survey of the condition of all branches of knowledge which had been cultivated up to that time. This work was composed with a view to that reform of the existing philosophy which Bacon always had before his eyes; and in the Latin edition of his works, forms the First Part of the *Instauratio Magna*. In the Second Part of the Instauratio, the *Novum Organon*, published in 1620, he more explicitly and confidently states his expectations on this subject. He points out how slightly and feebly the examination of nature had been pursued up to his time, and with what scanty fruit. He notes the indications of this in the very limited knowledge of the Greeks who had till then been the teachers of Europe, in the complaints of authors concerning the subtility and obscurity of the secrets of nature, in the dissensions of sects, in the absence of useful inventions resulting from theory, in the fixed form which the sciences had retained for two thousand years. Nor, he adds[*], is this wonderful; for how little of his thought and labour has man bestowed upon science! Out of twenty-five centuries scarce six have been favourable to the progress of knowledge. And even in those favoured times, natural philosophy received the

to

[*] Lib. i. Aphor. 78 *et seq.*

smallest share of man's attention; while the portion so given was marred by controversy and dogmatism; and even those who have bestowed a little thought upon this philosophy, have never made it their main study, but have used it as a passage or drawbridge to serve other objects. And thus, he says, the great Mother of the Sciences is thrust down with indignity to the offices of a handmaid; is made to minister to the labours of medicine or mathematics, or to give the first preparatory tinge to the immature minds of youth. From these and similar considerations of the errors of past time, he draws hope for the future, employing the same argument which Demosthenes uses to the Athenians. "That which is worst in the events of the past, is the best as a ground of trust in the future. For if you had done all that became you, and still had been in this condition, your case might be desperate; but since your failure is the result of your own mistakes, there is good hope that, correcting the error of your course, you may reach a prosperity yet unknown to you."

6. (II.) All Bacon's hope of improvement indeed was placed in an entire *change of the Method* by which science was pursued; and the boldness, and at the same time, (the then existing state of science being considered) the definiteness of his views of the change that was requisite are truly remarkable.

That all knowledge must begin with observation, is one great principle of Bacon's philosophy; but I hardly think it necessary to notice the inculcation of this maxim as one of his main services to the cause of sound knowledge, since it had, as we have seen, been fully insisted upon by others before him, and was growing rapidly into general acceptance without his aid. But if he was not the first to tell men that they must collect their knowledge from observation, he had no rival in his peculiar office of teaching them *how* science must thus be gathered from experience.

It appears to me that by far the most extraordinary parts of Bacon's works are those in which, with extreme earnestness and clearness, he insists upon a *graduated and successive induction*, as opposed to a hasty transit from special facts to the highest generalizations. The nineteenth Axiom of the First Book of the *Novum Organon* contains a view of the nature of true science most exact and profound; and, so far as I am aware, at the time perfectly new. "There are two ways, and can only be two, of seeking and finding truth. The one, from sense and particulars, takes a flight to the most general axioms, and from those principles and their truth, settled once for all, invents and judges of intermediate axioms. The other method collects axioms from sense and particulars, ascending *continuously and by degrees*, so that in the end it arrives at the most general axioms; this latter way is the true one, but hitherto untried."

It is to be remarked, that in this passage Bacon employs the term *axioms* to express any propositions collected from facts by induction, and thus fitted to become the starting-point of deductive reasonings. How far propositions so obtained may approach to the character of axioms in the more rigorous sense of the term, we have already in some measure examined; but that question does not here immediately concern us. The truly remarkable circumstance is to find this recommendation of a continuous advance from observation, by limited steps, through successive gradations of generality, given at a time when speculative men in general had only just begun to perceive that they must begin their course from experience in some way or other. How exactly this description represents the general structure of the soundest and most comprehensive physical theories, all persons who have studied the progress of science up to modern times can bear testimony; but perhaps this structure of science cannot in any other way be made so apparent as by those Tables of successive generalizations in which we have exhibited the history and constitution of some of the principal physical sciences, in the Chapter of the preceding Book which treats of the Logic of Induction. And the view which Bacon thus took of the true progress of science was not only new, but, so far as I am aware, has never been adequately illustrated up to the present day.

7. It is true, as I observed in the last Chapter, that Galileo had been led to see the necessity, not only of proceeding from experience in the pursuit of knowledge, but of proceeding cautiously and gradually; and he had exemplified this rule more than once, when, having made one step in discovery, he held back his foot, for a time, from the next step, however tempting. But Galileo had not reached this wide and commanding view of the successive subordination of many steps, all leading up at last to some wide and simple general truth. In catching sight of this principle, and in ascribing to it its due importance, Bacon's sagacity, so far as I am aware, wrought unassisted and unrivalled.

8. Nor is there any wavering or vagueness in Bacon's assertion of this important truth. He repeats it over and over again; illustrates it by a great number of the most lively metaphors and emphatic expressions. Thus he speaks of the successive *floors* (*tabulata*) of induction; and speaks of each science as a *pyramid** which has observation and experience for its basis. No images can better

---

* *Aug. Sc.*, lib. iii. c. 4. p. 194. So in other places, as *Nov. Org.*, i. Aphorism 104. "De scientiis tum demum bene sperandum est quando per scalam veram et per gradus continuos, et non intermissos aut hiulcos a particularibus ascendetur ad axiomata minora, et deinde ad media, alia aliis superiora, et postremo demum ad generalissima."

exhibit the relation of general and particular truths, as our own Inductive Tables may serve to show.

9. (III.) Again; not less remarkable is his contrasting this true Method of Science (while it was almost, as he says, yet untried) with the ancient and *vicious Method*, which began, indeed, with facts of observation, but rushed at once, and with no gradations, to the most general principles. For this was the course which had been actually followed by all those speculative reformers who had talked so loudly of the necessity of beginning our philosophy from experience. All these men, if they attempted to frame physical doctrines at all, had caught up a few facts of observation, and had erected a universal theory upon the suggestions which these offered. This process of illicit generalization, or, as Bacon terms it, Anticipation of Nature (*anticipatio naturæ*), in opposition to the Interpretation of Nature, he depicts with singular acuteness, in its character and causes. "These two ways," he says†, "both begin from sense and particulars; but their discrepancy is immense. The one merely skims over experience and particulars in a cursory transit; the other deals with them in a due and orderly manner. The one, at its very outset, frames certain general abstract principles, but useless; the other gradually rises to those principles which have a real existence in nature."

"The former path," he adds‡, "that of illicit and hasty generalization, is one which the intellect follows when abandoned to its own impulse; and this it does from the requisitions of logic. For the mind has a yearning which makes it dart forth to generalities, that it may have something to rest in; and after a little dallying with experience, becomes weary of it; and all these evils are augmented by logic, which requires these generalities to make a show with in its disputations."

† *Nov. Org.*, i. Aph. 22.   *Ib.*, Aph. 20.

" In a sober, patient, grave intellect," he further adds,
"the mind, by its own impulse, (and more especially if
it be not impeded by the sway of established opinions)
attempts in some measure that other and true way, of
gradual generalization; but this it does with small profit;
for the intellect, except it be regulated and aided, is a
faculty of unequal operation, and altogether unapt to
master the obscurity of things."

The profound and searching wisdom of these remarks
appears more and more, as we apply them to the various
attempts which men have made to obtain knowledge; when
they begin with the contemplation of a few facts, and pursue their speculations, as upon most subjects they have
hitherto generally done; for almost all such attempts have
led immediately to some process of illicit generalization,
which introduces an interminable course of controversy. In
the physical sciences, however, we have the further inestimable advantage of seeing the other side of the contrast
exemplified: for many of them, as our Inductive Tables
show us, have gone on according to the most rigorous
conditions of gradual and successive generalization; and
in consequence of this circumstance in their constitution,
possess, in each part of their structure, a solid truth,
which is always ready to stand the severest tests of reasoning and experiment.

We see how justly and clearly Bacon judged concerning the mode in which facts are to be employed in
the construction of science. This, indeed, has ever been
deemed his great merit: insomuch that many persons appear to apprehend the main substance of his doctrine to
reside in the maxim that facts of observation, and such
facts alone, are the essential elements of all true science.

10. (IV.) Yet we have endeavoured to establish the
doctrine that facts are but one of two ingredients of
knowledge both equally necessary;—that *Ideas* are no

less indispensable than facts themselves; and that except these be duly unfolded and applied, facts are collected in vain. Has Bacon then neglected this great portion of his subject? Has he been led by some partiality of view, or some peculiarity of circumstances, to leave this curious and essential element of science in its pristine obscurity? Was he unaware of its interest and importance?

We may reply that Bacon's philosophy, in its effect upon his readers in general, does *not* give due weight or due attention to the ideal element of our knowledge. He is considered as peculiarly and eminently the asserter of the value of experiment and observation. He is always understood to belong to the experiential, as opposed to the ideal school. He is held up in contrast to Plato and others who love to dwell upon that part of knowledge which has its origin in the intellect of man.

11. Nor can it be denied that Bacon has, in the finished part of his *Novum Organum*, put prominently forwards the necessary dependence of all our knowledge upon Experience, and said little of its dependence, equally necessary, upon the Conceptions which the intellect itself supplies. It will appear, however, on a close examination, that he was by no means insensible or careless of this internal element of all connected speculation. He held the balance, with no partial or feeble hand, between phenomena and ideas. He urged the Colligation of Facts, but he was not the less aware of the value of the Explication of Conceptions.

12. This appears plainly from some remarkable Aphorisms in the *Novum Organum*. Thus, in noticing the causes of the little progress then made by science, he states this:—" In the current Notions, all is unsound, whether they be logical or physical. *Substance, quality, action, passion,* even *being,* are not good Conceptions; still less are *heavy, light, dense, rare, moist, dry, generation,*

corruption, attraction, repulsion, element, matter, form, and
others of that kind; all are fantastical and ill-defined."
And in his attempt to exemplify his own system, he hesi-
tates* in accepting or rejecting the notions of *elementary,
celestial, rare,* as belonging to fire, since, as he says, they
are vague and ill-defined notions (*notiones vagæ nec bene
terminatæ*). In that part of his work which appears to
be completed, there is not, so far as I have noticed, any
attempt to fix and define any notions thus complained of
as loose and obscure. But yet such an undertaking appears to have formed part of his plan; and in the *Abece-
darium Naturæ*†, which consists of the heads of various
portions of his great scheme, marked by letters of the
alphabet, we find the titles of a series of dissertations
" On the Conditions of Beings," which must have had for
their object the elucidation of divers Notions essential to
science, and which would have been contributions to the
Explication of Conceptions, such as we have attempted
in a former part of this work. Thus some of the subjects
of these dissertations are;—Of Much and Little;—Of
Durable and Transitory;—Of Natural and Monstrous;—
Of Natural and Artificial. When the philosopher of
induction came to discuss these, considered as *conditions
of existence,* he could not do other than develop, limit,
methodize and define the Ideas involved in these Notions,
so as to make them consistent with themselves, and a fit
basis of demonstrative reasoning. His task would have
been of the same nature as ours has been, in that part of
this work which treats of the Fundamental Ideas of the
various classes of sciences.

13. Thus Bacon, in his speculative philosophy, took
firmly hold of both the handles of science; and if he had
completed his scheme, would probably have given due

* *Nov. Org.,* lib. ii. Aph. 19.
† *Inst. Mag.,* par. iii. (vol. viii. p. 244.)

attention to Ideas, no less than to Facts, as an element of our knowledge; while in his view of the general method of ascending from facts to principles, he displayed a sagacity truly wonderful. But we cannot be surprised, that in attempting to exemplify the method which he recommended, he should have failed. For the method could be exemplified only by some important discovery in physical science; and great discoveries, even with the most perfect methods, do not come at command. Moreover although the general structure of his scheme was correct, the precise import of some of its details could hardly be understood, till the actual progress of science had made men somewhat familiar with the kind of steps which it included.

14. (V.) Accordingly, Bacon's *Inquisition into the Nature of Heat*, which is given in the Second Book of the *Novum Organon* as an example of the mode of interrogating Nature, cannot be looked upon otherwise than as a complete failure. This will be evident if we consider that, although the exact nature of heat is still an obscure and controverted matter, the science of Heat now consists of many important truths; and that to none of these truths is there any approximation in Bacon's essay. From his process he arrives at this, as the "forma or true definition" of heat;—"that it is an expansive, restrained motion, modified in certain ways, and exerted in the smaller particles of the body." But the steps by which the science of Heat really advanced were, (as may be seen in the history[*] of the subject,) these;—The discovery of a *measure* of heat or temperature (the thermometer); The establishment of the *laws* of conduction and radiation; of the *laws* of specific heat, latent heat, and the like. Such steps have led to Ampère's *hypothesis*[†], that heat consists in the vibrations of an imponderable fluid;

and to Laplace's *hypothesis*, that temperature consists in
the internal radiation of such a fluid. These hypotheses
cannot yet be said to be even probable; but at least they
are so modified as to include some of the preceding laws
which are firmly established; whereas Bacon's hypothetical motion includes no laws of phenomena, explains
no process, and is indeed itself an example of illicit
generalization.

15. One main ground of Bacon's ill fortune in this
undertaking appears to be, that he was not aware of an
important maxim of inductive science, that we must first
obtain the *measure* and ascertain the *laws* of phenomena,
before we endeavour to discover their *causes*. The whole
history of thermotics up to the present time has been
occupied with the *former* step, and the task is not yet
completed: it is no wonder, therefore, that Bacon failed
entirely, when he so prematurely attempted the *second*.
His sagacity had taught him that the progress of science
must be gradual; but it had not led him to judge adequately how gradual it must be, nor of what different
kinds of inquiries, taken in due order, it must needs consist, in order to obtain success.

Another mistake, which could not fail to render it
unlikely that Bacon should really exemplify his precepts
by any actual advance in science, was, that he did not
justly appreciate the sagacity, the inventive genius, which
all discovery requires. He conceived that he could
supersede the necessity of such peculiar endowments.
"Our method of discovery in science," he says*, "is of
such a nature, that there is not much left to acuteness
and strength of genius, but all degrees of genius and
intellect are brought nearly to the same level." And he
illustrates this by comparing his method to a pair of
compasses, by means of which a person with no manual
skill may draw a perfect circle. In the same spirit he

* *Nov. Org.*, lib. i. Aph. 61.

speaks of proceeding by *due rejections;* and appears to imagine that when we have obtained a collection of facts, if we go on successively rejecting what is false, we shall at last find that we have, left in our hands, that scientific truth which we seek. I need not observe how far this view is removed from the real state of the case. The necessity of a *conception* which must be furnished by the mind in order to bind together the facts, could hardly have escaped the eye of Bacon, if he had cultivated more carefully the ideal side of his own philosophy. And any attempts which he could have made to construct such conceptions by mere rule and method, must have ended in convincing him that nothing but a peculiar inventive talent could supply that which was thus not contained in the facts, and yet was needed for the discovery.

16. (VI.) Since Bacon, with all his acuteness, had not divined circumstances so important in the formation of science, it is not wonderful that his attempt to reduce this process to a *Technical Form* is of little value. In the first place, he says*, we must prepare a natural and experimental history, good and sufficient; in the next place, the instances thus collected are to be arranged in Tables in some orderly way; and then we must apply a legitimate and true induction. And in his example†, he first collects a great number of cases in which heat appears under various circumstances, which he calls "a Muster of Instances before the intellect," (*comparentia instantiarum ad intellectum,*) or a *Table of the Presence* of the thing sought. He then adds a *Table of its Absence* in proximate cases, containing instances where heat does not appear; then a *Table of Degrees,* in which it appears with greater or less intensity. He then adds‡, that we must try to exclude several obvious suppositions, which

---

* *Nov. Org.,* lib. ii. Aph. 10.   † Aph. 11.

‡ Aph. 15, p. 105.

he does by reference to some of the instances he has collected; and this step he calls the *Exclusive*, or the *Rejection of Natures*. He then observes, (and justly,) that whereas truth emerges more easily from error than from confusion, we may, after this preparation, *give play to the intellect*, (fiat permissio intellectus,) and make an attempt at induction, liable afterwards to be corrected; and by this step, which he terms his *First Vindemiation*, or *Inchoate Induction*, he is led to the proposition concerning heat, which we have stated above.

17. In all the details of his example he is unfortunate. By proposing to himself to examine at once into the *nature* of heat, instead of the laws of special classes of phenomena, he makes, as we have said, a fundamental mistake; which is the less surprising since he had before him so few examples of the right course in the previous history of science. But further, his collection of instances is very loosely brought together; for he includes in his list the *hot* taste of aromatic plants, the *caustic* effects of acids, and many other facts which cannot be ascribed to heat without a studious laxity in the use of the word. And when he comes to that point where he permits his intellect its range, the conception of *motion* upon which it at once fastens, appears to be selected with little choice or skill, the suggestion being taken from flame*, boiling liquids, a blown fire, and some other cases. If from such examples we could imagine heat to be motion, we ought at least to have some gradation to cases of heat where no motion is visible, as in a red-hot iron. It would seem that, after a large collection of instances had been looked at, the intellect, even in its first attempts, ought not to have dwelt upon such an hypothesis as this.

18. After these steps, Bacon speaks of several classes of instances which, singling them out of the general and

* Page 110.

indiscriminate collection of facts, he terms *Instances with Prerogative;* and these he points out as peculiar aids and guides to the intellect in its task. These Instances with Prerogative have generally been much dwelt upon by those who have commented on the *Novum Organon.* Yet, in reality, such a classification, as has been observed by one of the ablest writers of the present day*, is of little service in the task of induction. For the instances are, for the most part, classed, not according to the ideas which they involve, or to any obvious circumstance in the facts of which they consist, but according to the extent or manner of their influence upon the inquiry in which they are employed. Thus we have Solitary Instances, Migrating Instances, Ostensive Instances, Clandestine Instances, so termed according to the degree in which they exhibit, or seem to exhibit, the property whose nature we would examine. We have Guide-Post Instances, (*Instantiæ Crucis,*) Instances of the Parted Road, of the Doorway, of the Lamp, according to the guidance they supply to our advance. Such a classification is much of the same nature as if, having to teach the art of building, we were to describe tools with reference to the amount and place of the work which they must do, instead of pointing out their construction and use :—as if we were to inform the pupil that we must have tools for lifting a stone up, tools for moving it sideways, tools for laying it square, tools for cementing it firmly. Such an enumeration of ends would convey little instruction as to the means. Moreover, many of Bacon's classes of instances are vitiated by the assumption that the " form," that is, the general law and cause of the property which is the subject of investigation, is to be looked for directly in the instances; which, as we have seen in his inquiry concerning heat, is a fundamental error.

* HERSCHEL, *On the Study of Nat. Phil.*, Art. 192.

19. Yet his phraseology in some cases, as in the *instantia cruois*, serves well to mark the place which certain experiments hold in our reasonings: and many of the special examples which he gives are full of acuteness and sagacity. Thus he suggests swinging a pendulum in a mine, in order to determine whether the attraction of the earth arises from the attraction of its parts; and observing the tide at the same moment in different parts of the world, in order to ascertain whether the motion of the water is expansive or progressive; with other ingenious proposals. These marks of genius may serve to counterbalance the unfavourable judgment of Bacon's aptitude for physical science which we are sometimes tempted to form, in consequence of his false views on other points; as his rejection of the Copernican system, and his undervaluing Gilbert's magnetical speculations. Most of these errors arose from a too ambitious habit of intellect, which would not be contented with any except very wide and general truths; and from an indistinctness of mechanical, and perhaps, in general, of mathematical ideas:—defects which Bacon's own philosophy was directed to remedy, and which, in the progress of time, it has remedied in others.

20. (VII.) Having thus freely given our judgment concerning the most exact and definite portion of Bacon's precepts, it cannot be necessary for us to discuss at any length the value of those more vague and general *Warnings* against prejudice and partiality, against intellectual indolence and presumption, with which his works abound. His advice and exhortations of this kind are always expressed with energy and point, often clothed in the happiest forms of imagery; and hence it has come to pass, that such passages are perhaps more familiar to the general reader than any other parts of his writings. Nor are Bacon's counsels without their importance, when we

have to do with those subjects in which prejudice and partiality exercise their peculiar sway. Questions of politics and morals, of manners, taste, or history, cannot be subjected to a scheme of rigorous induction; and though on such matters we venture to assert general principles, these are commonly obtained with some degree of insecurity, and depend upon special habits of thought, not upon mere logical connexion. Here, therefore, the intellect may be perverted, by mixing, with the pure reason, our gregarious affections, or our individual propensities; the false suggestions involved in language, or the imposing delusions of received theories. In these dim and complex labyrinths of human thought, *the Idol of the Tribe,* or *of the Den, of the Forum,* or *of the Theatre,* may occupy men's minds with delusive shapes, and may obscure or pervert their vision of truth. But in that Natural Philosophy with which we are here concerned, there is little opportunity for such influences. As far as a physical theory is completed through all the steps of a just induction, there is a clear daylight diffused over it which leaves no lurking-place for prejudice. Each part can be examined separately and repeatedly; and the theory is not to be deemed perfect till it will bear the scrutiny of all sound minds alike. Although, therefore, Bacon, by warning men against the idols or fallacious images above spoken of, may have guarded them from dangerous error, his precepts have little to do with Natural Philosophy: and we cannot agree with him when he says*, that the doctrine concerning these idols bears the same relation to the interpretation of nature as the doctrine concerning sophistical paralogisms bears to common logic.

21. (VIII.) There is one very prominent feature in Bacon's speculations which we must not omit to notice; it is a leading and constant object with him to apply his

---

* *Nov. Org.,* lib. i. Aph. 40.

knowledge to *Use.* The insight which he obtains into nature, he would employ in commanding nature for the service of man. He wishes to have not only principles but works. The phrase which best describes the aim of his philosophy is his own[*], "Ascendendo ad *axiomata,* descendendo ad *opera.*" This disposition appears in the first aphorism of the *Novum Organon,* and runs through the work. "Man, the *minister* and interpreter of nature, *does* and understands, so far as he has, in fact or in thought, observed the course of nature; and he cannot know or *do* more than this." It is not necessary for us to dwell much upon this turn of mind; for the whole of our present inquiry goes upon the supposition that an acquaintance with the laws of nature is worth our having for its own sake. It may be universally true, that knowledge is power; but we have to do with it not as power, but as knowledge. It is the formation of Science, not of Art, with which we are here concerned. It may give a peculiar interest to the history of science, to show how it constantly tends to provide better and better for the wants and comforts of the body; but *that* is not the interest which engages us in our present inquiry into the nature and course of philosophy. The consideration of the means which promote man's material well-being often appears to be invested with a kind of dignity, by the discovery of general laws which it involves; and the satisfaction which rises in our minds at the contemplation of such cases, men sometimes ascribe, with a false ingenuity, to the love of mere bodily enjoyment. But it is never difficult to see that this baser and coarser element is not the real source of our admiration. Those who hold that it is the main business of science to construct instruments for the uses of life, appear sometimes to be willing to accept the consequence which follows from

* *Nov. Org.,* lib. i. Ax. 103.

such a doctrine, that the first shoemaker was a philosopher worthy of the highest admiration*. But those who maintain such paradoxes, often, by a happy inconsistency, make it their own aim, not to devise some improved covering for the feet, but to delight the mind with acute speculations, exhibited in all the graces of wit and fancy.

It has been said† that the key of the Baconian doctrine consists in two words, Utility and Progress. With regard to the latter point, we have already seen that the hope and prospect of a boundless progress in human knowledge had sprung up in men's minds, even in the early times of imperial Rome ; and were most emphatically expressed by that very Seneca who disdained to reckon the worth of knowledge by its value in food and clothing. And when we say that Utility was the great business of Bacon's philosophy, we forget one-half of his characteristic phrase. "Ascendendo ad axiomata," no less than "descendendo ad opera," was, he repeatedly declared, the scheme of his path. He constantly spoke, we are told by his secretary‡, of two kinds of experiments, *experimenta fructifera*, and *experimenta lucifera*.

Again; when we are told by modern writers that Bacon merely recommended such induction as all men instinctively practise, we ought to recollect his own earnest and incessant declarations to the contrary. The induction hitherto practised is, he says, of no use for obtaining solid science. There are two ways§, "hæc via in usu est," "altera vera, sed intentata." Men have constantly been employed in *anticipation*; in illicit induction. The intellect left to itself rushes on in this road‖; the conclusions so obtained are persuasive¶ ; far more persuasive than inductions made with due caution**. But still

---

* *Edinb. Rev.*, No. cxxxii. p. 65.   † *Ib.*
‡ Pref. to the *Nat. Hist.*, i. 243.   § *Nov. Org.*, lib. i. Aph. 19.
‖ Aph. 20.   ¶ Aph. 27.   ** *Ib.*, 28.

this method must be rejected if we would obtain true knowledge. We shall then at length have ground of good hope for science when we proceed in another manner*. We must rise, not by a leap, but by small steps, by successive advances, by a gradation of ascents, trying our facts, and clearing our notions at every interval. The scheme of true philosophy, according to Bacon, is not obvious and simple, but long and technical, requiring constant care and self-denial to follow it. And we have seen that, in this opinion, his judgment is confirmed by the past history and present condition of science.

Again; it is by no means a just view of Bacon's character to place him in contrast to Plato. Plato's philosophy was the philosophy of Ideas; but it was not left for Bacon to set up the philosophy of Facts in opposition to that of Ideas. That had been done fully by the speculative reformers of the sixteenth century. Bacon had the merit of showing that Facts and Ideas must be combined; and not only so, but of divining many of the special rules and forms of this combination, when as yet there were no examples of them, with a sagacity hitherto quite unparalleled.

22. (IX.) With Bacon's unhappy political life we have here nothing to do. But we cannot but notice with pleasure how faithfully, how perseveringly, how energetically he discharged his great philosophical office of a Reformer of Methods. He had conceived the purpose of making this his object at an early period. When meditating the continuation of his *Novum Organon*, and speaking of his reasons for trusting that his work will reach some completeness of effect, he says†, "I am by two arguments thus persuaded. First, I think thus from the zeal and

---

* Aph. 104. So Aph. 105. "In constituendo axiomate forma *inductionis* alia quam adhuc in usu fuit excogitanda est," &c.

† *Ep. ad P. Fulgentium. Op.*, x. 330.

constancy of my mind, which has not waxed old in this
design, nor, after so many years, grown cold and indif-
ferent; I remember that about forty years ago I composed a juvenile work about these things, which with
great contrivance and a pompous title I called *temporis
partum maximum*, or the most considerable birth of time;
Next, that on account of its usefulness, it may hope the
Divine blessing." In stating the grounds of hope for
future progress in the sciences, he says*: "Some hope
may, we conceive, be ministered to men by our own
example: and this we say, not for the sake of boasting,
but because it is useful to be said. If any despond, let
them look at me, a man among all others of my age most
occupied with civil affairs, nor of very sound health,
(which brings a great loss of time;) also in this attempt
the first explorer, following the footsteps of no man, nor
communicating on these subjects with any mortal; yet,
having steadily entered upon the true road and made my
mind submit to things themselves, one who has, in this
undertaking, made, (as we think,) some progress." He
then proceeds to speak of what may be done by the
combined and more prosperous labours of others, in that
strain of noble hope and confidence, which rises again
and again, like a chorus, at intervals in every part of his
writings. In the *Advancement of Learning* he had said,
"I could not be true and constant to the argument I
handle, if I were not willing to go beyond others, but yet
not more willing than to have others go beyond me
again." In the Preface to the *Instauratio Magna*, he had
placed among his postulates those expressions which have
more than once warmed the breast of a philosophical
reformer†. "Concerning ourselves we speak not; but
as touching the matter which we have in hand, this we

* *Nov. Org.*, i. Aph. 113.
† See the motto to KANT's *Kritik der Reinen Vernunft*.

ask;—that men be of good hope, neither feign and imagine to themselves this our Reform as something of
infinite dimension and beyond the grasp of mortal man,
when in truth it is the end and true limit of infinite
error; and is by no means unmindful of the condition of
mortality and humanity, not confiding that such a thing
can be carried to its perfect close in the space of a single
age, but assigning it as a task to a succession of generations." In a later portion of the *Instauratio* he says:
"We bear the strongest love to the *human republic,* our
common country; and we by no means abandon the hope
that there will arise and come forth some man among
posterity, who will be able to receive and digest all that
is best in what we deliver; and whose care it will be to
cultivate and perfect such things. Therefore, by the
blessing of the Deity, to tend to this object, to open up
the fountains, to discover the useful, to gather guidance
for the way, shall be our task; and from this we shall
never, while we remain in life, desist."

23. (X.) We may add, that the spirit of piety as well
as of hope which is seen in this passage, appears to have
been habitual to Bacon at all periods of his life. We
find in his works several drafts of portions of his great
scheme, and several of them begin with a prayer. One
of these entitled, in the edition of his works, "The
Student's Prayer," appears to me to belong probably to
his early youth. Another, entitled "The Writer's Prayer,"
is inserted at the end of the Preface of the *Instauratio,* as
it was finally published. I will conclude my notice of
this wonderful man by inserting here these two prayers.

"To God the Father, God the Word, God the Spirit,
we pour forth most humble and hearty supplications;
that he, remembering the calamities of mankind, and the
pilgrimage of this our life, in which we wear out days
few and evil, would please to open to us new refresh-

ments out of the fountains of his goodness for the
alleviating of our miseries. This also we humbly and
earnestly beg, that human things may not prejudice such
as are divine; neither that, from the unlocking of the
gates of sense, and the kindling of a greater natural light,
anything of, incredulity, or intellectual night, may arise
in our minds towards divine mysteries. But rather, that
by our mind thoroughly cleansed and purged from fancy
and vanities, and yet subject and perfectly given up to
the Divine oracles, there may be given unto faith the
things that are faith's."

"Thou, O Father, who gavest the visible light as the
first-born of thy creatures, and didst pour into man the
intellectual light as the top and consummation of thy
workmanship, be pleased to protect and govern this work,
which coming from thy goodness, returneth to thy glory.
Thou, after thou hadst reviewed the works which thy
hands had made, beheldest that everything was very good,
and thou didst rest with complacency in them. But
man, reflecting on the works which he had made, saw that
all was vanity and vexation of spirit, and could by no
means acquiesce in them. Wherefore, if we labour in
thy works with the sweat of our brows, thou wilt make
us partakers of thy vision and thy Sabbath. We humbly
beg that this mind may be stedfastly in us; and that
thou, by our hands, and also by the hands of others on
whom thou shalt bestow the same spirit, wilt please to
convey a largess of new alms to thy family of mankind.
These things we commend to thy everlasting love, by our
Jesus, thy Christ, God with us. Amen."

CHAPTER XII.

FROM BACON TO NEWTON.

I. *Harvey*.—We have already seen that Bacon was by no means the first mover or principal author of the revolution in the method of philosophizing which took place in his time; but only the writer who proclaimed in the most impressive and comprehensive manner, the scheme, the profit, the dignity, and the prospects of the new philosophy. Those, therefore, who after him took up the same views are not to be considered as his successors, but as his fellow labourers; and the line of historical succession of opinions must be pursued without special reference to any. one leading character, as the principal figure of the epoch. I resume this line, by noticing a contemporary and fellow countryman of Bacon, Harvey, the discoverer of the circulation of the blood. This discovery was not published and generally accepted till near the end of Bacon's life; but the anatomist's reflections on the method of pursuing science, though strongly marked with the character of the revolution that was taking place, belong to a very different school from the Chancellor's. Harvey was a pupil of Fabricius of Acquapendente, whom we noticed among the practical reformers of the sixteenth century. He entertained, like his master, a strong reverence for the great names which had ruled in philosophy up to that time, Aristotle and Galen; and was disposed rather to recommend his own method by exhibiting it as the true interpretation of ancient wisdom, than to boast of its novelty. It is true, that he assigns, as his reason for publishing some of his researches*, "that by revealing the method I use in searching into

---

* *Anatomical Exercitations concerning the Generation of Living Creatures*, 1653. Preface.

things, I might propose to studious men, a new and (if I mistake not) a surer path to the attainment of know-ledge* ;" but he soon proceeds to fortify himself with the authority of Aristotle. In doing this, however, he has the very great merit of giving a living and practical cha-racter to truths which exist in the Aristotelian works, but which had hitherto been barren and empty professions. We have seen that Aristotle had asserted the importance of experience as one root of knowledge; and in this had been followed by the schoolmen of the middle ages : but this assertion came with very different force and effect from a man, the whole of whose life had been spent in obtaining, by means of experience, knowledge which no man had possessed before. In Harvey's general reflec-tions, the necessity of both the elements of knowledge, sensations and ideas, experience and reason, is fully brought into view, and rightly connected with the meta-physics of Aristotle. He puts the antithesis of these two elements with great clearness. "Universals are chiefly known to us, for science is begot by reasoning from

* He used similar expressions in conversation. George Ent, who edited his *Generation of Animals,* visited him, " at that time residing not far from the city; and found him very intent upon the perscruta-tion of nature's works, and with a countenance as cheerful, as mind imperturbed; Democritus like, chiefly searching into the cause of natural things." In the course of conversation the writer said, "It hath always been your choice, about the secrets of Nature, to consult Nature herself." " 'Tis true," replied he; " and I have constantly been of opinion that from thence we might acquire not only the knowledge of those less considerable secrets of Nature, but even a certain admiration of that Supreme Essence, the Creator. And though I have ever been ready to acknowledge, that many things have been discovered by learned men of former times; yet do I still believe that the number of those which remain yet concealed in the darkness of impervestigable Nature is much greater. Nay, I cannot forbear to wonder, and some-times smile at those, who persuade themselves, that all things were so consummately and absolutely delivered by Aristotle, Galen, or some other great name, as that nothing was left to the superaddition of any that succeeded."

universals to particulars; yet that very comprehension of
universals in the understanding springs from the perception
of singulars in our sense." Again, he quotes Aristotle's apparently opposite assertions:—that made in his
Physics*, "that we must advance from things which are
first known to us, though confusedly,' to things more distinctly intelligible in themselves; from the whole to the
part; from the universal to the particular;" and that
made in the *Analytics*†; that "Singulars are more known
to us and do first exist according to sense: for nothing is
in the understanding which was not before in the sense."
Both, he says, are true, though at first they seem to
clash: for "though in knowledge we begin with sense,
sensation itself is a universal thing." This he further
illustrates; and quotes Seneca, who says, that "Art
itself is nothing but the *reason* of the work, implanted in
the Artist's mind:" and adds, "the same way by which
we gain an Art, by the very same way we attain any kind
of science or knowledge whatever; for as Art is a habit
whose object is something to be done, so Science is a
habit whose object is something to be known; and as the
former proceedeth from the imitation of examples, so
this latter, from the knowledge of things natural. The
source of both is from sense and experience; since [but]
it is impossible that Art should be rightly purchased by
the one or Science by the other without a direction from
ideas." Without here dwelling on the relation of Art
and Science, (very justly stated by Harvey, except that
ideas exist in a very different form in the mind of the
Artist and the Scientist) it will be seen that this doctrine,
of science springing from experience with a direction
from ideas, is exactly that which we have repeatedly
urged, as the true view of the subject. From this view,
Harvey proceeds to infer the importance of a reference

* Lib. i. c. 2, 3.   † *Anal. Post.*, ii.

to sense in his own subject, not only for first discovering, but for receiving knowledge: " Without experience, not other men's but our own, no man is a proper disciple of any part of natural knowledge; without experimental skill in anatomy, he will no better apprehend what I shall deliver concerning generation, than a man born blind can judge of the nature and difference of colours, or one born deaf, of sounds." "If we do otherwise, we may get a humid and floating opinion, but never a solid and infallible knowledge: as is happenable to those who see foreign countries only in maps, and the bowels of men falsely described in anatomical tables. And hence it comes about, that in this rank age, we have many sophisters and bookwrights, but few wise men and philosophers." He had before declared "how unsafe and degenerate a thing it is, to be tutored by other men's commentaries, without making trial of the things themselves; especially since Nature's book is so open and legible." We are here reminded of Galileo's condemnation of the "paper philosophers." The train of thought thus expressed by the practical discoverers, spread rapidly with the spread of the new knowledge that had suggested it, and soon became general and unquestioned.

II. *Descartes.*—Such opinions are now among the most familiar and popular of those which are current among writers and speakers; but we should err much if we were to imagine that after they were once propounded they were never resisted or contradicted. Indeed, even in our own time, not only are such maxims very frequently practically neglected or forgotten, but the opposite opinions, and views of science quite inconsistent with those we have been explaining, are often promulgated and widely accepted. The philosophy of pure ideas has its commonplaces, as well as the philosophy of experience. And at the time of which we speak, the former philosophy, no less than the latter, had its great asserter and expounder; a man in his own time more admired than Bacon, regarded with more deference by a large body of disciples all over Europe, and more powerful in stirring up men's minds to a new activity of inquiry. I speak of Descartes, whose labours, considered as a philosophical system, were an endeavour to revive the method of obtaining knowledge by reasoning from our own ideas only, and to erect it in opposition to the method of observation and experiment. The Cartesian philosophy contained an attempt at a counter-revolution. Thus in this author's *Principia Philosophiæ**, he says that "he will give a short account of the principal phenomena of the world, not that he may use them as reasons to prove anything; for," adds he, "we desire to deduce effects from causes, not causes from effects; but only in order that out of the innumerable effects which we learn to be capable of resulting from the same causes, we may determine our mind to consider some rather than others." He had before said, "The principles which we have obtained [by pure *à priori* reasoning] are so vast and so fruitful, that many more consequences follow from them than we see contained in this visible world, and even many more than our mind can ever take a full survey of." And he professes to apply this method in detail. Thus in attempting to state the three fundamental laws of motion, he employs only *à priori* reasonings, and is in fact led into error in the third law which he thus obtains†. And in his *Dioptrics‡* he pretends to deduce the laws of reflection and refraction of light from certain comparisons (which are, in truth, arbitrary,) in which the radiation of light is represented by the motion of a ball impinging upon the reflecting or refracting body. It might be represented as a curious instance of the caprice of fortune,

* Pars iii. p. 45.   † See *Hist. Ind. Sci.*, ii. 50.   ‡ Cap. i. ii.

which appears in scientific as in other history, that Kepler, professing to derive all his knowledge from experience, and exerting himself with the greatest energy and perseverance, failed in detecting the law of refraction; while Descartes, who professed to be able to despise experiment, obtained the true law of sines. But as we have stated in the History*, Descartes appears to have learnt this law from Snell's papers. And whether this be so or not, it is certain that notwithstanding the profession of independence which his philosophy made, it was in reality constantly guided and instructed by experience. Thus in explaining the Rainbow (in which his portion of the discovery merits great praise) he speaks† of taking a globe of glass, allowing the sun to shine on one side of it, and noting the colours produced by rays after two refractions and one reflection. And in many other instances, indeed in all that relates to physics, the reasonings and explanations of Descartes and his followers were, consciously or unconsciously, directed by the known facts, which they had observed themselves or learnt from others.

But since Descartes thus, speculatively at least, set himself in opposition to the great reform of scientific method which was going on in his time, how, it may be asked, did he acquire so strong an influence over the most active minds of his time? How is it that he became the founder of a large and distinguished school of philosophers? How is it that he not only was mainly instrumental in deposing Aristotle from his intellectual throne, but for a time appeared to have established himself with almost equal powers, and to have rendered the Cartesian school as firm a body as the Peripatetic had been?

* *Hist. Ind. Sci.*, ii. 347.   † *Meteorum*, c. 8, p. 187.

The causes to be assigned for this remarkable result are, I conceive, the following. In the first place, the physicists of the Cartesian school did, as I have just stated, found their philosophy upon experiment; and did not practically, nor indeed, most of them, theoretically, assent to their master's boast of showing what the phenomena *must be*, instead of looking to see what they *are*. And as Descartes had really incorporated in his philosophy all the chief physical discoveries of his own and preceding times, and had delivered, in a more general and systematic shape than any one before him, the principles which he thus established, the physical philosophy of his school was in reality far the best then current; and was an immense improvement upon the Aristotelian doctrines, which had not yet been displaced as a system. Another circumstance which gained him much favour, was the bold and ostentatious manner in which he professed to begin his philosophy by liberating himself from all preconceived prejudice. The first sentence of his philosophy contains this celebrated declaration: "Since," he says, "we begin life as infants, and have contracted various judgments concerning sensible things before we possess the entire use of our reason, we are turned aside from the knowledge of truth by many prejudices: from which it does not appear that we can be any otherwise delivered, than if once in our life we make it our business to doubt of everything in which we discern the smallest suspicion of uncertainty." With this sweeping rejection or unhesitating scrutiny of all preconceived opinions, the power of the ancient authorities and masters in philosophy must obviously shrink away; and thus Descartes came to be considered as the great hero of the overthrow of the Aristotelian dogmatism. But in addition to these causes, and perhaps more powerful than all in procuring the assent of men to his doctrines, came the deductive and systematic character of his philosophy. For although all knowledge of the external world is in reality only to be obtained from observation, by inductive steps,—minute, perhaps, and slow, and many, as Galileo and Bacon had already taught;—the human mind conforms to these conditions reluctantly and unsteadily, and is ever ready to rush to general principles, and then to employ itself in deducing conclusions from these by synthetical reasonings; a task grateful, from the distinctness and certainty of the result, and the accompanying feeling of our own sufficiency. Hence men readily overlooked the precarious character of Descartes' fundamental assumptions, in their admiration of the skill with which a varied and complex universe was evolved out of them. And the complete and systematic character of this philosophy attracted men no less than its logical connexion. I may quote here what a philosopher* of our own time has said of another writer : " He owed his influence to various causes ; at the head of which may be placed that genius for system which, though it cramps the growth of knowledge, perhaps finally atones for that mischief by the zeal and activity which it rouses among followers and opponents, who discover truth by accident when in pursuit of weapons for their warfare. A system which attempts a task so hard as that of subjecting vast provinces of human knowledge to one or two principles, if it presents some striking instances of conformity to superficial appearances is sure to delight the framer ; and for a time to subdue and captivate the student too entirely for sober reflection and rigorous examination. In the first instance consistency passes for truth. When principles in some instances have proved sufficient to give an unexpected explanation of facts, the delighted reader is content to accept as true all other deductions from the principles. Specious

* MACKINTOSH, *Dissertation on Ethical Science.*

premises being assumed to be true, nothing more can be
required than logical inference.   Mathematical forms
pass current as the equivalent of mathematical certainty.
The unwary admirer is satisfied with the completeness
and symmetry of the plan of his house, unmindful of the
need of examining the firmness of the foundation and the
soundness of the materials.   The system-maker, like the
conqueror, long dazzles and overawes the world; but
when their sway is past, the vulgar herd, unable to
measure their astonishing faculties, take revenge by
trampling on fallen greatness."   Bacon had showed his
wisdom in his reflections on this subject, when he said
that "Method, carrying a show of total and perfect
knowledge, hath a tendency to generate acquiescence."

The main value of Descartes' physical doctrines consisted in their being arrived at in a way inconsistent with
his own professed method, namely, by a reference to
observation.   But though he did in reality begin from
facts, his system was nevertheless a glaring example
of that error which Bacon had called *Anticipation;* that
illicit generalization which leaps at once from special
facts to principles of the widest and remotest kind; such,
for instance, as the Cartesian doctrine, that the world is
an absolute *plenum,* every part being full of matter of
some kind, and that all natural effects depend on the
laws of motion.   Against this fault, to which the human
mind is so prone, Bacon had lifted his warning voice
in vain, so far as the Cartesians were concerned; as indeed,
to this day, one theorist after another pursues his course,
and turns a deaf ear to the Verulamian injunctions; perhaps even complacently boasts that he founds his theory
upon observation; and forgets that there are, as the
aphorism of the *Novum Organon* declares, two ways by
which this may be done;—the one hitherto in use and
suggested by our common tendencies, but barren and

worthless; the other almost untried, to be pursued only with effort and self-denial, but alone capable of producing true knowledge.

III. *Gassendi.*—Thus the lessons which Bacon taught were far from being generally accepted and applied at first. The amount of the influence of these two men, Bacon and Descartes, upon their age, has often been a subject of discussion. The fortunes of the Cartesian school, have been in some measure traced in the History of Science. But I may mention the notice taken of these two philosophers by Gassendi, a contemporary and countryman of Descartes. Gassendi, as I have elsewhere stated,* was associated with Descartes in public opinion, as an opponent of the Aristotelian dogmatism; but was not in fact a follower or profound admirer of that writer. In a Treatise on Logic, Gassendi gives an account of the Logic of various sects and authors; treating, in order, of the Logic of Zeno (the Eleatic), of Euclid (the Megarean), of Plato, of Aristotle, of the Stoics, of Epicurus, of Lullius, of Ramus; and to these he adds the Logic of Verulam, and the Logic of Cartesius. "We must not," he says, "on account of the celebrity it has obtained, pass over the Organon or Logic of Francis Bacon Lord Verulam, High Chancellor of England, whose noble purpose in our time it has been, to make an Instauration of the Sciences." He then gives a brief account of the *Novum Organon*, noticing the principal features in its rules, and especially the distinction between the vulgar induction which leaps at once from particular experiments to the more general axioms, and the chastised and gradual induction, which the author of the *Organon* recommends. In his account of the Cartesian Logic, he justly observes, that "He too imitated Verulam in this, that being about to build up a new philosophy from the foundation, he

* *Hist. Ind. Sci.*, ii. 137.

wished in the first place to lay aside all prejudice : and having then found some solid principle, to make that the ground-work of his whole structure. But he proceeds by a very different path from that which Verulam follows ; for while Verulam seeks aid from things, to perfect the cogitation of the intellect, Cartesius conceives, that when we have laid aside all knowledge of things, there is, in our thoughts alone, such a resource, that the intellect may by its own power arrive at a perfect knowledge of all, even the most abstruse things."

The writings of Descartes have been most admired, and his method most commended, by those authors who have employed themselves upon metaphysical rather than physical subjects of inquiry. Perhaps we might say that, in reference to such subjects, this method is not so vicious as at first, when contrasted with the Baconian induction, it seems to be : for it might be urged that the *thoughts* from which Descartes begins his reasonings are, in reality *experiments* of the kind which the subject requires us to consider: each such thought is a fact in the intellectual world; and of such facts, the metaphysician seeks to discover the laws. I shall not here examine the validity of this plea ; but shall turn to the consideration of the actual progress of physical science and its effect on men's minds.

IV. *Actual progress in Science.*—The practical discoverers were indeed very active and very successful during the seventeenth century which opened with Bacon's survey and exhortations. The laws of nature, of which men had begun to obtain a glimpse in the preceding century, were investigated with zeal and sagacity, and the consequence was that the foundations of most of the modern physical sciences were laid. That mode of research by experiment and observation, which had, a little time ago, been a strange, and to many, an unwelcome innovation, was now become the habitual course of philosophers. The revolution from the philosophy of tradition to the philosophy of experience was completed. The great discoveries of Kepler belonged to the preceding century. They are not, I believe, noticed, either by Bacon or by Descartes; but they gave a strong impulse to astronomical and mechanical speculators, by showing the necessity of a sound science of motion. Such a science Galileo had already begun to construct. At the time of which I speak, his disciples* were still labouring at this task, and at other problems which rapidly suggested themselves. They had already convinced themselves that air had weight; in 1643 Torricelli proved this practically by the invention of the Barometer; in 1647, Pascal proved it still further by sending the Barometer to the top of a mountain. Pascal and Boyle brought into clear view the fundamental laws of fluid equilibrium; Boyle and Mariotte determined the law of the compression of air as regulated by its elasticity. Otto Guericke invented the air pump, and by his "Magdeburg Experiments" on a vacuum, illustrated still further the effects of the air. Guericke pursued what Gilbert had begun, the observation of electrical phenomena; and these two physicists made an important step, by detecting repulsion as well as attraction in these phenomena. Gilbert had already laid the foundations of the science of Magnetism. The law of refraction, at which Kepler had laboured in vain, was, as we have seen, discovered by Snell (about 1621), and published by Descartes. Mersenne had discovered some of the more important parts of the theory of Harmonics. In sciences of a different kind, the same movement was visible. Chemical doctrines tended to assume a proper degree of generality, when Sylvius in 1679 taught the opposition of acid and alkali, and Stahl, soon after, the phlogistic theory of com-

* Castelli, Torricelli, Viviani, Baliani, Gassendi, Mersenne, Borelli, Cavalleri.

bustion. Steno had remarked the most important law of crystallography in 1669, that the angles of the same kind of crystals are always equal. In the sciences of classification, about 1680, Ray and Morison in England resumed the attempt to form a systematic botany, which had been interrupted for a hundred years, from the time of the memorable essay of Cæsalpinus. The grand discovery of the circulation of the blood by Harvey about 1619, was followed in 1651 by Pecquet's discovery of the course of the chyle. There could now no longer be any question whether science was progressive, or whether observation could lead to new truths.

Among these cultivators of science, such sentiments as have been already quoted became very familiar;—that knowledge is to be sought from nature herself by observation and experiment;—that in such matters tradition is of no force when opposed to experience, and that mere reasonings without facts cannot lead to solid knowledge. But I do not know that we find in these writers any more special rules of induction and scientific research which have since been confirmed and universally adopted. Perhaps too, as was natural in so great a revolution, the writers of this time, especially the second-rate ones, were somewhat too prone to disparage the labours and talents of Aristotle and the ancients in general, and to overlook the ideal element of our knowledge, in their zealous study of phenomena. They urged, sometimes in an exaggerated manner, the superiority of modern times in all that regards science, and the supreme and sole importance of facts in scientific investigations. There prevailed among them also a lofty and dignified tone of speaking of the condition and prospects of science, such as we are accustomed to admire in the Verulamian writings; for this, in a less degree, is epidemic among those who a little after his time speak of the new philosophy.

V. *Otto Guericke, &c.*—I need not illustrate these characteristics at any great length. I may as an example notice Otto Guericke's Preface to his *Experimenta Magdeburgica* (1670). He quotes a passage from Kircher's Treatise on the Magnetic Art, in which the author says, "Hence it appears how all philosophy, except it be supported by experiments, is empty, fallacious, and useless; what monstrosities philosophers, in other respects of the highest and subtlest genius, may produce in philosophy by neglecting experiment. Thus Experience alone is the Dissolver of Doubts, the Reconciler of Difficulties, the sole Mistress of Truth, who holds a torch before us in obscurity, unties our knots, teaches us the true causes of things." Guericke himself reiterates the same remark, adding that "philosophers, insisting upon their own thoughts and arguments merely, cannot come to any sound conclusion respecting the natural constitution of the world." Nor were the Cartesians slow in taking up the same train of reflection. Thus Gilbert Clark who, in 1660, published[*] a defence of Descartes' doctrine of a *plenum* in the universe, speaks in a tone which reminds us of Bacon, and indeed was very probably caught from him. "Natural philosophy formerly consisted entirely of loose and most doubtful controversies, carried on in high sounding words, fit rather to delude than to instruct men. But at last (by the favour of the Deity) there shone forth some more divine intellects, who taking as their counsellors reason and experience together, exhibited a new method of philosophizing. Hence has been conceived a strong hope that philosophers may embrace, not a shadow or empty image of Truth, but Truth herself: and that Physiology (Physics) scattering these controversies to the

---

[*] *De Plenitudine Mundi, in qua defenditur Cartesiana Philosophia contra sententias Francisci Baconi, Th. Hobbii et Sethi Wardi.*

winds, will contract an alliance with Mathematics. Yet this is hardly the work of one age; still less of one man. Yet let not the mind despond, or doubt not that, one party of investigators after another following the same method of philosophizing, at last, under good auguries, the mysteries of nature being daily unlocked as far as human feebleness will allow, Truth may at last appear in full, and these nuptial torches may be lighted."

As another instance of the same kind, I may quote the Preface to the First volume of the Transactions of the Academy of Sciences at Paris. " It is only since the present century," says the writer, " that we can reckon the revival of Mathematics and Physics. M. Descartes and other great men have laboured at this work with so much success, that in this department of literature, the whole face of things has been changed. Men have quitted a sterile system of physics, which for several generations had been always at the same point; the reign of words and terms is passed; men will have things; they establish principles which they understand, they follow those principles; and thus they make progress. Authority has ceased to have more weight than Reason: that which was received without contradiction because it had been long received, is now examined, and often rejected: and philosophers have made it their business to consult, respecting natural things, Nature herself rather than the Ancients." These had now become the commonplaces of those who spoke concerning the course and method of the Sciences.

VI. *Hooke.*—In England, as might be expected, the influence of Francis Bacon was more directly visible. We find many writers, about this time, repeating the truths which Bacon had proclaimed, and in almost every case showing the same imperfections in their views which we have noticed in him. We may take as an example of this Hooke's Essay, entitled " A General Scheme or Idea of the present state of Natural Philosophy, and how its defects may be remedied by a Methodical proceeding in the making Experiments and collecting Observations; whereby to compile a Natural History as a solid basis for the superstructure of true Philosophy." This Essay may be looked upon as an attempt to adapt the *Novum Organon* to the age which succeeded its publication. We have in this imitation, as in the original, an enumeration of various mistakes and impediments which had in preceding times prevented the progress of knowledge; exhortations to experiment and observation as the only solid basis of Science; very ingenious suggestions of trains of inquiry, and modes of pursuing them; and a promise of obtaining scientific truths when facts have been duly accumulated. This last part of his scheme the author calls *a Philosophical Algebra;* and he appears to have imagined that it might answer the purpose of finding unknown causes from known facts, by means of certain regular processes, in the same manner as Common Algebra finds unknown from known quantities. But this part of the plan appears to have remained unexecuted. The suggestion of such a method was a result of the Baconian notion that invention in a discoverer might be dispensed with. We find Hooke adopting the phrases in which this notion is implied: thus he speaks of the understanding as "being very prone to run into the affirmative way of judging, and wanting patience to follow and prosecute the negative way of inquiry, by rejection of disagreeing natures." And he follows Bacon also in the error of attempting at once to obtain from the facts the discovery of a " nature," instead of investigating first the measures and the laws of phenomena. I return to more general notices of the course of men's thoughts on this subject.

VII. *Royal Society.*—Those who associated themselves

together for the prosecution of science quoted Bacon as
their leader, and exulted in the progress made by the philosophy which proceeded upon his principles.   Thus in
Oldenburg's Dedication of the Transactions of the Royal
Society of London for 1670, to Robert Boyle, he says;
" I am informed by such as well remember the best and
worst days of the famous Lord Bacon, that though he
wrote his *Advancement of Learning* and his *Instauratio
Magna* in the time of his greatest power, yet his greatest
reputation rebounded first from the most intelligent
foreigners in many parts of Christendom:" and after
speaking of his practical talents and his public employments, he adds, " much more justly still may we wonder
how, without any great skill in Chemistry, without much
pretence to the Mathematics or Mechanics, without optic
aids or other engines of late invention, he should so much
transcend the philosophers then living, in judicious
and clear instructions, in so many useful observations
and discoveries, I think I may say beyond the records of
many ages."   And in the end of the Preface to the same
volume, he speaks with great exultation of the advance
of science all over Europe, referring undoubtedly to facts
then familiar.   " And now let envy snarl, it cannot stop
the wheels of active philosophy, in no part of the known
world ;—not in France, either in Paris or in Caen :—not
in Italy, either in Rome, Naples, Milan, Florence, Venice,
Bononia or Padua ;—in none of the Universities either on
this or on that side of the seas, Madrid and Lisbon, all
the best spirits in Spain and Portugal, and the spacious
and remote dominions to them belonging ;—the Imperial
Court and the Princes of Germany ; the Northern Kings
and their best luminaries ; and even the frozen Moscovite
and Russian have all ɩtaken the operative ferment : and
it works high and prevails every way, to the encourage-
ment of all sincere lovers of knowledge and virtue."

Again, in the Preface for 1672, he pursues the same thought into detail. "We must grant that in the last age, when operative philosophy began to recover ground, and to tread on the heels of triumphant Philology; emergent adventures and great successes were encountered by dangerous oppositions and strong obstructions. Galilæus and others in Italy suffered extremities for their celestial discoveries; and here in England Sir Walter Raleigh, when he was in his greatest lustre, was notoriously slandered to have erected a school of atheism, because he gave countenance to chemistry, to practical arts, and to curious mechanical operations, and designed to form the best of them into a college. And Queen Elizabeth's Gilbert was a long time esteemed extravagant for his magnetisms; and Harvey for his diligent researches in pursuance of the circulation of the blood. But when our renowned Lord Bacon had demonstrated the methods for a perfect restoration of all parts of real knowledge; and the generous and philosophical Peireskius had, soon after, agitated in all parts to redeem the most instructive antiquities, and to excite experimental essays and fresh discoveries; the success became on a sudden stupendous; and effective philosophy began to sparkle, and even to flow into beams of shining light all over the world."

The formation of the Royal Society of London and of the Academy of Sciences of Paris, from which proceeded the declamations just quoted, were among many indications, belonging to this period, of the importance which states as well as individuals had by this time begun to attach to the cultivation of science. The English Society was established almost immediately when the restoration of the monarchy appeared to give a promise of tranquillity to the nation (in 1660), and the French Academy very soon afterwards (in 1666). These measures were very soon followed by the establishment of the Observatories of Paris and Greenwich (in 1667 and 1675); which may be considered to be a kind of public recognition of the astronomy of observation, as an object on which it was the advantage and the duty of nations to bestow their wealth.

VIII. *Bacon's New Atalantis.*—When philosophers had their attention turned to the boundless prospect of increase to the knowledge and powers and pleasures of man which the cultivation of experimental philosophy seemed to promise, it was natural that they should think of devising institutions and associations by which such benefits might be secured. Bacon had drawn a picture of a society organized with a view to such purpose, in his fiction of the " New Atalantis." The imaginary teacher who explains this institution to the inquiring traveller, describes it by the name of *Solomon's House;* and says *, " The end of our foundation is the knowledge of causes and secret motions of things; and the enlarging the bounds of the human empire to effecting of things possible." And, as parts of this House, he describes caves and wells, chambers and towers, baths and gardens, parks and pools, dispensatories and furnaces, and many other contrivances, provided for the purpose of making experiments of many kinds. He describes also the various employments of the Fellows of this College, who take a share in its researches. There are *merchants of light,* who bring books and inventions from foreign countries ; *depredators,* who gather the experiments which exist in books ; *mystery-men,* who collect the experiments of the mechanical arts ; *pioneers* or *miners,* who invent new experiments ; and *compilers,* " who draw the experiments of the former into titles and tables, to give the better light for the drawing of observations and axioms out of them." There are

* BACON'S *Works,* vol. ii. 111.

also *dowry-men or benefactors*, that cast about how to draw out of the experiments of their fellows things of use and practice for man's life; *lamps*, that direct new experiments of a more penetrating light than the former; *inoculators*, that execute the experiments so directed. Finally, there are the *interpreters of nature*, that raise the former discoveries by experiments into greater observations (that is, more general truths) axioms and aphorisms. Upon this scheme we may remark, that fictitious as it undisguisedly is, it still serves to exhibit very clearly some of the main features of the author's philosophy:—namely, his steady view of the necessity of ascending from facts to the most general truths by several stages;—an exaggerated opinion of the aid that could be derived in such a task from technical separation of the phenomena and a distribution of them into tables;—a belief, probably incorrect, that the offices of experimenter and interpreter may be entirely separated, and pursued by different persons with a certainty of obtaining success;—and a strong determination to make knowledge constantly subservient to the uses of life.

IX. *Cowley.*—Another project of the same kind, less ambitious but apparently more directed to practice, was published a little later (1657) by another eminent man of letters in this country. I speak of Cowley's "Proposition for the Advancement of Experimental Philosophy." He suggests that a College should be established at a short distance from London, endowed with a revenue of four thousand pounds, and consisting of twenty professors with other members. The objects of the labours of these professors he describes to be, first, to examine all knowledge of nature delivered to us from former ages and to pronounce it sound or worthless; second, to recover the lost inventions of the ancients; third, to improve all arts that we now have; lastly, to discover

others that we yet have not.   In this proposal we cannot
help marking the visible declension from Bacon's more
philosophical view.   For we have here only a very vague
indication of improving old arts and discovering new,
instead of the two clear Verulamian antitheses, Experiments and Axioms deduced from them, on the one
hand, and on the other an ascent to general Laws, and a
derivation, from these, of Arts for daily use.   Moreover
the prominent place which Cowley has assigned to the
verifying the knowledge of former ages and recovering
" the lost inventions and drowned laws of the ancients,"
implies a disposition to think too highly of traditionary
knowledge; a weakness which Bacon's scheme shows
him to have fully overcome.   And thus it has been up
to the present day, that with all Bacon's mistakes, in the
philosophy of scientific method few have come up to
him, and perhaps none have gone beyond him.

Cowley exerted himself to do justice to the new philosophy in verse as well as prose, and his Poem to the
Royal Society expresses in a very noble manner those
views of the history and prospects of philosophy which
prevailed among the men by whom the Royal Society
was founded.   The fertility and ingenuity of comparison
which characterise Cowley's poetry are well known ; and
these qualities are in this instance largely employed for
the embellishment of his subject.   Many of the comparisons which he exhibits are apt and striking.   Philosophy is a ward whose estate (human knowledge) is, in
his nonage, kept from him by his guardians and tutors;
(a case which the ancient rhetoricians were fond of
taking as a subject of declamation;) and these wrong-
doers retain him in unjust tutelage and constraint for
their own purposes ; until

> Bacon at last, a mighty man, arose,
> (Whom a wise King, and Nature, chose
> Lord Chancellor of both their laws,)
> And boldly undertook the injured pupil's cause.

Again, Bacon is one who breaks a scarecrow Priapus which stands in the garden of knowledge. Again, Bacon is one who, instead of a picture of painted grapes, gives us real grapes from which we press "the thirsty soul's refreshing wine." Again, Bacon is like Moses, who led the Hebrews forth from the barren wilderness, and ascended Pisgah;—

> Did on the very border stand
> Of the blest promised land,
> And from the mountain's top of his exalted wit
> Saw it himself and showed us it.

The poet however adds, that Bacon discovered, but did not conquer this new world; and that the men whom he addresses must subdue these regions. These "champions" are then ingeniously compared to Gideon's band:

> Their old and empty pitchers first they brake
> And with their hands then lifted up the light.

There were still at this time some who sneered at or condemned the new philosophy; but the tide of popular opinion was soon strongly in its favour. I have elsewhere * noticed a pasquinade of the poet Boileau in 1682, directed against the Aristotelians. At this time, and indeed for long afterwards, the philosophers of France were Cartesians. The English men of science, although partially and for a time they accepted some of Descartes' opinions, for the most part carried on the reform independently, and in pursuance of their own views. And they very soon found a much greater leader than Descartes to place at their head, and to take as their authority, so far as they acknowledged authority, in their speculations. I speak of Newton, whose influence upon the philosophy of science I must now consider.

---

* *Hist. Ind. Sci.*, ii. 137.

Chapter XIII.

NEWTON.

1. Bold and extensive as had been the anticipations of those whose minds were excited by the promise of the new philosophy, the discoveries of Newton respecting the mechanics of the universe, brought into view truths more general and profound than those earlier philosophers had hoped or imagined. With these vast accessions to human knowledge, men's thoughts were again set in action; and philosophers made earnest and various attempts to draw, from these extraordinary advances in science, the true moral with regard to the conduct and limits of the human understanding. They not only endeavoured to verify and illustrate, by these new portions of science, what had recently been taught concerning the methods of obtaining sound knowledge; but they were also led to speculate concerning many new and more interesting questions relating to this subject. They saw, for the first time, or at least far more clearly than before, the distinction between the inquiry into the *laws*, and into the *causes* of phenomena. They were tempted to ask, how far the discovery of causes could be carried; and whether it would soon reach, or clearly point to, the ultimate cause. They were driven to consider whether the properties which they discovered were essential properties of all matter, necessarily and primarily involved in its essence, though revealed to us at a late period by their derivative effects. These questions even now agitate the thoughts of speculative men. Some of them have already, in this work, been discussed, or arranged in the places which our view of the philosophy of these subjects assigns to them. But we must here notice them

as they occurred to Newton himself and his immediate followers.

2. The general Baconian notion of the method of philosophizing, that it consists in ascending from phenomena, through various stages of generalization, to truths of the highest order, received, in Newton's discovery of the universal mutual gravitation of every particle of matter, that pointed actual exemplification, for want of which it had hitherto been almost overlooked, or at least very vaguely understood. That great truth, and the steps by which it was established, afford, even now, by far the best example of the successive ascent, from one scientific truth to another,—of the repeated transition from less to more general propositions,—which we can yet produce ; as may be seen in the Table which exhibits the relation of these steps in Book XI. Newton himself did not fail to recognize this feature in the truths which he exhibited. Thus, he says*, " By the way of Analysis we proceed from compounds to ingredients, as from motions to the forces producing them; and in general, from effects to their causes. and from particular causes to more general ones, till the argument end in the most general." And in like manner in another Query† : " The main business of natural philosophy is to argue from phenomena without feigning hypotheses, and to deduce causes from effects, till we come to the First Cause, which is certainly not mechanical."

3. Newton appears to have had a horror of the term *hypothesis,* which probably arose from his acquaintance with the rash and illicit general assumptions of Descartes. Thus in the passage just quoted, after declaring that gravity must have some other cause than matter, he says, " Later philosophers banish the consideration of such a cause out of Natural Philosophy, feigning hypo-

---

* *Optics,* Qv. 31, near the end.　　　　　† Qu. 28.

theses for explaining all things mechanically, and referring other causes to metaphysics." In the celebrated Scholium at the end of the *Principia*, he says, "Whatever is not deduced from the phenomena, is to be termed *hypothesis;* and hypotheses, whether metaphysical or physical, or occult causes, or mechanical, have no place in experimental philosophy. In this philosophy, propositions are deduced from phenomena, and rendered general by induction." And in another place, he arrests the course of his own suggestions, saying, "Verum hypotheses non fingo." I have already attempted to show that this is, in reality, a superstitious and self-destructive spirit of speculation. Some hypotheses are necessary, in order to connect the facts which are observed; some new principle of unity must be applied to the phenomena, before induction can be attempted. What is requisite is, that the hypothesis should be close to the facts, and not connected with them by other arbitrary and untried facts; and that the philosopher should be ready to resign it as soon as the facts refuse to confirm it. We have seen in the History, that it was by such a use of hypotheses, that both Newton himself, and Kepler, on whose discoveries those of Newton were based, made their discoveries. The suppositions of a force tending to the sun and varying inversely as the square of the distance; of a mutual force between all the bodies of the solar system; of the force of each body arising from the attraction of all its parts; not to mention others, also propounded by Newton,—were all hypotheses before they were verified as theories. It is related that when Newton was asked how it was that he saw into the laws of nature so much further than other men, he replied, that if it were so, it resulted from his keeping his thoughts steadily occupied upon the subject which was to be thus penetrated. But what is this occupation of the thoughts, if it be not the process of keeping the phenomena clearly in view, and trying, one after another, all the plausible hypotheses which seem likely to connect them, till at last the true law is discovered? Hypotheses so used are a necessary element of discovery.

4. With regard to the details of the process of discovery, Newton has given us some of his views, which are well worthy of notice, on account of their coming from him; and which are real additions to the philosophy of this subject. He speaks repeatedly of the *analysis* and *synthesis* of observed facts; and thus marks certain steps in scientific research, very important, and not, I think, clearly pointed out by his predecessors. Thus he says*, "As in Mathematics, so in Natural Philosophy, the investigation of difficult things by the method of analysis ought ever to precede the method of composition. This analysis consists in making experiments and observations, and in drawing general conclusions from them by induction, and admitting of no objections against the conclusions, but such as are taken from experiments or other certain truths. And although the arguing from experiments and observations by induction be no demonstration of general conclusions; yet it is the best way of arguing which the nature of things admits of, and may be looked upon as so much the stronger, by how much the induction is more general." And he then observes, as we have quoted above, that by this way of analysis we proceed from compounds to ingredients, from motions to forces, from effects to causes, and from less to more general causes. The *analysis* here spoken of includes the steps which in this work we call the *decomposition* of facts, the exact *observation* and *measurement* of the phenomena, and the *colligation* of facts; the necessary intermediate step, the *selection* and *explication* of the appropriate conception, being passed over, in the fear of

* *Op.*, Qu. 31.

seeming to encourage the fabrication of hypotheses. The *synthesis* of which Newton here speaks consists of those steps of *deductive reasoning*, proceeding from the conception once assumed, which are requisite for the comparison of its consequences with the observed facts. This statement of the process of research, is, as far as it goes, perfectly exact.

5. In speaking of Newton's precepts on the subject, we are naturally led to the celebrated "Rules of Philosophizing," inserted in the second edition of the *Principia*. These rules have generally been quoted and commented on with an almost unquestioning reverence. Such Rules, coming from such an authority, cannot fail to be highly interesting to us; but at the same time, we cannot here evade the necessity of scrutinizing their truth and value, according to the principles which our survey of this subject has brought into view. The Rules stand at the beginning of that part of the *Principia* (the Third Book) in which he infers the mutual gravitation of the sun, moon, planets, and all parts of each. They are as follows:

"Rule I. We are not to admit other causes of natural things than such as both are true, and suffice for explaining their phenomena.

"Rules II. Natural effects of the same kind are to be referred to the same causes, as far as can be done.

"Rule III. The qualities of bodies which cannot be increased or diminished in intensity, and which belong to all bodies in which we can institute experiments, are to be held for qualities of all bodies whatever.

"Rule IV. In experimental philosophy, propositions collected from phenomena by induction, are to be held as true either accurately or approximately, notwithstanding contrary hypotheses; till other phenomena occur by which they may be rendered either more accurate or liable to exception."

In considering these Rules, we cannot help remarking, in the first place, that they are constructed with an intentional adaptation to .the case with which Newton has to deal,—the induction of Universal Gravitation; and are intended to protect the reasonings before which they stand. Thus the first Rule is designed to strengthen the inference of gravitation from the celestial phenomena, by describing it as a *vera causa*, a true cause; the second countenances the doctrine that the planetary motions are governed by mechanical forces, as terrestrial motions are; the third rule appears intended to justify the assertion of gravitation, as a *universal* quality of bodies; and the fourth contains, along with a general declaration of the authority of induction, the author's usual protest against hypotheses, levelled at the Cartesian hypotheses especially.

6. *Of the First Rule.*—We, however, must consider these Rules in their general application, in which point of view they, have often been referred to, and have had very great authority allowed them. One of the points which has been most discussed, is that maxim which requires that the causes of phenomena which we assign should be true causes, *veræ causæ*. Of course this does not mean that they should be *the* true or right cause; for although it is the philosopher's aim to discover such causes, he would be little aided in his search of truth, by being told that it is truth which he is to seek. The rule has generally been understood to prescribe that in attempting to account for any class of phenomena, we must assume such causes only, as *from other considerations*, we know to exist. Thus gravity, which was employed in explaining the motions of the moon and planets, was already known to exist and operate at the earth's surface.

Now the Rule thus interpreted is, I conceive, an injurious limitation of the field of induction. For it forbids us to look for a cause, except among the causes

with which we are already familiar.   But if we follow this
rule, how shall we ever become acquainted with any new
cause?   Or how do we know that the phenomena which
we contemplate do really arise from some cause which
we already truly know?   If they do not, must we still
insist upon making them depend upon some of our known
causes; or must we abandon the study of them altogether?   Must we, for example, resolve to refer the action
of radiant heat to the air, rather than to any peculiar
fluid or ether, because the former is known to exist, the
latter is merely assumed for the purpose of explanation?
But why should we do this?   Why should we not endeavour to learn the cause from the effects, even if it be not
already known to us?   We can infer causes, which are
new when we first became acquainted with them.   Chemical Forces, Optical Forces, Vital Forces, are known to
us only by chemical and optical and vital phenomena; must
we, therefore, reject their existence or abandon their
study?   They do not conform to the double condition,
that they shall be sufficient and *also* real : they are true,
only so far as they explain the facts, but are they, therefore, unintelligible or useless?   Are they not highly important and instructive subjects of speculation?   And if
the gravitation which rules the motions of the planets
had not existed at the earth's surface;—if it had been
there masked and concealed by the superior effect of
magnetism, or some other extraneous force, might not
Newton still have inferred, from Kepler's laws, the tendency of the planets to the sun ; and from their perturbations, their tendency to each other?   His discoveries
would still have been immense, if the cause which he
assigned had not been a *vera causa* in the sense now
contemplated.

7. But what do we mean by calling gravity a " true
cause?"   How do we learn its reality?   Of course, by its

effects, with which we are familiar;—by the weight and fall of bodies about us. These strike even the most careless observer. No one can fail to see that all bodies which we come in contact with are heavy;—that gravity acts in our neighbourhood here upon earth. Hence, it may be said, this cause is at any rate a true cause, whether it explains the celestial phenomena or not.

But if this be what is meant by a *vera causa*, it appears strange to require that in all cases we should find such a one to account for all classes of phenomena. Is it reasonable or prudent to demand that we shall reduce every set of phenomena, however minute, or abstruse, or complicated, to causes so obviously existing as to strike the most incurious, and to be familiar among men? How can we expect to find *such veræ causæ* for the delicate and recondite phenomena which an exact and skilful observer detects in chemical, or optical, or electrical experiments? The facts themselves are too fine for vulgar apprehension; their relations, their symmetries, their measures require a previous discipline to understand them. How then can their causes be found among those agencies with which the common unscientific herd of mankind are familiar? What likelihood is there that causes held for real by such persons, shall explain facts which such persons cannot see or cannot understand?

Again: if we give authority to such a rule, and require that the causes by which science explains the facts which she notes and measures and analyses, shall be causes which men, without any special study, have already come to believe in, from the effects which they casually see around them, what is this, except to make our first rude and unscientific persuasions the criterion and test of our most laborious and thoughtful inferences? What is it, but to give to ignorance and thoughtlessness the right of pronouncing upon the convictions of intense study and long disciplined thought? "Electrical atmospheres' surrounding electrized bodies, were at one time held to be a "true cause" of the effects which such bodies produce. These atmospheres, it was said, are obvious to the senses; we feel them like a spider's web on the hands and face. Æpinus had to answer such persons, by proving that there are no atmospheres, no effluvia, but only repulsion. He thus, for a *true cause* in the vulgar sense of the term, substituted an *hypothesis*; yet who doubts that what he did was an advance in the science of electricity?

8. Perhaps some persons may be disposed to say, that Newton's Rule does not enjoin us to take those causes only which we clearly know, or suppose we know, to be really existing and operating, but only causes *of such kinds* as we have already satisfied ourselves do exist in nature. It may be urged that we are entitled to infer that the planets are governed in their motions by an attractive force, because we find, in the bodies immediately subject to observation and experiment, that such motions are produced by attractive forces, for example by that of the earth. It may be said that we might on similar grounds infer forces which unite particles of chemical compounds, or deflect particles of light, because we see adhesion and deflection produced by forces.

But it is easy to show that the Rule, thus laxly understood, loses all significance. It prohibits no hypothesis; for all hypotheses suppose causes *such as*, in some case or other, we have seen in action. No one would think of explaining phenomena by referring them to forces and agencies altogether different from any which are known; for on this supposition, how could he pretend to reason about the effects of the assumed causes, or undertake to prove that they would explain the facts? Some close similarity with some known kind of cause is requisite, in order that the hypothesis may have the appearance of an

explanation.   No forces, or virtues, or sympathies, or
fluids, or ethers, would be excluded by *this* interpretation
of *veræ causæ*.   Least of all, would such an interpretation reject the Cartesian hypothesis of vortices; which
undoubtedly, as I conceive, Newton intended to condemn
by his Rule.   For that *such* a case as a whirling fluid,
carrying bodies round a centre in orbits, does occur, is too
obvious to require proof.   Every eddying stream, or blast
that twirls the dust in the road, exhibits examples of such
action, and would justify the assumption of the vortices
which carry the planets in their courses; as indeed, without doubt, such facts suggested the Cartesian explanation
of the solar system.   The vortices, in this mode of considering the subject, are at the least as *real* a cause of
motion as gravity itself.

9. Thus the Rule which enjoins "true causes," is nuga-
tory, if we take *veræ causæ* in the extended sense of any
causes of a real *kind*, and unphilosophical if we understand the term of *those very* causes which we familiarly
suppose to exist.   But it may be said that we are to
designate as "true causes," not those which are collected
in a loose, confused and precarious manner, by undisciplined minds, from obvious phenomena, but those which
are justly and rigorously inferred.   Such a cause, it may
be added, gravity is; for the facts of the downward
pressures and downward motions of bodies at the earth's
surface lead us, by the plainest and strictest induction, to
the assertion of such a force.   Now to this interpretation
of the Rule there is no objection; but then, it must be
observed, that on this view, terrestrial gravity is inferred
by the same process as celestial gravitation; and the
cause is no more entitled to be called "true," because it is
obtained from the former, than because it is obtained from
the latter class of facts.   We thus obtain an intelligible
and tenable explanation of a *vera causa ;* but then, by this

explanation its *verity* ceases to be distinguishable from its other condition, that it " suffices for the explanation of the phenomena." The assumption of universal gravitation accounts for the fall of a stone ; it also accounts for the revolutions of the Moon or of Saturn ; but since both these explanations are of the same kind, we cannot with justice make the one a criterion or condition of the admissibility of the other.

10. But still, the Rule, so understood, is so far from being unmeaning or frivolous, that it expresses one of the most important tests which can be given of a sound physical theory,   It is true, the explanation of one set of facts may be of the same nature as the explanation of the other class : but then, that the cause explains *both* classes, gives it a very different claim upon our attention and assent from that which it would have if it explained one class only.   The very circumstance that the two explanations coincide, is a most weighty presumption in their favour.   It is the testimony of two witnesses in behalf of the hypothesis ; and in proportion as these two witnesses are separate and independent, the conviction produced by their agreement is more and more complete.   When the explanation of two kinds of phenomena, distinct and not apparently connected, leads us to the same cause, such a coincidence does give a reality to the cause, which it has not while it merely accounts for those appearances which suggested the supposition.   This coincidence of propositions inferred from separate classes of facts, is exactly what we noticed in the last Book, as one of the most decisive characteristics of a true theory, under the name of the *Consilience of Inductions.*

That Newton's First Rule of Philosophizing, so understood, authorizes the inferences which he himself made, is really the ground on which they are so firmly believed by philosophers.   Thus when the doctrine of a gravity varying inversely as the square of the distance from the body, accounted at the same time for the relations of times and distances in the planetary orbits and for the amount of the moon's deflection from the tangent of her orbit, such a doctrine became most convincing : or again, when the doctrine of the universal gravitation of all parts of matter, which explained so admirably the inequalities of the moon's motions, also gave a satisfactory account of a phenomenon utterly different, the precession of the equinoxes. And of the same kind is the evidence in favour of the undulatory theory of light, when the assumption of the length of an undulation, to which we are led by the colours of thin plates, is found to be identical with that length which explains the phenomena of diffraction; or when the hypothesis of transverse vibrations, suggested by the facts of polarization, explains also the laws of double refraction. When such a convergence of two trains of induction points to the same spot, we can no longer suspect that we are wrong. Such an accumulation of proof really persuades us that we have to do with a *vera causa*. And if this kind of proof be multiplied ;—if we again find other facts of a sort uncontemplated in framing our hypothesis, but yet clearly accounted for when we have adopted the supposition ;—we are still further confirmed in our belief; and by such accumulation of proof we may be so far satisfied, as to believe without conceiving it possible to doubt. In this case, when the validity of the opinion adopted by us has been repeatedly confirmed by its sufficiency in unforeseen cases, so that all doubt is removed and forgotten, the theoretical cause takes its place among the realities of the world, and becomes *a true cause*.

11. Newton's Rule then, to avoid mistakes, might be thus expressed; That " we may, provisorily, assume such hypothetical cause as will account for any given class of

natural phenomena; but that when two different classes of facts lead us to the same hypothesis, we may hold it to be a *true cause*." And this Rule will rarely or never mislead us. There are no instances, in which a doctrine recommended in this manner has afterwards been discovered to be false. There have been hypotheses which have explained many phenomena, and kept their ground long, and have afterwards been rejected. But these have been hypotheses which explained only one class of phenomena; and their fall took place when another kind of facts was examined and brought into conflict with the former. Thus the system of eccentrics and epicycles accounted for all the observed *motions* of the planets, and was the means of expressing and transmitting all astronomical knowledge for two thousand years. But then, how was it overthrown? By considering the *distances* as well as motions of the heavenly bodies. Here was a second class of facts; and when the system was adjusted so as to agree with the one class, it was at variance with the other. These cycles and epicycles could not be true, because they could not be made a just representation of the facts. But if the measures of distance as well as of position had conspired in pointing out the cycles and epicycles, as the paths of the planets, the paths so determined could not have been otherwise than their real paths; and the epicyclical theory would have been, at least geometrically, true.

12. *Of the Second Rule.*—Newton's Second Rule directs that "natural events of the *same kind* are to be referred to the *same causes*, so far as can be done." Such a precept at first appears to help us but little; for all systems, however little solid, profess to conform to such a rule. When any theorist undertakes to explain a class of facts, he assigns causes which according to him, will by their natural action, as seen in other cases, produce the effects in question. The events which he accounts for by

his hypothetical cause, are, he holds, of the same kind as those which such a cause is known to produce. Kepler, in ascribing the planetary motions to magnetism, Descartes, in explaining them by means of vortices, held that they were referring celestial motions to the causes which give rise to terrestrial motions of the same kind. The question is, *Are* the effects of the same kind? This once settled, there will be no question about the propriety of assigning them to the same cause. But the difficulty is, to determine *when* events are of the same kind, Are the motions of the planets of the same kind with the motion of a body moving freely in a curvilinear path, or do they not rather resemble the motion of a floating body swept round by a whirling current? The Newtonian and the Cartesian answered this question differently. How then can we apply this Rule with any advantage?

13. To this we reply, that there is no way of escaping this uncertainty and ambiguity, but by obtaining a clear possession of the ideas which our hypothesis involves, and by reasoning rigorously from them. Newton asserts that the planets move in free paths, acted on by certain forces. The most exact calculation gives the closest agreement of the results of this hypothesis with the facts. Descartes asserts that the planets are carried round by a fluid. The more rigorously the conceptions of force and the laws of motion are applied to this hypothesis, the more signal is its failure in reconciling the facts to one another. Without such calculation we can come to no decision between the two hypotheses. If the Newtonian hold that the motions of the planets are *evidently* of the *same kind* as those of a body describing a curve in free space, and therefore, like that, to be explained by a force acting upon the body; the Cartesian denies that the planets do move in free space. They are, he maintains, immersed in a plenum. It is only when it appears that comets

pass through this plenum in all directions with no impediment, and that no possible form and motion of its whirlpools can explain the forces and motions which are observed in the solar system, that he is compelled to allow the Newtonian's classification of events of the *same kind*.

Thus it does not appear that this Rule of Newton can be interpreted in any distinct and positive manner, otherwise than as enjoining that, in the task of induction, we employ clear ideas, rigorous reasoning, and close and fair comparison of the results of the hypothesis with the facts. These are, no doubt, important and fundamental conditions of a just induction; but in this injunction we find no peculiar or technical criterion by which we may satisfy ourselves that we are right, or detect our errors. Still, of such general prudential rules, none can be more wise than one which thus, in the task of connecting facts by means of ideas, recommends that the ideas be clear, the facts correct, and the chain of reasoning which connects them without a flaw.

14. *Of the Third Rule.*—The Third Rule, that " qualities which are observed without exception be held to be universal, " as I have already said, seems to be intended to authorize the assertion of gravitation as a universal attribute of matter. We formerly stated, in treating of Mechanical Ideas*, that this application of such a Rule appears to be a mode of reasoning far from conclusive. The assertion of the universality of any property of bodies must be grounded upon the reason of the case, and not upon any arbitrary maxim. Is it intended by this Rule to prohibit any further examination how far gravity is an original property of matter, and how far it may be resolved into the result of other agencies? We know perfectly well that this was not Newton's intention;

---

* Book iii. c. 10.

since the cause of gravity was a point which he proposed to himself as a subject of inquiry. It would certainly be very unphilosophical to pretend, by this Rule of Philosophizing, to prejudge the question of such hypotheses as that of Mosotti, That gravity is the excess of the electrical attraction over electrical repulsion: and yet to adopt this hypothesis, would be to suppose electrical forces more truly universal than gravity; for according to the hypothesis, gravity, being the inequality of the attraction and repulsion, is only an accidental and partial relation of these forces. Nor would it be allowable to urge this Rule as a reason of assuming that double stars are attracted to each other by a force varying according to the inverse square of the distance; without examining, as Herschel and others have done, the orbits which they really describe. But if the Rule is not available in such cases, what is its real value and authority? and in what cases are they exemplified?

15. In a former part of this work*, it was shown that the fundamental laws of motion, and the properties of matter which these involve, are, after a full consideration of the subject, unavoidably assumed as universally true. It was further shown, that although our knowledge of these laws and properties be gathered from experience, we are strongly impelled, some philosophers think authorized, to look upon these as not only universally, but necessarily true. It was also stated, that the law of gravitation, though its universality may be deemed probable, does not apparently involve the same necessity as the fundamental laws of motion. But it was pointed out that these are some of the most abstruse and difficult questions of the whole of philosophy; involving the profound, perhaps insoluble, problem of the identity or diversity of ideas and things. It cannot, therefore, be deemed

* Book iii. c. 9, 10, 11.

philosophical to cut these Gordian knots by peremptory maxims, which encourage us to decide without rendering a reason. Moreover, it appears clear that the reason which is rendered for this Rule by the Newtonians is quite untenable; namely, that we know extension, hardness, and inertia, to be universal qualities of bodies by experience alone, and that we have the same evidence of experience for the universality of gravitation. We have already observed that we cannot, with any propriety, say that we *find* by experience all bodies are extended. This could not be a just assertion, except we could conceive the possibility of our finding the contrary. But who can conceive our finding by experience some bodies which are not extended? It appears, then, that the reason given for the Third Rule of Newton involves a mistake respecting the nature and authority of experience. And the Rule itself cannot be applied without attempting to decide by the casual limits of observation, questions which necessarily depend upon the relations of ideas.

16. *Of the Fourth Rule.*—Newton's Fourth Rule is, that "Propositions collected from phenomena by induction, shall be held to be true, notwithstanding contrary hypotheses; but shall be liable to be rendered more accurate, or to have their exceptions pointed out, by additional study of phenomena." This Rule contains little more than a general assertion of the authority of induction, accompanied by Newton's usual protest against hypotheses.

The really valuable part of the Fourth Rule is that which implies that a constant verification, and, if necessary, rectification, of truths discovered by induction, should go on in the scientific world. Even when the law is, or appears to be, most certainly exact and universal, it should be constantly exhibited to us afresh in the form of experience and observation. This is necessary, in order to

discover exceptions and modifications if such exist; and
if the law be rigorously true, the contemplation of it, as
exemplified in the world of phenomena, will best give us
that clear apprehension of its bearings which may lead us
to see the ground of its truth.

The concluding clause of this Fourth Rule appears, at
first, to imply that all inductive propositions are to be
considered as merely provisional and limited, and never
secure from exception. But to judge thus would be to
underrate the stability and generality of scientific truths;
for what man of science can suppose that we shall hereafter discover exceptions to the universal gravitation of
all parts of the solar system? And it is plain that the
author did not intend the restriction to be applied so
rigorously; for in the Third Rule, as we have just seen,
he authorizes us to infer universal properties of matter
from observation, and carries the liberty of inductive
inference to its full extent. The Third Rule appears to
encourage us to assert a law to be universal, even in cases
in which it has not been tried; the Fourth Rule seems
to warn us that the law may be inaccurate, even in cases
in which it has been tried. Nor is either of these suggestions erroneous; but both the universality and the
rigorous accuracy of our laws are proved by reference to
Ideas rather than to Experience; a truth which, perhaps,
the philosophers of Newton's time were somewhat disposed to overlook.

17. The disposition to ascribe all our knowledge to
Experience, appears in Newton and the Newtonians by
other indications; for instance, it is seen in their extreme
dislike to the ancient expressions by which the principles
and causes of phenomena were described, as the *occult
causes* of the Schoolmen, and the *forms* of the Aristotelians,
which had been adopted by Bacon. Newton says*, that

* *Optics*, Qu. 31.

the particles of matter not only possess inertia, but also active principles, as gravity, fermentation, cohesion; he adds, " These principles I consider not as Occult Qualities, supposed to result from the Specific Forms of things, but as General Laws of Nature, by which the things themselves are formed: their truth appearing to us by phenomena, though their causes be not yet discovered. For these are manifest qualities, and their causes only are occult. And the Aristotelians gave the name of *occult qualities*, not to manifest qualities, but to such qualities only as they supposed to lie hid in bodies, and to the unknown causes of manifest effects: such as would be the causes of gravity, and of magnetick and electrick attractions, and of fermentations, if we should suppose that these forces or actions arose from qualities unknown to us, and incapable of being discovered and made manifest. Such occult qualities put a stop to the improvement of Natural Philosophy, and therefore of late years have been rejected. To tell us that every species of things is endowed with an occult specific quality by which it acts and produces manifest effects, is to tell us nothing: but to derive two or three general principles of motion from phenomena, and afterwards to tell us how the properties and actions of all corporeal things follow from these manifest principles, would be a great step in philosophy, though the causes of those principles were not yet discovered: and therefore I scruple not to propose the principles of motion above maintained, they being of very general extent, and leave their causes to be found out."

18. All that is here said is highly philosophical and valuable; but we may observe that the investigation of *specific forms*, in the sense in which some writers had used the phrase, was no means a frivolous or unmeaning object of inquiry. Bacon and others had used *form* as

equivalent to *law**. If we could ascertain that arrangement of the particles of a crystal from which its external crystalline form and other properties arise, this arrangement would be the *internal form* of the crystal. If the undulatory theory be true, the *form* of light is transverse vibrations: if the emission theory be maintained, the *form* of light is particles moving in straight lines, and deflected by various forces. Both the terms, *form* and *law*, imply an ideal connexion of sensible phenomena; form supposes matter which is moulded to the form; law supposes objects which are governed by the law. The former term refers more precisely to existences, the latter to occurrences. The latter term is now the more familiar, and is, perhaps, the better metaphor: but the former also contains the essential antithesis which belongs to the subject, and might be used in expressing the same conclusions.

But occult causes, employed in the way in which Newton describes, had certainly been very prejudicial to the progress of knowledge, by stopping inquiry with a mere word. The absurdity of such pretended explanations had not escaped ridicule. The pretended physician in the comedy gives an example of an occult cause or virtue.

> Mihi demandatur
> A doctissimo Doctore
> *Quare* Opium facit dormire:

---

* *Nov. Org.*, lib. ii. Aph. 2. Licet enim in natura nihil existet præter corpora individua, edentia actus puros individuos ex lege; in doctrinis tamen illa ipsa lex, ejusque inquisitio, et inventio, et explicatio, pro fundamento est tam ad sciendum quam ad operandum. Eam autem *legem*, ejusque *paragraphos, formarum* nomine intelligimus; præsertim cum hoc vocabulum invaluerit, et familiter occurrat.

Aph. 17. Eadem res est *forma* calidi vel *forma* luminis, et *lex* calidi aut *lex* luminis.

Et ego respondeo,
Quia est in eo
Virtus dormitiva,
Cujus natura est sensus assoupire.

19. But the most valuable part of the view presented to us in the quotation just given from Newton is the distinct separation, already noticed as peculiarly brought into prominence by him, of the determination of the *laws* of phenomena, and the investigation of their *causes.* The maxim, that the former inquiry must precede the latter, and that if the general laws of facts be discovered, the result is highly valuable, although the causes remain unknown, is extremely important; and had not, I think, ever been so strongly and clearly stated, till Newton both repeatedly promulgated the precept, and added to it the weight of the most striking examples.

We have seen that Newton, along with views the most just and important concerning the nature and methods of science, had something of the tendency, prevalent in his time, to suspect or reject, at least speculatively, all elements of knowledge except observation. This tendency was, however, in him so corrected and restrained by his own wonderful sagacity and mathematical habits, that it scarcely led to any opinion which we might not safely adopt. But we must now consider the cases in which this tendency operated in a more unbalanced manner, and led to the assertion of doctrines which, if consistently followed, would destroy the very foundations of all general and certain knowledge.

Chapter XIV.

LOCKE AND HIS FRENCH FOLLOWERS.

1. In the constant opposition and struggle of the schools of philosophy, which consider our Senses and our Ideas, respectively, as the principal sources of our knowledge, we have seen that at the period of which we now treat, the tendency was to exalt the external and disparage the internal element. The disposition to ascribe our knowledge to observation alone, had already, in Bacon's time, led him to dwell to a disproportionate degree upon that half of his subject; and had tinged Newton's expressions, though it had not biassed his practice. But this partiality soon assumed a more prominent shape, becoming extreme in Locke, and extravagant in those who professed to follow him.

Indeed Locke appears to owe his popularity and influence as a popular writer mainly to his being one of the first to express, in a plain and unhesitating manner, opinions which had for some time been ripening in the minds of a large portion of the cultivated public. Hobbes had already promulgated the main doctrines, which Locke afterwards urged, on the subject of the origin and nature of our knowledge: but in him these doctrines were combined with offensive opinions on points of morals, government, and religion, so that their access to general favour was impeded: and it was to Locke that they were indebted for the extensive influence which they soon after obtained. Locke owed this authority mainly to the intellectual circumstances of the time. Although a writer of great merit, he by no means possesses such metaphysical acuteness or such philosophical largeness of view, or such a charm of writing, as to give him the high place he has held in the literature of Europe. But he

came at a period when the reign of Ideas was tottering
to its fall. All the most active and ambitious spirits had
gone over to the new opinions, and were prepared to
follow the fortunes of the Philosophy of Experiment,
then in the most prosperous and brilliant condition, and
full of still brighter promise. There were, indeed, a few
learned and thoughtful men who still remained faithful
to the empire of Ideas; partly, it may be, from a too
fond attachment to ancient systems; but partly, also,
because they knew that there were subjects of vast importance in which experience did not form the whole
foundation of our knowledge. They knew, too, that
many of the plausible tenets of the new philosophy were
revivals of fallacies which had been discussed and refuted
in ancient times. But the advocates of mere experience
came on with a vast store of weighty truth among their
artillery, and with the energy which the advance usually
bestows. The ideal system of philosophy could, for the
present, make no effectual resistance; Locke, by putting
himself at the head of the assault, became the hero of his
day: and his name has been used as the watchword of
those who adhere to the philosophy of the senses up to
our own times.

2. Locke himself did not assert the exclusive authority of the senses in the extreme unmitigated manner in
which some who call themselves his disciples have done.
But this is the common lot of the leaders of revolutions,
for they are usually bound by some ties of affection and
habit to the previous state of things, and would not
destroy all traces of that condition: while their followers
attend, not to their inconsistent wishes, but to the meaning of the revolution itself; and carry out, to their genuine and complete results, the principles which won the
victory, and which have been brought out more sharp
from the conflict. Thus Locke himself does not assert

that all our ideas are derived from sensation, but from sensation *and reflection*. But it was easily seen that, in this assertion, two very heterogeneous elements were conjoined: that while to pronounce Sensation the origin of ideas is a clear decided tenet, the acceptance or rejection of which determines the general character of our philosophy; to make the same declaration concerning Reflection is in the highest degree vague and ambiguous, since reflection may either be resolved into a mere modification of sensation, as was done by one school, or may mean all that the opposite school oppose to sensation, under the name of Ideas. Hence the clear and strong impression which fastened upon men's minds, and which does in fact represent all the systematic and consistent part of Locke's philosophy, was, that in it all our ideas are represented as derived from Sensation.

3. We need not spend much time in pointing out the inconsistencies into which Locke fell; as all must fall into inconsistences who recognize no source of knowledge except the senses. Thus he maintains that our Idea of Space is derived from the senses of sight and touch; our Idea of Solidity from the touch alone. Our Notion of Substance is an unknown support of unknown qualities, and is illustrated by the Indian fable of the tortoise which supports the elephant, which supports the world. Our Notion of Power or Cause is in like manner got from the senses. And yet, though these ideas are thus mere fragments of our experience, Locke does not hesitate to ascribe to them necessity and universality when they occur in propositions. Thus he maintains the necessary truth of geometrical properties: he asserts that the resistance arising from solidity is absolutely insurmountable*; he conceives that nothing short of Omnipotence can annihilate a particle of matter†; and he has

---

* Book xi. c. **4**, sec. 3.   † *Ib.*, c. 13, sec. 22.

no misgivings in arguing upon the axiom that Every thing must have a cause. He does not perceive that, upon his own account of the origin of our knowledge, we can have no right to make any of these assertions. If our knowledge of the truths which concern the external world were wholly derived from experience, all that we could venture to say would be,—that geometrical properties of figures are true *as far as we have tried them;*—that we have seen *no example* of a solid body being reduced to occupy less space by pressure, or of a material substance annihilated by natural means;—and that *wherever we have examined,* we have found that every change has had a cause. Experience can never entitle us to declare that what she has not seen is impossible; still less, that things which she can not see are certain. Locke himself intended to throw no doubt upon the certainty of either human or divine knowledge; but his principles, when men discarded the temper in which he applied them, and the checks to their misapplication which he conceived that he had provided, easily led to a very comprehensive scepticism. His doctrines tended to dislodge from their true bases the most indisputable parts of knowledge; as, for example, pure and mixed mathematics. It may well be supposed, therefore, that they shook the foundations of many other parts of knowledge in the minds of common thinkers.

It was not long before these consequences of the overthrow of ideas showed themselves in the speculative world. I have already in a previous part of this work * mentioned Hume's sceptical inferences from Locke's maxim, that we have no ideas except those which we acquire by experience; and the doctrines set up in opposition to this by the metaphysicians of Germany. I might trace the progress of the sensational opinions in

* Book iii. c. 3. Modern Opinions respecting the Idea of Cause.

Britain till the reaction took place here also : but they were so much more clearly and decidedly followed out in France, that I shall pursue their history in that country.

4. *The French Followers of Locke, Condillac, &c.*—Most of the French writers who adopted Locke's leading doctrines, rejected the "Reflection," which formed an anomalous part of his philosophy, and declared that sensation alone was the source of ideas. Among these writers Condillac was the most distinguished. He expressed the leading tenet of their school in a clear and pointed manner by saying that " All ideas are tranformed sensations." We have already considered this point*, and need not here longer dwell upon it.

Opinions such as these tend to annihilate, as we have seen, one of the two co-ordinate elements of our knowledge. Yet they were far from being so prejudicial to the progress of science, or even of the philosophy of science, as might have been anticipated. One reason of this was, that they were practically corrected, especially among the cultivators of Natural Philosophy, by the study of mathematics ; for that study did really supply all that was requisite on the ideal side of science, so far as the ideas of space, time, and number, were concerned, and partly also with regard to the idea of cause and others. And the methods of discovery, though the philosophy of them made no material advance, were practically employed with so much activity, and in so many various subjects, that a certain kind of prudence and skill in this employment was very widely diffused.

5. *Importance of Language.*—In one respect this school of metaphysicians rendered a very valuable service to the philosophy of science. They brought into prominent notice the great importance of *words* and *terms*

* B. i. c. 4.

in the formation and progress of knowledge, and pointed out that the office of language is not only to convey and preserve our thoughts, but to perform the analysis in which reasoning consists. They were led to this train of speculation, in a great measure, by taking pure mathematical science as their standard example of substantial knowledge. Condillac, rejecting, as we have said, almost all those ideas on which universal and demonstrable truths must be based, was still not at all disposed to question the reality of human knowledge; but was, on the contrary, a zealous admirer of the evidence and connexion which appear in those sciences which have the ideas of space and number for their foundation, especially the latter. He looked for the grounds of the certainty and reality of the knowledge which these sciences contain; and found them, as he conceived, in the nature of the *language* which they employ. The *Signs* which are used in arithmetic and algebra enable us to keep steadily in view the identity of the same quantity under all the forms which, by composition and decomposition, it may be made to assume; and these Signs also not only express the operations which are performed, but suggest the extension of the operations according to analogy. Algebra, according to him, is only a very perfect language; and language answers its purpose of leading us to truth, by possessing the characteristics of algebra. Words are the symbols of certain groups of impressions or facts; they are so selected and applied as to exhibit the analogies which prevail among these facts; and these analogies are the truths of which our knowledge consists. " Every language is an analytical method; every analytical method is a language*;" these were the truths " alike new and simple," as he held, which he conceived that he had demonstrated. "The art of speaking, the art of writing,

* *Langue des Calculs*, p. 1.

the art of reasoning, the art of thinking, are only, at bottom, one and the same art*." Each of these operations consists in a succession of analytical operations; and words are the marks by which we are able to fix our minds upon the steps of this analysis.

7. The analysis of our impressions and notions does in reality lead to truth, not only in virtue of the identity of the whole with its parts, as Condillac held, but also in virtue of certain Ideas which govern the synthesis of our sensations, and which contain the elements of universal truths, as we have all along endeavoured to show. But although Condillac overlooked or rejected this doctrine, the importance of words, as marking the successive steps of this synthesis and analysis, is not less than he represented it to be. Every truth, once established by induction from facts, when it is become familiar under a brief and precise form of expression, becomes itself a fact; and is capable of being employed, along with other facts of a like kind, as the materials of fresh inductions. In this successive process, the term, like the cord of a fagot, both binds together the facts which it includes, and makes it possible to manage the assemblage as a single thing. On occasion of most discoveries in science, the selection of a technical term is an essential part of the proceeding. In the History of Science, we have had numerous opportunities of remarking this; and the List of technical terms given as an Index to that work, refers us, by almost every word, to one such occasion. And these terms, which thus have had so large a share in the formation of science, and which constitute its language, do also offer the means of analysing its truths, each into *its* constituent truths; and these into facts more special, till the original foundations of our most general propositions are clearly exhibited. The relations of general

* *Grammaire*, p. xxxvi.

and particular truths are most evidently represented by
the Inductive Tables given in Book XI. But each step
in each of these Tables has its proper form of expression,
familiar among the cultivators of science; and the analysis which our Tables display, is commonly performed in
men's minds, when it becomes necessary, by fixing the
attention successively upon a series of words, not upon
the lines of a Table. Language offers to the mind such
a scale or ladder as the Table offers to the eye; and as
such Tables present to us, as we have said, the Logic of
Induction, that is, the formal conditions of the soundness
of our reasoning from facts, we may with propriety say
that a just analysis of the meaning of words is an essential portion of Inductive Logic.

In saying this, we must not forget that a decomposition of general truths into ideas, as well as into facts,
belongs to our philosophy; but the point we have here
to remark, is the essential importance of words to the
latter of these processes. And this point had not ever
had its due weight assigned to it till the time of Condillac and other followers of Locke, who pursued their
speculations in the spirit I have just described. The
doctrine of the importance of terms is the most considerable addition to the philosophy of science which has been
made since the time of Bacon*.

8. *The French Encyclopædists.*—The French *Encyclo-
pédie*, published in 1751, of which Diderot and Dalembert were the editors, may be considered as representing
the leading characters of European philosophy during the
greater part of the eighteenth century. The writers in

---

* Since the selection and construction of terms is thus a matter of
so much consequence in the formation of science, it is proper that
systematic rules, founded upon sound principles, should be laid down
for the performance of this operation. Some such rules have accordingly been suggested in another part of this work.

this work belong for the most part to the school of Locke and Condillac; and we may make a few remarks upon them, in order to bring into view one or two points in addition to what we have already said of that school. The *Discours Preliminaire*, written by Dalembert, is celebrated as containing a view of the origin of our knowledge, and the connexion and classification of the sciences.

A tendency of the speculations of the Encyclopedists, as of the School of Locke in general, is to reject all ideal principles of connexion among facts, as something which experience, the only source of true knowledge, does not give. Hence all certain knowledge consists only in the recognition of the same thing under different aspects, or different forms of expression. Axioms are not the result of an original relation of ideas, but of the use, or it may be the abuse*, of words. In like manner, the propositions of Geometry are a series of modifications,—of distortions so to speak,—of one original truth; much as if the proposition were stated in the successive forms of expression presented by a language which was constantly growing more and more artificial. Several of the sciences which rest upon physical principles, that is, (says the writer) truths of experience or simple hypotheses, have only an experimental or hypothetical certainty. Impenetrability added to the idea of extent is a mystery in addition: the nature of motion is a riddle for philosophers: the metaphysical principle of the laws of percussion is equally concealed from them. The more profoundly they study the idea of matter and of the properties which represent it, the more obscure this idea becomes; the more completely does it escape them.

9. This is a very common style of reflection, even down to our own times. I have endeavoured to show that concerning the Fundamental Ideas of space, of force and

* *Disc. Prelim.*, p. viii.

resistance, of substance, external quality, and the like, we know enough to make these Ideas the grounds of certain and universal truths;—enough to supply us with axioms from which we can demonstratively reason.   If men wish for any other knowledge of the nature of matter than that which ideas, and facts conformable to ideas, give them, undoubtedly their desire will be frustrated, and they will be left in a mysterious vacancy; for it does not appear how such knowledge as they ask for could be knowledge at all.   But in reality, this complaint of our ignorance of the real nature of things proceeds from the rejection of ideas, and the assumption of the senses alone as the ground of knowledge.   " Observation and calculation are the only sources of truth :"—this is the motto of the school of which we now speak.   And its import amounts to this :—that they reject all ideas except the idea of number, and recognize the modifications which parts undergo by addition and subtraction as the only modes in which true propositions are generated.   The laws of nature are assemblages of facts : the truths of science are assertions of the identity of things which are the same. " By the avowal of almost all philosophers," says a writer of this school*, " the most sublime truths, when once simplified and reduced to their lowest terms, are converted into facts, and thenceforth present to the mind only this proposition ; the white is white, the black is black."

These statements are true in what they positively assert, but they involve error in the denial which by implication they convey.   It is true that observation and demonstration are the only sources of scientific truth ; but then, demonstration may be founded on other grounds besides the elementary properties of number.   It is true that the theory of gravitation is but the assertion of a general fact ; but this is so, not because a sound theory

* HELVETIUS *Sur l'Homme,* c. xxiii.

does not involve ideas, but because our apprehension of a fact does.

10. Another characteristic indication of the temper of the Encyclopedists and of the age to which they belong, is the importance by them assigned to those practical *Arts* which minister to man's comfort and convenience. Not only in the body of the Encyclopedia are the Mechanical Arts placed side by side with the Sciences, and treated at great length; but in the Preliminary Discourse, the preference assigned to the liberal over the mechanical Arts is treated as a prejudice*, and the value of science is spoken of as measured by its utility. "The discovery of the Mariner's Compass is not less advantageous to the human race than the explanation of its properties would be to physics.—Why should we not esteem those to whom we owe the fusee and the escapement of watches as much as the inventors of Algebra?" And in the classification of sciences which accompanies the Discourse, the labours of artisans of all kinds have a place.

This classification of the various branches of science contained in the Dissertation is often spoken of. It has for its basis the classification proposed by Bacon, in which the parts of human knowledge are arranged according to the faculties of the mind in which they originate; and these faculties are taken, both by Bacon and by Dalembert, as Memory, Reason, and Imagination. The insufficiency of Bacon's arrangement as a scientific classification is so glaring, that the adoption of it, with only superficial modifications, at the period of the Encyclopedia, is a remarkable proof of the want of original thought and real philosophy at the time of which we speak.

11. We need not trace further the opinion which derives all our knowledge from the senses in its application

* p. xiii.

to the philosophy of Science. Its declared aim is to reduce all knowledge to the knowledge of Facts; and it rejects all inquiries which involve the Idea of Cause, and similar Ideas, describing them as "metaphysical," or in some other damnatory way. It professes, indeed, to discard all Ideas; but, as we have long ago seen, some Ideas or other are inevitably included even in the simplest Facts. Accordingly the speculations of this school are compelled to retain the relations of Position, Succession, Number and Resemblance, which are rigorously ideal relations. The philosophy of Sensation, in order to be consistent, ought to reject these Ideas along with the rest, and to deny altogether the possibility of general knowledge.

When the opinions of the Sensational School had gone to an extreme length, a Reaction naturally began to take place in men's minds. Such have been the alternations of opinion, from the earliest ages of human speculation. Man may perhaps have existed in an original condition in which he was only aware of the impressions of Sense; but his first attempts to analyse his perceptions brought under his notice Ideas as a separate element, essential to the existence of knowledge. Ideas were thenceforth almost the sole subject of the study of philosophers; of Plato and his disciples, professedly; of Aristotle, and still more of the followers and commentators of Aristotle, practically. And this continued till the time of Galileo, when the authority of the Senses again began to be asserted; for it was shown by the great discoveries which were then made, that the Senses had at least some share in the promotion of knowledge. As discoveries more numerous and more striking were supplied by Observation, the world gradually passed over to the opinion that the share which had been ascribed to Ideas in the formation of real knowledge was altogether a delusion, and that Sen-

sation alone was true.   But when this was asserted as a
general doctrine, both its manifest falsity and its alarming
consequences roused men's minds, and made them recoil
from the extreme point to which they were approaching.
Philosophy again oscillated back towards Ideas; and over
a great part of Europe, in the clearest and most comprehensive minds, this regression from the dogmas of the
Sensational School is at present the prevailing movement.
We shall conclude our review by noticing a few indications of this state of things.

CHAPTER XV.

THE REACTION AGAINST THE SENSATIONAL SCHOOL.

1. WHEN Locke's *Essay* appeared, it was easily seen
that its tendency was to urge, in a much more rigorous
sense than had previously been usual, the ancient maxim
of Aristotle, adopted by the schoolmen of the middle ages,
that " nothing exists in the intellect but what has entered
by the senses."   Leibnitz expressed in a pointed manner
the limitation with which this doctrine had always been
understood.   " Nihil est in intellectu quod non prius
fuerit in sensu;—*nempe*," he added, "*nisi intellectus ipse.*"
To this it has been objected*, that we cannot say that
the intellect is *in* the intellect.   But this remark is
obviously frivolous; for the faculties of the understanding
(which are what the argument against the Sensational
School requires us to reserve) may be said to be in the
understanding, with as much justice as we may assert
there are in it the impressions derived from sense.   And
when we take account of these faculties, and of the Ideas

* See Mr. SHARPE'S *Essays*.

to which, by their operation, we necessarily subordinate our apprehension of phenomena, we are led to a refutation of the philosophy which makes phenomena, unconnected by Ideas, the source of all knowledge. The succeeding opponents of the Lockian school insisted upon and developed in various ways this remark of Leibnitz, or some equivalent view.

2. It was by inquiries into the foundations of morals that English philosophers were led to question the truth of Locke's theory. Dr. Price, in his *Review of the Principal Questions in Morals*, first published in 1757, maintained that we cannot with propriety assert all our ideas to be derived from sensation and reflection. He pointed out, very steadily, the other source *. "The power, I assert, that *understands*, or the faculty within us that discerns *truth*, and that compares all the objects of thought and *judges* of them, is a spring of new ideas." And he exhibits the antithesis in various forms. "Were not *sense* and *knowledge* entirely different, we should rest satisfied with sensible impressions, such as light, colours and sounds, and inquire no further about them, at least when the impressions are strong and vigorous: whereas on the contrary we necessarily desire some further acquaintance with them, and can never be satisfied till we have subjected them to the survey of reason. Sense presents *particular* forms to the mind, but cannot rise to any *general* ideas. It is the intellect that examines and compares the presented forms, that rises above individuals to universal and abstract ideas; and thus looks downward upon objects, takes in at one view an infinity of particulars, and is capable of discovering general truths. Sense sees only the outside of things, reason acquaints itself with their natures. Sensation is only a mode of feeling in the mind; but knowledge implies an active and vital energy in the mind."

* P. 16.   † P. 18.

3. The necessity of refuting Hume's inferences from the mere-sensation system led other writers to limit, in various ways, their assent to Locke. Especially was this the case with a number of intelligent metaphysicians in Scotland, as Reid, Beattie, Dugald Stewart and Thomas Brown. Thus Reid asserts*, "that the account which Mr. Locke himself gives of the Idea of Power cannot be reconciled to his favourite doctrine, that all our simple ideas have their origin from sensation or reflection." Reid remarks, that our memory and our reasoning power come in for a share in the origin of this idea: and in speaking of reasoning, he obviously assumes the axiom that every event must have a cause. By succeeding writers of this school, the assumption of the fundamental principles, to which our nature in such cases irresistibly directs us, is more clearly pointed out. Thus Stewart defends the form of expression used by Price†. "A variety of intuitive judgments might be mentioned, involving simple ideas, which it is impossible to trace to any origin but to the power which enables us to form these judgments. Thus it is surely an intuitive truth that the sensations of which I am conscious, and all those I remember, belong to one and the same being, which I call *myself*. Here is an intuitive judgment involving the simple idea of *Identity*. In like manner, the changes which I perceive in the universe impress me with a conviction that some cause must have operated to produce them. Here is an intuitive judgment involving the simple Idea of *Causation*. When we consider the adjacent angles made by a straight line standing upon another, and perceive that their sum is equal to two right angles, the judgment we form involves a simple idea of *Equality*. To say, therefore, that the Reason or the Understanding

* *Essays on the Powers of the Human Mind*, i. 31.
† *Outlines of Moral Phil.*, p. 138.

is a source of new ideas, is not so exceptionable a mode
of speaking as has been sometimes supposed.   According
to Locke, *Sense* furnishes our ideas, and Reason perceives
their agreements and disagreements.   But the truth is,
that these agreements and disagreements are in many
instances, simple ideas, of which no analysis can be given ;
and of which the origin must, therefore be referred to
Reason, according to Locke's own doctrine."   This view,
according to which the Reason or Understanding is the
source of certain simple ideas, such as Identity, Causation,
Equality, which ideas are necessarily involved in the
intuitive judgments which we form, when we recognize
fundamental truths of science, approaches very near in
effect to the doctrine which in this work we have presented, of Fundamental Ideas belonging to each science,
and manifesting themselves in the axioms of the science.
It may be observed, however, that by attempting to enumerate these ideas and axioms, so as to lay the foundations of the whole body of physical science; and by
endeavouring, as far as possible, to simplify and connect
each group of such Ideas; we have at least given a more
systematic form to this doctrine.   We have, moreover,
traced it into many consequences to which it necessarily
leads, but which do not appear to have been contemplated
by the metaphysicians of the Scotch school.   But I
gladly acknowledge my obligations to the writers of that
school; and I trust that in the near agreement of my
views on such points with theirs, there is ground for
believing the system of philosophy which I have in this
work presented, to be that to which the minds of thoughtful men, who have meditated on such subjects, are generally tending.

4. As a further instance that such a tendency is at
work, I may make a quotation from an eminent English
philosophical writer of another school.   "If you will be

at the pains," says Archbishop Whately*, " carefully to analyse the simplest description you hear of any transaction or state of things, you will find that the process which almost invariably takes place is, in logical language, this : that each individual *has in his mind* certain major premises or principles relative to the subject in question ;—that observation of what actually presents itself to the senses, supplies minor premises ; and that the statement given (and which is reported as a thing experienced) consists, in fact, of the *conclusions* drawn from the combinations of these premises." The major premises here spoken of are the Fundamental Ideas, and the Axioms and Propositions to which they lead ; and whatever is regarded as a fact of observation is necessarily a conclusion in which these propositions are assumed ; for these contain, as we have said, the conditions of our experience. Our experience conforms to these axioms and their consequences, whether or not the connexion be stated in a logical manner, by means of premises and a conclusion.

5. The same persuasion is also suggested by the course which the study of metaphysics has taken of late years in France. In that country, as we have seen, the Sensational System, which was considered as the necessary consequence of the revolution begun by Locke, obtained a more complete ascendancy than it did in England ; and in that country too, the reaction, among metaphysical and moral writers, when its time came, was more decided and rapid than it was among Locke's own countrymen. It would appear that M. Laromiguière was one of the first to give expression to this feeling, of the necessity of a modification of the sensational philosophy. He began by professing himself the disciple of Condillac, even while he was almost unconsciously subverting the fundamental principles of that writer. And thus, as M. Cousin justly

observes*, his opinions had the more powerful effect from being presented, not as thwarting and contradicting, but as sharing and following out the spirit of his age. M. Laromiguière's work, entitled *Essai sur les Facultés de l'Ame*, consists of lectures given to the Faculty of Letters of the Academy of Paris, in the years 1811, 1812 and 1813. In the views which these lectures present, there is much which the author has in common with Condillac. But he is led by his investigation to assert†, that it is not true that sensation is the sole fundamental element of our thoughts and our understanding. *Attention* also is requisite: and here we have an element of quite another kind. For sensation is passive; attention is active. Attention does not spring out of sensation; the passive principle is not the reason of the active principle. Activity and passivity are two facts entirely different. Nor can this activity be defined or derived; being, as the author says, a fundamental idea. The distinction is manifest by its own nature; and we may find evidence of it in the very forms of language. To *look* is more than to *see*; to *hearken* is more than to *hear*. The French language marks this distinction with respect to other senses also. "On *voit*, et l'on *regarde*; on *entend*, et l'on *ecoute*; on *sent*, et l'on *flaire*; on *goûte*, et l'on *savoure*." And thus the mere sensation, or capacity of feeling, is only the occasion on which the attention is exercised; while the attention is the foundation of all the operations of the understanding.

The reader of the former part of this work, will have seen how much we have insisted upon the activity of the mind, as the necessary basis of all knowledge. In all observation and experience, the mind is active, and by its activity apprehends all sensations in subordination to its own ideas; and thus it becomes capable of collecting

---

* *Fragmens Philosophiques*, i. 53. † *Ib.*, i. 67.

knowledge from phenomena, since ideas involve general relations and connexions, which sensations of themselves cannot involve. And thus we see that, in this respect also, our philosophy stands at that point to which the speculations of the most reflective men have of late constantly been verging.

6. M. Cousin himself, from whom we have quoted the above account of Laromiguière, shares in this tendency, and has argued very energetically and successfully against the doctrines of the Sensational School. He has made it his office once more to bring into notice among his countrymen, the doctrine of ideas as the sources of knowledge ; and has revived the study of Plato, who may still be considered as one of the great leaders of the ideal school. The large portion of M. Cousin's works which refers to questions out of the reach of our present review makes it suitable not to dwell longer upon them in this place.

7. We turn to speculations more closely connected with our present subject. M. Ampère, a French man of science, well entitled by his extensive knowledge, and large and profound views, to deal with the philosophy of the sciences, published in 1834, his *Essai sur la Philosophie des Sciences, ou Exposition analytique d'un Classification Naturelle de toutes les Connaissances Humaines.* In this remarkable work we see strong evidence of the progress of the reaction against the system which derives our knowledge from sensation only. The author starts from a maxim, that in classing the sciences, we must not only regard the nature of the objects about which each science is concerned, but also the point of view under which it considers them : that is, the *ideas* which each science involves. M. Ampère also gives briefly his views of the intellectual constitution of man ; a subject on which he had long and sedulously employed his thoughts ; and

these views are far from belonging to the Sensational School. Human thought, he says, is composed of phenomena and of conceptions. Phenomena are external, or *sensitive;* and internal, or *active.* Conceptions are of four kinds; *primitive,* as space and motion, duration and cause; *objective,* as our idea of matter and substance; *onomatic,* or those which we associate with the general terms which language presents to us; and *explicative,* by which we ascend to causes after a comparative study of phenomena. He teaches further, that in deriving ideas from sensation, the mind is not passive; but exerts an action which, when voluntary, is called *attention,* but when it is, as it often is, involuntary, may be termed *reaction.*

I shall not dwell upon the examination of these opinions*; but I may remark, that both in the recognition of conceptions as an original and essential element of the mind, and in giving a prominent place to the active function of the mind, in the origin of our knowledge, this view approaches to that which I have presented in the preceding part of this work; although undoubtedly with considerable differences.

8. The classification of the sciences which M. Ampère proposes, is founded upon a consideration of the sciences themselves; and is, the author conceives, in accordance with the conditions of natural classifications, as exhibited in Botany and other sciences. It is of a more symmetrical kind, and exhibits more steps of subordination, than that to which I have been led; it includes also practical Art as well as theoretical Science; and it is extended to moral and political as well as physical Sciences. It will not be necessary for me here to examine it in detail: but I may remark, that it is throughout a *dichotomous* division, each higher number being subdivided into two

---

* See also the vigorous critique of Locke's *Essay,* by Lemaistre, *Soirées de St. Petersbourg.*

lower ones, and so on. In this way, M. Ampère obtains sciences of the First Order, each of which is divided into two sciences of the Second, and four of the Third Order. Thus Mechanics is divided into *Cinematics, Statics, Dynamics,* and *Molecular Mechanics;* Physics is divided into *Experimental Physics, Chemistry, Stereometry,* and *Atomology;* Geology is divided into *Physical Geography, Mineralogy, Geonomy,* and *Theory of the Earth.* Without here criticizing these divisions or their principle, I may observe that *Cinematics,* the doctrine of motion without reference to the force which produces it, is a portion of knowledge which our investigation has led us also to see the necessity of erecting into a separate science; and which we have termed *Pure Mechanism.* Of the divisions of Geology, *Physical Geography,* especially as explained by M. Ampère, is certainly a part of the subject, both important and tolerably distinct from the rest. *Geonomy* contains what we have termed in the History, *Descriptive Geology;*— the exhibition of the facts separate from the inquiry into their causes; while our *Physical Geology* agrees with M. Ampère's *Theory of the Earth. Mineralogy* appears to be placed by him in a different place from that which it occupies in our scheme: but in fact, he uses the term for a different science;—he applies it to the classification not of *simple minerals,* but of *rocks,* which is a science auxiliary to geology, and which has sometimes been called *Petralogy.* What we have termed *Mineralogy,* M. Ampère unites with *Chemistry.* "It belongs," he says*, "to Chemistry, and not to Mineralogy, to inquire how many atoms of silicium and of oxygen compose silica; to tell us that its primitive form is a rhombohedron of certain angles, that it is called *quartz,* &c.: leaving, on one hand, to Molecular Geometry the task of explaining the different secondary forms which may result from this primitive

* P. 210.

form; and on the other hand, leaving to Mineralogy the office of describing the different varieties of quartz, and the rocks in which they occur, according as the quartz is crystallized, transparent, coloured, amorphous, solid, or in sand." But we may remark, that by adopting this arrangement, we separate from Mineralogy, almost all the knowledge, and absolutely all the general knowledge, which books professing to treat of that science have usually contained. The consideration of Mineralogical Classifi- cations, which, as may be seen in the History of Science, is so curious and instructive, is forced into the domain of Chemistry, although many of the persons who figure in it were not at all properly chemists. And we lose, in this way, the advantage of that peculiar office which, in our arrangement, Mineralogy fills; of forming a rigorous transition from the sciences of classification to those which consider the mathematical properties of bodies; and connecting the external characters and the internal constitution of bodies by means of a system of important general truths. I conceive, therefore, that our disposition of this science, and our mode of applying the name, are far more convenient than those of M. Ampère.

9. We have seen the reaction against the pure sensa- tional doctrines operating very powerfully in England and in France. But it was in Germany that these doc- trines were most decidedly rejected; and systems in extreme opposition to these put forth with confidence, and received with applause. Of the authors who gave this impulse to opinions in that country, Kant was the first, and by far the most important. I have already endeavoured to explain how he was roused, by the scep- ticism of Hume, to examine wherein the fallacy lay which appeared to invalidate all reasonings from effect to cause; and how this inquiry terminated in a conviction that the foundations of our reasonings on this and similar

points were to be sought in the mind, and not in the phenomena ;—in the *subject* and not in the *object*. The revolution in the customary mode of contemplating human knowledge which Kant's opinions involved, was most complete. He himself, with no small justice, compares* it with the change produced by Copernicus's theory of the solar system. "Hitherto," he says, "men have assumed that all our knowledge must be regulated by the objects of it; yet all attempts to make out anything concerning objects *à priori* by means of our conceptions," (as for instance their geometrical properties) "must, on this foundation, be unavailing. Let us then try whether we cannot make out something more in the problems of metaphysics, by assuming that objects must be regulated by our knowledge, since this agrees better with that supposition, which we are prompted to make, that we can know something of them *à priori*. This thought is like that of Copernicus, who, when he found that nothing was to be made of the phenomena of the heavens so long as everything was supposed to turn about the spectator, tried whether the matter might not be better explained if he made the spectator turn, and left the stars at rest. We may make the same essay in metaphysics, as to what concerns our intuitive knowledge respecting objects. If our apprehension of objects must be regulated by the properties of the objects, I cannot comprehend how we can possibly know anything about them *à priori*. But if the object, as apprehended by us, be regulated by the constitution of our faculties of apprehension, I can readily conceive this possibility." From this he infers that our experience must be regulated by our conceptions.

10. This view of the nature of knowledge soon superseded entirely the doctrines of the Sensational School

* *Kritik der Reinen Vernunft*, Pref., p. xv.

amòng the metaphysicians of Germany. These philosophers did not gradually modify and reject the dogmas of
Locke and Condillac, as was done in England and
France*; nor did they endeavour to ascertain the extent
of the empire of Ideas by a careful survey of its several
provinces, as we have been doing in the previous part of
the present work. The German metaphysicians saw at
once that Ideas and Things, the Subjective and the Objective elements of our knowledge, were, by Kant's system,
brought into opposition and correlation, as equally real
and equally indispensable. Seeing this, they rushed at
once to the highest and most difficult problem of philosophy,—to determine what this correlation is;—to discover how Ideas and Things are at the same time opposite
and identical;—how the world, while it is distinct from
and independent of us, is yet, as an object of our knowledge, governed by the conditions of our thoughts. The
attempts to solve this problem, taken in the widest sense,
including the forms which it assumes in Morals, Politics,
the Arts, and Religion, as well as in the Material Sciences,
have, since that time, occupied the most profound speculators of Germany; and have given rise to a number of
systems, which, rapidly succeeding each other, have, each
in its day, been looked upon as a complete solution of
the problem. To trace the characters of these various
systems, does not belong to the business of the present
Book: my task at present is ended when I have shown,
as I have now done, how the progress of thought in the
philosophical world, followed from the earliest up to the
present time, has led to that recognition of the co-existence and joint necessity of the two opposite elements of

---

* The sensational system never acquired in Germany the ascendancy which it obtained in England and France; but I am compelled
here to pass over the history of philosophy in Germany, except so far
as it affects ourselves.

our knowledge ; and when I have pointed out processes adapted to the extension of our knowledge, which a true view of its nature has suggested or may suggest.

In the latter portion· of my task something still remains to be done, which will be the subject of the ensuing Book.

BOOK XIII.

OF METHODS EMPLOYED IN THE FORMATION OF SCIENCE.

CHAPTER I.

INTRODUCTION.

1. IN the last Book but one of this work, we pointed out certain general Characters of scientific knowledge which may often serve to distinguish it from opinions of a looser or vaguer kind. In the last Book we traced the steps by which men were led to a perception, more or less clear, of those characteristics; and in the course of this review, we had to consider various precepts and maxims offered by philosophers as fitted to guide us in the pursuit of exact and general truths. Other contributions of the same kind to the philosophy of science might be noticed, and some which contain more valuable suggestions, and indicate a more practical acquaintance with the subject than any which have yet been quoted. Among these, I must especially distinguish Sir John Herschel's *Discourse on the Study of Natural Philosophy.* But my object in this work is not so much to relate the history, as to present the really valuable results of preceding labours. I shall, therefore, proceed no further with the criticism of other authors; but shall endeavour to collect, both from them and from my own researches and reflections, such views and such rules as seem best

adapted to assist us in the discovery and recognition of scientific truth; or, at least, such as may enable us to understand the process by which this truth is obtained. We would present to the reader the Philosophy and, if possible, the Art, of Discovery.

2. But, in truth, we must acknowledge, before we proceed with this subject, that, speaking with strictness, an *Art of Discovery* is not possible;—that we can give no Rules for the pursuit of truth which shall be universally and peremptorily applicable;—and that the helps which we can offer to the inquirer in such cases are limited and precarious. Still we trust it will be found that aids may be pointed out which are neither worthless nor uninstructive. The mere classification of examples of successful inquiry, to which our rules give occasion, is full of interest for the philosophical speculator. And if our maxims direct the discoverer to no operations which might not have occurred of themselves, they may still concentrate our attention on that which is most important and characteristic in these operations, and may direct us to the best mode of insuring their success. I shall, therefore, attempt to resolve the Process of Discovery into its parts, and to give an account as distinct as may be of Rules and Methods which belong to each portion of the process.

3. In the Eleventh Book we considered the three main parts of the process by which science is constructed: namely, the Decomposition and Observation of Complex Facts; the Explication of our Ideal Conceptions; and the Colligation of Elementary Facts by means of those Conceptions. The first and last of these three steps are capable of receiving additional accuracy by peculiar processes. They may further the advance of science in a more effectual manner when directed by special technical *Methods*, of which in the present Book we must give a

brief view. In this more technical form, the observation of facts involves the *Measurement of Phenomena;* and the Colligation of Facts includes all arts and rules by which the process of Induction can be assisted. Hence we shall have here to consider *Methods of Observation,* and *Methods of Induction,* using these phrases in the widest sense. The second of the three steps above mentioned, the Explication of our Conceptions, does not admit of being much assisted by methods, although something may be done by Education and Discussion.

4. The Methods of Induction, of which we have to speak, apply only to the first step in our ascent from phenomena to laws of nature;—the discovery of *Laws of Phenomena.* A higher and ulterior step remains behind, and follows in natural order the discovery of Laws of Phenomena; namely, the *Discovery of Causes;* and this must be stated as a distinct and essential process in a complete view of the course of science. Again, when we have thus ascended to the causes of phenomena and of their laws, we can often reason downwards from the cause so discovered; and we are thus led to suggestions of new phenomena, or to new explanations of phenomena already known. Such proceedings may be termed *Applications* of our Discoveries; including in the phrase, *Verifications* of our Doctrines by such an application of them to observed facts. Hence we have the following series of processes concerned in the formation of science.

    (1.) Decomposition of Facts;

    (2.) Measurement of Phenomena;

    (3.) Explication of Conceptions;

    (4.) Induction of Laws of Phenomena;

    (5.) Induction of Causes;

    (6.) Application of Inductive Discoveries.

5. Of these six processes, the methods by which the second and fourth may be assisted are here our peculiar

object of attention. The treatment of these subjects in the present work must necessarily be scanty and imperfect, although we may perhaps be able to add something to what has hitherto been systematically taught on these heads. Methods of Observation and of Induction might of themselves form an abundant subject for a treatise, and hereafter probably will do so, in the hands of future writers. A few remarks, offered as contributions to this subject, may serve to show how extensive it is, and how much more ready it now is than it ever before was, for a systematic discussion.

Of the above steps of the formation of science, the first, the Decomposition of Facts, has already been sufficiently explained in the Eleventh Book: for if we pursue it into further detail and exactitude, we find that we gradually trench upon some of the succeeding parts. I, therefore, proceed to treat of the second step, the Measurement of Phenomena;—of *methods* by which this work, in its widest sense, is executed, and these I shall term Methods of Observation.

CHAPTER II.

OF METHODS OF OBSERVATION.

1. I SHALL speak, in this chapter, of Methods of exact and systematic observation, by which such facts are collected as form the materials of precise scientific propositions. These Methods are very various, according to the nature of the subject inquired into, and other circumstances: but a great portion of them agree in being processes of measurement. These I shall peculiarly consider: and in the first place those referring to Number, Space, and Time, which are at the same time objects and instruments of measurement.

2. But though we have to explain how observations may be made as perfect as possible, we must not forget that in most cases complete perfection is unattainable. *Observations are never perfect.* For we observe phenomena by our senses, and measure their relations in time and space; but our senses and our measures are all, from various causes, inaccurate. If we have to observe the exact place of the moon among the stars, how much of instrumental apparatus is necessary! This apparatus has been improved by many successive generations of astronomers, yet it is still far from being perfect. And the senses of man, as well as his implements, are limited in their exactness. Two different observers do not obtain precisely the same measures of the time and place of a phenomenon; as, for instance, of the moment at which the moon occults a star, and the point of her *limb* at which the occultation takes place. Here, then, is a source of inaccuracy and error, even in astronomy, where the means of exact observation are incomparably more complete than they are in any other department of human research. In other cases, the task of obtaining accurate measures is far more difficult. If we have to observe the tides of the ocean when rippled with waves, we can see the average level of the water first rise and then fall; but how hard is it to select the exact moment when it is at its greatest height, or the exact highest point which it reaches! It is very easy, in such a case, to err by many minutes in time, and by several inches in space.

Still, in many cases, good Methods can remove very much of this inaccuracy, and to these we now proceed.

3. (I.) *Number.*—Number is the first step of measurement, since it measures itself, and does not, like space and time, require an arbitrary standard. Hence the first exact observations, and the first advances of rigorous knowledge, appear to have been made by means of number; as for

example,—the number of days in a month and in a year;
—the cycles according to which eclipses occur;—the
number of days in the revolutions of the planets; and
the like. All these discoveries, as we have seen in the
History of Astronomy, go back to the earliest period of
the science, anterior to any distinct tradition; and these
discoveries presuppose a series, probably a very long series,
of observations, made principally by means of number.
Nations so rude as to have no other means of exact measurement, have still systems of numeration by which they
can reckon to a considerable extent. Very often, such
nations have very complex systems, which are capable of
expressing numbers of great magnitude. Number supplies
the means of measuring other quantities, by the assumption of a *unit* of measure of the appropriate kind: but
where nature supplies the unit, number is applicable directly and immediately. Number is an important element
in the Classificatory as well as in the Mathematical
Sciences. The History of those Sciences shows how the
formation of botanical systems was effected by the adoption of number as a leading element by Cæsalpinus; and
how afterwards the Reform of Linnæus in classification depended in a great degree on his finding, in the
pistils and stamens, a better numerical basis than those
before employed. In like manner, the number of rays
in the membrane of the gills*, and the number of rays in
the fins of fish, were found to be important elements in
ichthyological classification by Artedi and Linnæus.
There are innumerable instances, in all parts of Natural
History, of the importance of the observation of number.
And in this observation, no instrument, scale or standard
is needed, or can be applied; except the scale of natural
numbers, expressed either in words or in figures, can be
considered as an instrument.

* *Hist. Ind. Sci.*, iii. 364—365.

4. (II.) *Measurement of Space.*—Of quantities admitting of *continuous* increase and decrease, (for number is discontinuous,) space is the most simple in its mode of measurement, and requires most frequently to be measured. The obvious mode of measuring space is by the repeated application of a material measure, as when we take a foot-rule and measure the length of a room. And in this case the foot-rule is the *unit* of space, and the length of the room is expressed by the number of such units which it contains: or, as it may not contain an exact number, by a number with a *fraction*. But besides this measurement of linear space, there is another kind of space which, for purposes of science, it is still more important to measure, namely, angular space. The visible heavens being considered as a sphere, the portions and paths of the heavenly bodies are determined by drawing circles on the surface of this sphere, and are expressed by means of the parts of these circles thus intercepted: by such measures the doctrines of astronomy were obtained in the very beginning of the science. The arcs of circles thus measured, are not like linear spaces, reckoned by means of an *arbitrary* unit; for there is a *natural unit*, the total circumference, to which all arcs may be referred. For the sake of convenience, the whole circumference is divided into 360 parts or *degrees ;* and by means of these degrees and their parts, all arcs are expressed. The *arcs* are the measures of the *angles at the centre,* and the degrees may be considered indifferently as measuring the one or the other of these quantities.

5. In the History of Astronomy*, I have described the method of observation of celestial angles employed by the Greeks. They determined the lines in which the heavenly bodies were seen, by means either of Shadows, or of Sights; and measured the angles between such lines

* *Hist. Ind. Sci.,* i. 197.

by arcs or rules properly applied to them. The Armill, Astrolabe, Dioptra, and Parallactic Instrument of the ancients were some of the instruments thus constructed. Tycho Brahe greatly improved the methods of astronomical observation by giving steadiness to the frame of his instruments, (which were large *quadrants*,) and accuracy to the divisions of the *limb**. But the application of the *telescope* to the astronomical quadrant and the fixation of the centre of the field by a *cross* of fine wires placed in the focus, was an immense improvement of the instrument, since it substituted a precise visual ray, pointing to the star, instead of the coarse coincidence of Sights. The accuracy of observation was still further increased by applying to the telescope a *micrometer* which might subdivide the smaller divisions of the arc.

6. By this means, the precision of astronomical observation was made so great, that very minute angular spaces could be measured : and it then became a question whether discrepancies which appeared at first as defects in the theory, might not arise sometimes from a bending or shaking of the instrument, and from the degrees marked on the limb being really somewhat unequal, instead of being rigorously equal. Accordingly, the framing and balancing of the instrument, so as to avoid all possible tremor or flexure, and the exact division of an arc into equal parts, became great objects of those who wished to improve astronomical observations. The observer no longer gazed at the stars from a lofty tower, but placed his telescope on the solid ground, and braced and balanced it with various contrivances. Instead of a quadrant, an entire circle was introduced (by Ramsden ;) and various processes were invented for the dividing of instruments. Among these we may notice Troughton's method of dividing ; in which the visual ray of a microscope was substi-

* *Hist. Ind. Sci.*, ii. 267.

tuted for the points of a pair of compasses, and, by *stepping* round the circle, the partial arcs were made to bear their exact relation to the whole circumference.

7. Astronomy is not the only science which depends on the measurement of angles. Crystallography also requires exact measures of this kind; and the *goniometer*, especially that devised by Wollaston, supplies the means of obtaining such measures. The science of Optics also, in many cases, requires the measurement of angles.

8. In the measurement of linear space, there is no natural standard which offers itself. Most of the common measures appear to be taken from some part of the human body; as a *foot*, a *cubit*, a *fathom*; but such measures cannot possess any precision, and are altered by convention: thus there were in ancient times many kinds of cubits; and in modern Europe, there are a great number of different standards of the foot, as the Rhenish foot, the Paris foot, the English foot. It is very desirable that, if possible, some permanent standard, founded in nature, should be adopted; for the conventional measures are lost in the course of ages; and thus, dimensions expressed by means of them become unintelligible. Two different natural standards have been employed in modern times: the French have referred their measures of length to the total circumference of a meridian of the earth; a quadrant of this meridian consists of ten million units or *metres*. The English have fixed their linear measure by reference to the length of a pendulum which employs an exact second of time in its small osillation. Both these methods occasion considerable difficulties in carrying them into effect; and are to be considered mainly as means of recovering the standard if it should ever be lost. For common purposes, some material standard is adopted as authority for the time: for example, the standard which in England possessed legal authority up to the year 1835 was pre-

served in the House of Parliament; and was lost in the conflagration which destroyed that edifice. The standard of length now generally referred to by men of science in England is that which is in the possession of the Astronomical Society of London.

9. A standard of length being established, the artifices for applying it, and for subdividing it in the most accurate manner, are nearly the same as in the case of measures of arcs: as for instance, the employment of the visual rays of microscopes instead of the legs of compasses and the edges of rules; the use of micrometers for minute measurements; and the like. Many different modes of avoiding error in such measurements have been devised by various observers, according to the nature of the cases with which they had to deal*.

10. (III.) *Measurement of Time.*—The methods of measuring Time are not so obvious as the methods of measuring space; for we cannot apply one portion of time to another, so as to test their equality. We are obliged to begin by assuming some change as the measure of time. Thus the motion of the sun in the sky, or the length and position of the shadows of objects, were the first modes of measuring the parts of the day. But what assurance had men, or what could they have, that the motion of the sun or of the shadow was uniform? They could have no such assurance, till they had adopted some measure of smaller times; which smaller times, making up larger times by repetition, they took as the standard of uniformity;— for example, an hour-glass, or a clepsydra which answered the same purpose among the ancients. There is no apparent reason why the successive periods measured by the emptying of the hour-glass should be unequal; they are

* On the precautions employed in astronomical instruments for the measure of space, see SIR J. HERSCHEL's *Astronomy*, (in the *Cabinet Cyclopædia*,) Art. 103—110.

implicitly accepted as equal; and by reference to these, the uniformity of the sun's motion may be verified. But the great improvement in the measurement of time was the use of a pendulum for the purpose by Galileo, and the application of this device to clocks by Huyghens in 1656. For the successive oscillations of a pendulum are rigorously equal, and a clock is only a train of machinery employed for the purpose of counting these oscillations. By means of this invention, the measure of time in astronomical observations became as accurate as the measure of space.

11. What is the *natural unit* of time? It was assumed from the first by the Greek astronomers, that the sidereal days, measured by the revolution of a star from any meridian to the same again, are exactly equal; and all improvements in the measure of time tended to confirm this assumption. The sidereal day is therefore the natural standard of time. But the solar day, determined by the diurnal revolution of the sun, although not rigorously invariable as the sidereal day is, undergoes scarcely any perceptible variation; and since the course of daily occurrences is regulated by the sun, it is far more convenient to seek the basis of our unit of time in *his* motions. Accordingly the solar day (the *mean* solar day) is divided into 24 hours, and these, into minutes and seconds; and this is our scale of time. Of such time, the sidereal day has 23 hours 56 minutes 4·09 seconds. And it is plain that by such a statement the length of the hour is fixed, with reference to a sidereal day. The *standard* of time (and the standard of space in like manner) equally answers its purpose, whether or not it coincides with any *whole number* of units.

12. Since the sidereal day is thus the standard of our measures of time, it becomes desirable to refer to it, constantly and exactly, the instruments by which time is

measured, in order that we may secure ourselves against error. For this purpose in astronomical observatories, observations are constantly made of the transit of stars across the meridian; the *transit instrument* with which this is done being adjusted with all imaginable regard to accuracy*.

13. When exact measures of time are required in other than astronomical observations, the same instruments are still used, namely, clocks and chronometers. In chronometers, the regulating part is an oscillating body; not, as in clocks, a pendulum oscillating by the force of gravity, but a wheel swinging to and fro on its centre, in consequence of the vibrations of a slender coil of elastic wire. To divide time into still smaller portions than these vibrations, other artifices are used; some of which will be mentioned under the next head.

14. (IV.) *Conversion of Space and Time.*—Space and time agree in being extended quantities, which are made up and measured by the repetition of homogeneous parts. If a body move uniformly, whether in the way of revolving or otherwise, the *space* which any point describes, is *proportional* to the *time* of its motion; and the space and the time may each be taken as a measure of the other. Hence in such cases, by taking space instead of time, or time instead of space, we may often obtain more convenient and precise measures, than we can by measuring directly the element with which we are concerned.

The most prominent example of such a conversion, is the measurement of the Right Ascension of stars, (that is, their angular distance from a standard meridian† on the celestial sphere,) by means of the time employed in their coming to the meridian of the place of observation.

---

* On the precautions employed in the measure of time by astronomers, see HERSCHEL's *Astron.*, Art. 115—127.

† A *meridian* is a circle passing through the poles about which the celestial sphere revolves. The meridian *of any place* on the earth is that meridian which is exactly over the place.

Since, as we have already stated, the visible celestial sphere, carrying the fixed stars, revolves with perfect uniformity about the pole; if we observe the stars as they come in succession to a fixed circle passing through the poles, the intervals of time between these observations will be proportional to the angles which the meridian circles passing through these stars make at the poles where they meet; and hence, if we have the means of measuring time with great accuracy, we can, by watching the *times* of the transits of successive stars across some visible mark in our own meridian, determine the *angular distances* of the meridian circles of all the stars from one another.

Accordingly, now that the pendulum clock affords astronomers the means of determining time exactly, a measurement of the Right Ascensions of heavenly bodies by means of a clock and a transit instrument, is a part of the regular business of an observatory. If the sidereal clock be so adjusted that it marks the beginning of its scale of time when the first point of Right Ascension is upon the visible meridian of our observatory, the point of the scale at which the clock points when any other star is in our meridian, will truly represent the Right Ascension of the star.

Thus as the motion of the stars is our measure of time, we employ time, conversely, as our measure of the places of the stars. The celestial machine and our terrestrial machines correspond to each other in their movements; and the star steals silently and steadily across our meridian line, just as the pointer of the clock steals past the mark of the hour. We may judge of the scale of this motion by considering that the full moon employs about two minutes of time in sailing across any fixed line seen against the sky, transverse to her path: and all the celestial bodies, carried along by the revolving sphere, travel at the same rate.

15. In this case, up to a certain degree, we render our measures of astronomical angles more exact and convenient by substituting time for space; but when, in the very same kind of observation, we wish to proceed to a greater degree of accuracy, we find that it is best done by substituting space for time. In observing the transit of a star across the meridian, if we have the clock within hearing, we can count the beats of the pendulum by the noise which they make, and tell exactly at which second of time the passage of the star across the visible thread takes place; and thus we measure Right Ascension by means of time. But our perception of time does not allow us to divide a second into ten parts, and to pronounce whether the transit takes place three-tenths, six-tenths, or seven-tenths of a second after the preceding beat of the clock. This, however, can be done by the usual mode of observing the transit of a star. The observer, listening to the beat of his clock, fastens his attention upon the star at each beat, and especially at the one immediately before and the one immediately after the passage of the thread: and by this means he has these two positions and the positions of the thread so far present to his intuition at once, that he can judge in what proportion the thread is nearer to one position than than the other, and can thus divide the intervening second in its due proportion. Thus if he observe that at the beginning of the second the star is on one side of the thread, and at the end of the second on the other side; and that the two distances from the thread are as two to three, he knows that the transit took place at two-fifths (or four-tenths) of a second after the former beat. In this way a second of time in astronomical observations may, by a skilful observer, be divided into ten equal parts; although when time is observed as time, a tenth of a second appears almost to escape our senses. From the above explanation, it will be seen that the reason why

the subdivision is possible in the way thus described, is this:—that the moment of time thus to be divided is so small, that the eye and the mind can retain, to the end of this moment, the impression of position which it received at the beginning. Though the two positions of the star, and the intermediate thread, are seen successively, they can be contemplated by the mind as if they were seen simultaneously : and thus it is precisely the smallness of this portion of time which enables us to subdivide it by means of space.

16. There is another case, of somewhat a different kind, in which time is employed in measuring space; namely, when space, or the standard of space, is defined by the length of a pendulum oscillating in a given time. We might in this way define any space by the time which a pendulum of such a length would take in oscillating; and thus we might speak, as was observed by those who suggested this device, of five minutes of cloth, or a rope half an hour long. We may observe, however, that in this case, the space is *not proportional* to the time. And we may add, that though we thus appear to avoid the arbitrary standard of space (for as we have seen, the standard of measures of time is a natural one,) we do not do so in fact : for we assume the invariableness of gravity, which really varies (though very slightly,) from place to place.

17. (V.) *The Method of Repetition in Measurement.* —In many cases we can give great additional accuracy to our measurements by repeatedly adding to itself the quantity which we wish to measure. Thus if we wished to ascertain the exact breadth of a thread, it might not be easy to determine whether it was one-ninetieth, or one-ninety-fifth, or one-hundredth part of an inch; but if we find that ninety-six such threads placed side by side occupy exactly an inch, we have the precise measure of the breadth of the thread. In the same manner, if two

clocks are going nearly at the same rate, we may not be able to distinguish the excess of an oscillation of one of the pendulums over an oscillation of the other: but when the two clocks have gone for an hour, one of them may have gained ten seconds upon the other; thus showing that the proportion of their times of oscillation is 3610 to 3600.

In the latter of these instances, we have the principle of repetition truly exemplified, because (as has been justly observed by Sir J. Herschel*,) there is then "a juxtaposition of units without error,"—"one vibration commences exactly where the last terminates, no part of time being lost or gained in the addition of the units so counted." In space, this juxtaposition of units without error cannot be rigorously accomplished, since the units must be added together by material contact (as in the case of the threads,) or in some equivalent manner. Yet the principle of repetition has been applied to angular measurement with considerable success in Borda's Repeating Circle. In this instrument, the angle between two objects which we have to observe, is repeated along the graduated limb of the circle by turning the telescope from one object to the other, alternately fastened to the circle (by its *clamp*) and loose from it (by unclamping). In this manner the errors of graduation may (theoretically) be entirely got rid of: for if an angle repeated *nine* times be found to go twice round the circle, it must be *exactly* eighty degrees: and where the repetition does not give an exact number of circumferences, it may still be made to subdivide the error to any required extent.

18. Connected with the principle of repetition, is the *Method of coincidences* or *interferences*. If we have two Scales, on one of which an inch is divided into 10, and on the other into 11 equal parts; and if, these Scales

* *Disc. Nat. Phil.*, Art. 121.

being placed side by side, it appear that the beginning of the latter Scale is between the 2nd and 3rd division of the former, it may not be apparent what fraction added to 2 determines the place of beginning of the second Scale as measured on the first. But if it appear also that the 3rd division of the second Scale *coincides* with a certain division of the first, (the 5th,) it is certain that 2 and *three-tenths* is the *exact* place of the beginning of the second Scale, measured on the first Scale. The 3rd division of the 11 Scale will coincide (or interfere with) a division of the 10 Scale, when the beginning or *zero* of the 11 divisions is three-tenths of a division beyond the preceding line of the 10 Scale; as will be plain on a little consideration. And if we have two Scales of equal units, in which each unit is divided into nearly, but not quite, the same number of equal parts (as 10 and 11, 19 and 20, 29 and 30,) and one sliding on the other, it will always happen that some one or other of the division lines will coincide, or very nearly coincide; and thus the exact position of the beginning of one unit, measured on the other scale, is determined. A sliding scale, thus divided for the purpose of subdividing the units of that on which it slides, is called a *Vernier*, from the name of its inventor.

19. The same Principle of Coincidence or Interference is applied to the exact measurement of the length of time occupied in the oscillation of a pendulum. If a detached pendulum, of such a length as to swing in little less than a second, be placed before the seconds' pendulum of a clock, and if the two pendulums begin to move together, the former will gain upon the latter, and in a little while their motions will be quite discordant. But if we go on watching, we shall find them, after a time, to agree again exactly; namely, when the detached pendulum has gained one complete oscillation (back and forwards,) upon the clock pendulum, and again coincides

with it in its motion. If this happen after 5 minutes, we know that the times of oscillation of the two pendulums are in the proportion of 300 to 302, and therefore the detached pendulum oscillates in $\frac{150}{151}$ of a second. The accuracy which can be obtained in the measure of an oscillation by this means is great; for the clock can be compared (by observing transits of the stars or otherwise) with the natural standard of time, the sidereal day. And the moment of coincidence of the two pendulums may, by proper arrangements, be very exactly determined.

We have hitherto spoken of methods of measuring time and space, but other elements also may be very precisely measured by various means.

20. (VI.) *Measurement of Weight.*—Weight, like space and time, is a quantity made up by addition of parts, and may be measured by similar methods. The principle of repetition is applicable to the measurement of weight; for if two bodies be put in the same pan of a balance, their weights are exactly added.

There may be difficulties of practical workmanship in carrying into effect the mathematical conditions of a perfect balance; for example, in securing an exact equality of the effective arms of the beam in all positions. These difficulties are evaded by the *Method of double weighing;* according to which the standard weights, and the body which is to be weighed, are successively put in the *same* pan, and made to balance by a third body in the opposite scale. By this means the different lengths of the arms of the beam, and other imperfections of the balance, become of no consequence*.

21. There is no natural *Standard* of weight. The conventional weight taken as the standard, is the weight of a given bulk of some known substance; for instance,

---

* For other methods of measuring weights accurately, see FARA-DAY's *Chemical Manipulation*, p. 25.

a *cubic foot of water*. But in order that this may be definite, the water must not contain any portion of heterogeneous substance: hence it is required that the water be *distilled* water.

22. (VII.) *Measurement of Secondary Qualities.*—We have already seen[*] that secondary qualities are estimated by means of conventional Scales, which refer them to space, number, or some other definite expression. Thus the Thermometer measures heat; the Musical Scale, with or without the aid of number, expresses the pitch of a note; and we may have an exact and complete Scale of Colours, pure and impure. We may remark, however, that with regard to sound and colour, the estimates of the ear and the eye are not superseded, but only assisted: for if we determine what a note is, by comparing it with an instrument known to be in tune, we still leave the ear to decide when the note is *in unison* with one of the notes of the instrument. And when we compare a colour with our chromatometer, we judge by the eye which division of the chromatometer it *matches*. Colour and sound have their Natural Scales, which the eye and ear habitually apply; what science requires is, that those scales should be systematized. We have seen that several conditions are requisite in such scales of qualities: the observer's skill and ingenuity are mainly shown in devising such scales and methods of applying them.

23. The Method of Coincidences is employed in harmonics: for if two notes are nearly, but not quite, in unison, the coincidences of the vibrations produce an audible undulation in the note, which is called the *howl;* and the exactness of the unison is known by this howl vanishing.

24. (VIII.) *Manipulation.*—The process of applying practically methods of experiment and observation, is

---

[*] Book iii. c. 2. Of the Measure of Secondary Qualities.

termed Manipulation; and the value of observations depends much upon the proficiency of the observer in this art. This skill appears, as we have said, not only in devising means and modes of measuring results, but also in inventing and executing arrangements by which elements are subjected to such conditions as the investigation requires: in finding and using some material combination by which nature shall be asked the question which we have in our minds. To do this in any subject may be considered as a peculiar Art, but especially in Chemistry; where "many experiments, and even whole trains of research, are essentially dependent for success on mere manipulation*." The changes which the chemist has to study,—compositions, decompositions, and mutual actions, affecting the internal structure rather than the external form and motion of bodies,—are not familiarly recognized by common observers, as those actions are which operate upon the total mass of a body: and hence it is only when the chemist has become, to a certain degree, familiar with his science, that he has the power of observing. He must learn to interpret the effects of mixture, heat, and other chemical agencies, so as to see in them those facts which chemistry makes the basis of her doctrines. And in learning to interpret this language, he must also learn to call it forth;—to place bodies under the requisite conditions, by the apparatus of his own laboratory and the operations of his own fingers. To do this with readiness and precision, is, as we have said, an Art, both of the mind and of the hand, in no small degree recondite and difficult. A person may be well acquainted with all the doctrines of chemistry, and may yet fail in the simplest experiment. How many precautions and observances, what resource and invention, what delicacy and vigilance,

* FARADAY's *Chemical Manipulation*, p. 3.

are requisite in chemical manipulation, may be seen by reference to Dr. Faraday's work on that subject.

25. The same qualities in the observer are requisite in some other departments of science; for example, in the researches of Optics: for in these, after the first broad facts have been noticed, the remaining features of the phenomena are both very complex and very minute; and require both ingenuity in the invention of experiments, and a keen scrutiny of their results. We have instances of the application of these qualities in most of the optical experimenters of recent times, and certainly in no one more than Sir David Brewster. Omitting here all notice of his succeeding labours, his *Treatise on New Philosophical Instruments*, published in 1813, is an excellent model of the kind of resource and skill of which we now speak. I may mention as an example of this skill, his mode of determining the refractive power of an *irregular* fragment of any transparent substance. At first this might appear an impossible problem; for it would seem that a regular and smooth surface are requisite, in order that we may have any measurable refraction. But Sir David Brewster overcame the difficulty by immersing the fragment in a combination of fluids, so mixed, that they had the same refractive power as the specimen. The question, *when* they had this power, was answered by noticing when the fragment became so transparent that its surface could hardly be seen; for this happened when, the refractive power within and without the fragment being the same, there was no refraction at the surface. And this condition being obtained, the refractive power of the fluid, and therefore of the fragment, was easily ascertained.

26. (IX.) *The Education of the Senses.*—Colour and Musical Tone are, as we have seen, determined by means of the Senses, whether or not Systematical Scales are

used in expressing the observed fact. Systematical Scales
of sensible qualities, however, not only give precision to
the record, but to the observation. But for this purpose
such an Education of the Senses is requisite as may
enable us to apply the scale immediately. The memory
must retain the sensation or perception to which the
technical term or degree of the scale refers. Thus with
regard to colour, as we have said already*, when we find
such terms as *tin-white* or *pinchbeck-brown*, the metallic
colour so denoted ought to occur at once to our recollec-
tion without delay or search. The observer's senses,
therefore, must be educated, at first by an actual exhibition of the standard, and afterwards by a familiar use of
it, to understand readily and clearly each phrase and
degree of the scales which in his observations he has to
apply. This is not only the best, but in many cases the
only way in which the observation can be expressed.
Thus *glassy lustre, fatty lustre, adamantine lustre,* denote
certain kinds of shining in minerals, which appearances
we should endeavour in vain to describe by periphrasis;
and which the terms, if considered as terms in common
language, would by no means clearly discriminate : for who,
in common language, would say that coal has a fatty
lustre? But these terms, in their conventional sense, are
perfectly definite; and when the eye is once familiarized
with this application of them, are easily and clearly intelligible.

27. The education of the senses, which is thus requisite in order to understand well the terminology of any
science, must be acquired by an inspection of the objects
which the science deals with; and is, perhaps, best promoted by the practical study of Natural History. In the
different departments of Natural History, the descriptions
of species are given by means of an extensive technical

* Book viii. c. 3. Terminology.

terminology: and that education of which we now speak, ought to produce the effect of making the observer as familiar with each of these terms as we are with the words of our common language. The terms have a much more precise meaning than other terms, since they are defined by express convention, and not learnt by common usage merely. Yet though they are thus defined, not the definition, but the perception itself, is that which the term suggests to the proficient.

In order to use the terminology to any good purpose, the student must possess it, not as a dictionary, but as a language. The terminology of his sciences must be the natural historian's most familiar tongue. He must learn to think in such language. And when this is achieved, the terminology, as I have elsewhere said*, though to an uneducated eye cumbrous and pedantical, is felt to be a useful implement, not an oppressive burden. The impatient schoolboy looks upon his grammar and vocabulary as irksome and burdensome; but the accomplished student who has learnt the language by means of them, knows that they have given him the means of expressing what he thinks, and even of thinking more precisely. And as the study of language thus gives precision to the thoughts, the study of Natural History, and especially of the descriptive part of it, gives precision to the senses.

The Education of the Senses is also greatly promoted by the practical pursuit of any science of experiment and observation, as chemistry or astronomy. The methods of manipulating, of which we have just spoken, in chemistry, and the methods of measuring extremely minute portions of space and time which are employed in astronomy, and which are described in the former part of this chapter, are among the best modes of educating the senses for purposes of scientific observation.

* *Hist. Ind. Sci.*, iii. 307.

28. By the various Methods of precise observation which we have thus very briefly described, facts are collected, of an exact and definite kind; they are then bound together in general laws, by the aid of general ideas and of such methods as we have now to consider. It is true, that the ideas which enable us to combine facts into general propositions, do commonly operate in our minds while we are still engaged in the office of observing. Ideas of one kind or other are requisite to connect our phenomena into facts, and to give meaning to the terms of our descriptions: and it frequently happens, that long before we have collected all the facts which induction requires, the mind catches the suggestion which some of these ideas offer, and leaps forwards to a conjectural law while the labour of observation is yet unfinished. But though this actually occurs, it is easy to see that the process of combining and generalizing facts is, in the order of nature, posterior to, and distinct from, the process of observing facts. Not only is this so, but there is an intermediate step which, though inseparable from all successful generalization, may be distinguished from it in our survey; and may, in some degree, be assisted by peculiar methods. To the consideration of such methods we now proceed.

Chapter III.

OF METHODS OF ACQUIRING CLEAR SCIENTIFIC IDEAS; *and first* OF INTELLECTUAL EDUCATION.

THE ways in which men become masters of those clear and yet comprehensive conceptions which the formation and reception of science require, are mainly two; which, although we cannot reduce them to any exact scheme, we may still, in a loose use of the term, call *Methods* of acquiring clear Ideas. These two ways are Education and Discussion.

1. (I.) *Idea of Space.*—It is easily seen that Education may do at least something to render our ideas distinct and precise. To learn Geometry in youth, tends, manifestly, to render our idea of space clear and exact. By such an education, all the relations, all the consequences of this idea, come to be readily and steadily apprehended; and thus it becomes easy for us to understand portions of science which otherwise we should by no means be able to comprehend. The conception of *similar triangles* was to be mastered, before the disciples of Thales could see the validity of his method of determining the height of lofty objects by the length of their shadows. The conception of *the sphere with its circles* had to become familiar, before the annual motion of the sun and its influence upon the lengths of days could be rightly traced. The properties of circles, combined with the *pure* doctrine of motion*, were required as an introduction to the theory of Epicycles: the properties of *conic sections* were needed, as a preparation for the discoveries of Kepler. And not only was it necessary that men should possess a *knowledge* of certain figures and their properties; but it was equally necessary that they should have the *habit of reasoning* with perfect steadiness, precision, and conclusiveness concerning the relations of space. No small discipline of the mind is requisite, in most cases, to accustom it to go, with complete insight and security, through the demonstrations respecting intersecting planes and lines, dihedral and trihedral angles, which occur in solid geometry. Yet how absolutely necessary is a perfect mastery of such reasonings, to him who is to explain the motions of the moon in latitude and longitude! How necessary, again,

* See b. ii. c. 12.

is the same faculty to the student of crystallography! Without mathematical habits of conception and of thinking, these portions of science are perfectly inaccessible. But the early study of plane and solid geometry gives to all tolerably gifted persons, the habits which are thus needed. The discipline of following the reasonings of didactic works on this subject, till we are quite familiar with them, and of devising for ourselves reasonings of the same kind, (as, for instance, the solutions of problems proposed,) soon gives the mind the power of *discoursing* with perfect facility concerning the most complex and multiplied relations of space, and enables us to refer to the properties of all plane and solid figures as surely as to the visible forms of objects. Thus we have here a signal instance of the efficacy of education in giving to our Conceptions that clearness, which the formation and existence of science indispensably require.

2. It is not my intention here to enter into the details of the form which should be given to education, in order that it may answer the purposes now contemplated. But I may make a remark, which the above examples naturally suggest, that in a mathematical education, considered as a preparation for furthering or understanding physical science, Geometry is to be cultivated, far rather than Algebra:—the properties of space are to be studied and reasoned upon as they are in themselves, not as they are replaced and disguised by symbolical representations. It is true, that when the student is become quite familiar with elementary geometry, he may often enable himself to deal in a more rapid and comprehensive manner with the relations of space, by using the language of symbols and the principles of symbolical calculation: but this is an ulterior step, which may be added to, but can never be substituted for, the direct cultivation of geometry. The method of symbolical reasoning employed upon sub-

jects of geometry and mechanics, has certainly achieved some remarkable triumphs in the treatment of the theory of the universe. These successful applications of symbols in the highest problems of physical astronomy appear to have made some teachers of mathematics imagine that it is best to *begin* the pupil's course with such symbolical generalities. But this mode of proceeding will be so far from giving the student clear ideas of mathematical relations, that it will involve him in utter confusion, and probably prevent his ever obtaining a firm footing in geometry. To commence mathematics in such a way, would be much as if we should begin the study of a language by reading the highest strains of its lyrical poetry.

3. (II.) *Idea of Number, &c.*—The study of mathematics, as I need hardly observe, develops and renders exact, our conceptions of the relations of number, as well as of space. And although, as we have already noticed, even in their original form the conceptions of number are for the most part very distinct, they may be still further improved by such discipline. In complex cases, a methodical cultivation of the mind in such subjects is needed: for instance, questions concerning cycles, and intercalations, and epacts, and the like, require very great steadiness of arithmetical apprehension in order that the reasoner may deal with them rightly. In the same manner, a mastery of problems belonging to the science of Pure Motion, or, as I have termed it, Mechanism, requires either great natural aptitude in the student, or a mind properly disciplined by suitable branches of mathematical study.

4. Arithmetic and Geometry have long been standard portions of the education of cultured persons throughout the civilized world; and hence all such persons have been able to accept and comprehend those portions of science which depend upon the idea of space: for instance

the doctrine of the globular form of the earth, with its consequences, such as the measures of latitude and longitude;—the heliocentric system of the universe in modern, or the geocentric in ancient times;—the explanation of the rainbow; and the like. In nations where there is no such education, these portions of science cannot exist as a part of the general stock of the knowledge of society, however intelligently they may be pursued by single philosophers dispersed here and there in the community.

5. (III.) *Idea of Force.*—As the idea of Space is brought out in its full evidence by the study of Geometry, so the idea of Force is called up and developed by the study of the science of Mechanics. It has already been shown, in our scrutiny of the Ideas of the Mechanical Sciences, that Force, the Cause of motion or of equilibrium, involves an independent Fundamental Idea, and is quite incapable of being resolved into any mere modification of our conceptions of space, time, and motion. And in order that the student may possess this idea in a precise and manifest shape, he must pursue the science of Mechanics in the mode which this view of its nature demands;—that is, he must study it as an independent science, resting on solid elementary principles of its own, and not built upon some other unmechanical science as its substructure. He must trace the truths of Mechanics from their own axioms and definitions; these axioms and definitions being considered as merely means of bringing into play the Idea on which the science depends. The conceptions of force and matter, of action and reaction, of momentum and inertia, with the reasonings in which they are involved, cannot be evaded by any substitution of lines or symbols for the conceptions. Any attempts at such substitution would render the study of Mechanics useless as a preparation of the mind for physical science; and would, indeed, except counteracted by great natural clearness of

thought on such subjects, fill the mind with confused and vague notions, quite unavailing for any purposes of sound reasoning. But, on the other hand, the study of Mechanics, in its genuine form, as a branch of education, is fitted to give a most useful and valuable precision of thought on such subjects; and is the more to be recommended, since, in the general habits of most men's minds, the mechanical conceptions are tainted with far greater obscurity and perplexity than belongs to the conceptions of number, space, and motion.

6. As habitually distinct conceptions of space and motion were requisite for the reception of the doctrines of formal astronomy, (the Ptolemaic and Copernican system,) so a clear and steady conception of force is indispensably necessary for understanding the Newtonian system of physical astronomy. It may be objected that the study of Mechanics as a science has not commonly formed part of a liberal education in Europe, and yet that educated persons have commonly accepted the Newtonian system. But to this we reply, that although most persons of good intellectual culture have professed to assent to the Newtonian system of the universe, yet they have, in fact, entertained it in so vague and perplexed a manner as to show very clearly that a better mental preparation than the usual one is necessary, in order that such persons may really understand the doctrine of universal attraction. I have already spoken of the prevalent indistinctness of mechanical conceptions*; and need not here dwell upon the indications, constantly occurring in conversation and in literature, of the utter inaccuracy of thought on such subjects which may often be detected; for instance, in the mode in which many men speak of centrifugal and centripetal forces;—of projectile and central forces;—of the effect of the moon

* Vol. i. p. 257.

upon the waters of the ocean ; and the like. The incoherence of ideas which we frequently witness on such points, shows us clearly that, in the minds of a great number of men, well educated according to the present standard, the acceptance of the doctrine of universal gravitation is a result of traditional prejudice, not of rational conviction. And those who are Newtonians on such grounds, are not at all more intellectually advanced by being Newtonians in the nineteenth century, than they would have been by being Ptolemaics in the fifteenth.

7. It is undoubtedly in the highest degree desirable that all great advances in science should become the common property of all cultivated men. And this can only be done by introducing into the course of a liberal education such studies as unfold and fix in men's minds the fundamental ideas upon which the new discovered truths rest. The progress made by the ancients in geography, astronomy, and other sciences, led them to assign, wisely and well, a place to arithmetic and geometry among the steps of an ingenuous education. The discoveries of modern times have rendered these steps still more indispensable ; for we cannot consider a man as cultivated up to the standard of his times, if he is not only ignorant of, but incapable of comprehending, the greatest achievements of the human intellect. And as innumerable discoveries of all ages have thus secured to Geometry her place as a part of good education, so the great discoveries of Newton make it proper to introduce Elementary Mechanics as a part of the same course. If the education deserve to be called good, the pupil will not remain ignorant of those discoveries, the most remarkable extensions of the field of human knowledge which have ever occurred. Yet he cannot by possibility comprehend them, except his mind be previously disciplined

by mechanical studies. The period appears now to be arrived when we may venture, or rather when we are bound to endeavour, to include a new class of fundamental ideas in the elementary discipline of the human intellect. This is indispensable, if we wish to educe the powers which we know that it possesses, and to enrich it with the wealth which lies within its reach*.

8. By the view which is thus presented to us of the nature and objects of intellectual education, we are led to consider the mind of man as undergoing a progress from age to age. By the discoveries which are made, and by the clearness and evidence which, after a time, (not suddenly nor soon,) the truths thus discovered acquire, one portion of knowledge after another becomes elementary; and if we would really secure this progress, and make men share in it, these new portions must be treated as elementary in the constitution of a liberal education. Even in the rudest forms of intelligence, man is immeasurably elevated above the unprogressive brute, for the idea of number is so far developed that he can count his flock or his arrows. But when number is contemplated in a speculative form, he has made a vast additional progress; when he steadily apprehends the relations of space, he has again advanced; when in thought he carries these relations into the vault of the sky, into the expanse of the universe, he reaches a higher intellectual position. And when he carries into these wide regions, not only the relations of space and time, but of cause and effect, of force and reaction, he has again made an intellectual advance; which, wide as it is at first, is accessible to all; and with which all should acquaint themselves, if they really desire to prosecute with energy the ascending path of truth and knowledge which lies

* The University of Cambridge has, by a recent law, made an examination in Elementary Mechanics requisite for the Degree of B.A.

before them. This should be an object of exertion to all ingenuous and hopeful minds. For that exertion is necessary,—that after all possible facilities have been afforded, it is still a matter of toil and struggle to appropriate to ourselves the acquisitions of great discoverers, is not to be denied. Elementary mechanics, like elementary geometry, is a study accessible to all; but like that too, or perhaps more than that, it is a study which requires effort and contention of mind,—a forced steadiness of thought. It is long since one complained of this labour in geometry; and was answered that in that region there is no *Royal Road.* The same is true of Mechanics, and must be true of all branches of solid education. But we should express the truth more appropriately in our days by saying that there is no *Popular Road* to these sciences. In the mind, as in the body, strenuous exercise alone can give strength and activity. The art of exact thought can be acquired only by the labour of close thinking.

9. (IV.) *Chemical Ideas.*—We appear then to have arrived at a point of human progress in which a liberal education of the scientific intellect should include, besides arithmetic, elementary geometry and mechanics. The question then occurs to us, whether there are any other Fundamental Ideas, among those belonging to other sciences, which ought also to be made part of such an education;—whether, for example, we should strive to develop in the minds of all cultured men the ideas of *polarity,* mechanical and chemical, of which we spoke in a former part of this work.

The views to which we have been conducted by the previous inquiry lead us to reply that it would not be well to make *chemical* polarities, at any rate, a subject of elementary instruction. For even the most profound and acute philosophers who have speculated upon this subject,—they who are leading the van in the march of discovery, do not seem yet to have reduced their thoughts on this subject to a consistency, or to have taken hold of this idea of polarity in a manner quite satisfactory to their own minds. This part of the subject is, therefore, by no means ready to be introduced into a course of general elementary education ; for, with a view to such a purpose, nothing less than the most thoroughly luminous and transparent condition of the idea' will suffice. Its whole efficacy, as a means and object of disciplinal study, depends upon there being no obscurity, perplexity, or indefiniteness with regard to it, beyond that transient deficiency which at first exists in the learner's mind, and is to be removed by his studies. The idea of chemical polarity is not yet in this condition ; and therefore is not yet fit for a place in education. Yet since this idea of polarity is the most general idea which enters into chemistry, and appears to be that which includes almost all the others, it would be unphilosophical, and inconsistent with all sound views of science, to introduce into education some chemical conceptions, and to omit those which depend upon this idea : indeed such a partial adoption of the science could hardly take place without not only omitting, but misrepresenting, a great part of our chemical knowledge. The conclusion to which we are necessarily led, therefore, is this :—that at present chemistry cannot with any advantage, form a portion of a general intellectual education*.

10. (V.) *Natural-History Ideas.*—But there remains still another class of Ideas, with regard to which we may very properly ask whether they may not advantageously form a portion of a liberal education : I mean the Ideas

* I do not here stop to prove that an education (if it be so called) in which the memory only retains the verbal expression of results, while the mind does not apprehend the piinciples of the subject, and therefore cannot even understand the words in which its doctrines are expiessed, is of no value whatever to the intellect, but rather, is highly hurtful to the habits of thinking and reasoning.

of definite Resemblance and Difference, and of one set
of resemblances subordinate to another, which form the
bases of the classificatory sciences.   These Ideas are
developed by the study of the various branches of Natural
History, as Botany, and Zoology; and beyond all doubt,
those pursuits, if assiduously followed, very materially
affect the mental habits.   There is this obvious advantage
to be looked for from the study of Natural History, considered as a means of intellectual discipline:—that it
gives us, in a precise and scientific form, examples of
the classing and naming of objects; which operations the
use of common language leads us constantly to perform
n aloose and inexact way.   In the usual habits of our
minds and tongues, things are distinguished or brought
together, and names are applied, in a manner very indefinite, vacillating, and seemingly capricious: and we may
naturally be led to doubt whether such defects can be
avoided;—whether exact distinctions of things, and rigorous use of words be possible.   Now upon this point we
may receive the instruction of Natural History; which
proves to us, by the actual performance of the task, that
a precise classification and nomenclature are attainable,
at least for a mass of objects of the same kind.   Further,
we also learn from this study, that there may exist not
only an exact distinction of kinds of things, but a series
of distinctions, one set subordinate to another, and the
more general including the more special, so as to form a
system of classification.   All these are valuable lessons.
If by the study of Natural History we evolve, in a clear
and well defined form, the conceptions of *genus, species,*
and of *higher* and *lower steps* of classification, we communicate precision, clearness, and method to the intellect,
through a great range of its operations.

11. It must be observed, that in order to attain the
disciplinal benefit which the study of Natural History is

fitted to bestow, we must teach the *natural* not the artificial *classifications*; or at least the natural as well as the artificial. For it is important for the student to perceive that there are classifications, not merely arbitrary, founded upon some *assumed* character, but natural, recognized by some *discovered* character; he ought to see that our classes being collected according to one mark, are confirmed by many marks not originally stated in our scheme; and are thus found to be grouped together, not by a single resemblance, but by a mass of resemblances, indicating a natural affinity. That objects may be collected into such groups, is a highly important lesson, which Natural History alone, pursued as the science of *natural classes*, can teach.

12. Natural History has not unfrequently been made a portion of education: and has in some degree produced such effects as we have pointed out. It would appear however, that its lessons have in general been very imperfectly learnt or understood by persons of general education: and that there are perverse intellectual habits very generally prevalent in the cultivated classes, which ought ere now to have been corrected by the general teaching of Natural History. We may detect among speculative men many prejudices respecting the nature and rules of reasoning, which arise from pure mathematics having been so long and so universally the instrument of intellectual cultivation. Pure Mathematics reasons from definitions: whatever term is introduced into her pages, as a *circle*, or a *square*, its definition comes along with it: and this definition is supposed to supply all that the reasoner needs to know, respecting the term. If there be any doubt concerning the validity of the conclusion, the doubt is resolved by recurring to the definitions. Hence it has come to pass that in other subjects also, men seek for and demand definitions as the most secure foundation of reasoning. The definition and the term defined

are conceived to be so far identical, that in all cases the one may be substituted for the other; and such a substitution is held to be the best mode of detecting fallacies.

13. It has been already shown that even geometry is not founded upon definitions alone: and we shall not here again analyse the fallacy of this belief in the supreme value of definitions. But we may remark that the study of Natural History appears to be the proper remedy for this erroneous habit of thought. For in every department of Natural History the object of our study is *kinds* of things, not one of which kinds can be rigorously defined, yet all of them are sufficiently definite. In these cases we may indeed give a specific description of one of the kinds, and may call it a definition; but it is clear that such a definition does not contain the essence of the thing. We say* that the Rose Tribe are "Polypetalous dicotyledons, with lateral styles, superior simple ovaria, regular perigynous stamens, exalbuminous definite seeds, and alternate stipulate leaves." But no one would say that this was our essential conception of a rose, to be substituted for it in all cases of doubt or obscurity, by way of making our reasonings perfectly clear. Not only so; but as we have already seen†, the definition does not even apply to all the tribe. For the stipulæ are absent in Lowea: the albumen is present in Neillia: the fruit of Spiræa sorbifolia is capsular. If, then, we can possess any certain knowledge in Natural History, (which no cultivator of the subject will doubt,) it is evident that our knowledge cannot depend on the possibility of laying down exact definitions and reasoning from them.

14. But it may be asked, if we cannot define a word, or a class of things which a word denotes, how can we distinguish what it does mean from what it does not mean? How can we say that it signifies one thing

* LINDLEY's *Nat. Syst. Bot.*, p. 81.　　† B. viii., c. 2., p. 475.

rather than another, except we declare what is its signification?

The answer to this question involves the general principle of a natural method of classification, which has already been stated* and need not here be again dwelt on. It has been shown that names of *kinds* of things (*genera*) associate them according to total resemblances, not partial characters. The principle which connects a group of objects in natural history is not a *definition*, but a *type*. Thus we take as the type of the Rose family, it may be, the common wild rose; all species which resemble this more than they resemble any other group of species are also roses, and form one genus. All genera which resemble Roses more than they resemble any other group of genera are of the same family. And thus the Rose family is collected about some one species, which is the type or central point of the group.

In such an arrangement, it may readily be conceived that though the nucleus of each group may cohere firmly together, the outskirts of contiguous groups may approach, and may even be intermingled, so that some species may doubtfully adhere to one group or another. Yet this uncertainty does not all affect the truths which we find ourselves enabled to assert with regard to the general mass of each group. And thus we are taught that there may be very important differences between two groups of objects, although we are unable to tell where the one group ends and where the other begins; and that there may be propositions of indisputable truth, in which it is impossible to give unexceptionable definitions of the terms employed.

15. These lessons are of the highest value with regard to all employments of the human mind; for the mode in which words in common use acquire their mean-

* B. viii., c. 2., p. 476.

ing, approaches far more nearly to the *Method of Type* than to the method of definition. The terms which belong to our practical concerns, or to our spontaneous and unscientific speculations, are rarely capable of exact definition. They have been devised in order to express assertions, often very important, yet very vaguely conceived: and the signification of the word is extended, as far as the assertion conveyed by it can be extended, by apparent connexion or by analogy. And thus in all the attempts of man to grasp at knowledge, we have an exemplification of that which we have stated as the rule of induction, that Definition and Proposition are mutually dependent, each adjusted so as to give value and meaning to the other: and this is so, even when both the elements of truth are defective in precision: the Definition being replaced by an incomplete description or a loose reference to a Type; and the Proposition being in a corresponding degree insecure.

16. Thus the study of Natural History, as a corrective of the belief that definitions are essential to substantial truth, might be of great use; and the advantage which might thus be obtained is such as well entitles this study to a place in a liberal education. We may further observe, that in order that Natural History may produce such an effect, it must be studied by inspection of the *objects* themselves, and not by the reading of books only. Its lesson is, that we must in all cases of doubt or obscurity refer, not to words or definitions, but to things. The Book of Nature is its dictionary: it is there that the natural historian looks, to find the meaning of the words which he uses*. So long as a plant, in its most essential

* It is a curious example of the influence of the belief in definitions, that elementary books have been written in which Natural History is taught in the way of question and answer, and consequently by means of words alone. In such a scheme, of course all objects are *defined :* and

parts, is more like a rose than anything else, it is a rose. He knows no other definition.

17. (VI.) *Well-established Ideas alone to be used.*— We may assert in general what we have above stated specially with reference to the fundamental principles of chemistry:—no Ideas are suited to become the elements of elementary education, till they have not only become perfectly distinct and fixed in the minds of the leading cultivators of the science to which they belong; but till they have been so for some considerable period. The entire clearness and steadiness of view which is essential to sound science, must have time to extend itself to a wide circle of disciples. The views and principles which are detected by the most profound and acute philosophers, are soon appropriated by all the most intelligent and active minds of their own and of the following generations; and when this has taken place, (and not till then,) it is right, by a proper constitution of our liberal education, to extend a general knowledge of such principles to all cultivated persons. And it follows, from this view of the matter, that we are by no means to be in haste to adopt, into our course of education, all new discoveries as soon as they are made. They require some time, in order to settle into their proper place and position in men's minds, and to show themselves under their true aspects; and till this is done, we confuse and disturb, rather than enlighten and unfold, the ideas of learners, by introducing the discoveries into our elementary instruction. Hence it was perhaps reasonable that a century should elapse from the time of Galileo before the rigor-

we may easily anticipate the value of the knowledge thus conveyed. Thus, " Iron is a well-known hard metal, of a darkish gray colour, and very elastic:" " Copper is an orange-coloured metal, more sonorous than any other, and the most elastic of any except iron." This is to pervert the meaning of education, and to make it a business of mere words.

ous teaching of mechanics became a general element of intellectual training; and the doctrine of universal gravitation was hardly ripe for such an employment till the end of the last century. We must not direct the unformed youthful mind to launch its little bark upon the waters of speculation, till all the agitation of discovery, with its consequent fluctuation and controversy, has well subsided.

18. But it may be asked, How is it that time operates to give distinctness and evidence to scientific ideas? In what way does it happen that views and principles, obscure and wavering at first, after a while become luminous and steady? Can we point out any process, any intermediate steps, by which this result is produced? If we can, this process must be an important portion of the subject now under our consideration.

To this we reply, that the transition from the hesitation and contradiction with which true ideas are first received, to the general assent and clear apprehension which they afterwards obtain, takes place through various arguments for and against them, and various modes of presenting and testing them, all which we may include under the term Discussion, which we have already mentioned as the second of the two ways by which scientific views are developed into full maturity.

Chapter IV.

OF METHODS OF ACQUIRING CLEAR SCIENTIFIC IDEAS, *continued.*—OF THE DISCUSSION OF IDEAS.

1. It is easily seen that in every part of science, the establishment of a new set of ideas has been accompanied with much of doubt and dissent. And by means of dis-

cussions so occasioned, the new conceptions, and the opinions which involve them, have gradually become definite and clear. The authors and asserters of the new opinions, in order to make them defensible, have been compelled to make them consistent: in order to recommend them to others, they have been obliged to make them more entirely intelligible to themselves. And thus the terms which formed the main points of the controversy, although applied in a loose and vacillating manner at first, have in the end become perfectly definite and exact. The opinions discussed have been, in their main features, the same throughout the debate; but they have at first been dimly, and at last clearly apprehended: like the objects of a landscape, at which we look through a telescope ill adjusted, till, by sliding the tube backwards and forwards, we at last bring it into focus, and perceive every feature of the prospect sharp and bright.

2. We have in the last Book but one* fully exemplified this gradual progress of conceptions from obscurity to clearness by means of Discussion. We have seen, too, that this mode of treating the subject has never been successful, except when it has been associated with an appeal to facts as well as to reasonings. A combination of experiment with argument, of observation with demonstration, has always been found requisite in order that men should arrive at those distinct conceptions which give them substantial truths. The arguments used led to the rejection of undefined, ambiguous, self-contradictory notions; but the reference to facts led to the selection, or at least to the retention, of the conceptions which were both true and useful. The two correlative processes, definition and true assertion, the formation of clear ideas and the induction of laws, went on together.

* B. xi. c. 2, Of the Explication of Conceptions.

Thus those discussions by which scientific conceptions are rendered ultimately quite distinct and fixed, include both reasonings from principles and illustrations from facts. At present we turn our attention more peculiarly to the former part of the process; according to the distinction already drawn, between the explication of conceptions and the colligation of facts. The Discussions of which we here speak, are the Method (if they may be called a *method*) by which the Explication of Conceptions is carried to the requisite point among philosophers.

3. In the scrutiny of the Fundamental Ideas of the Sciences which forms the previous Part of this work, and in the *History of the Inductive Sciences*, I have, in several instances, traced the steps by which, historically speaking, these Ideas have obtained their ultimate and permanent place in the minds of speculative men. I have thus exemplified the reasonings and controveries which constitute such Discussion as we now speak of. I have stated, at considerable length, the various attempts, failures, and advances, by which the ideas which enter into the science of mechanics were evolved into their present evidence. In like manner we have seen the conception of refracted rays of light, obscure and confused in Seneca, growing clearer in Roger Bacon, more definite in Descartes, perfectly distinct in Newton. The polarity of light, at first contemplated with some perplexity, became very distinct to Malus, Young, and Fresnel; yet the phenomena of circular polarization, and still more, the circular polarization of fluids, leave us, even at present, some difficulty in fully mastering this conception. The related polarities of electricity and magnetism are not yet fully comprehended, even by our greatest philosophers. One of Mr. Faraday's late papers (the Fourteenth Series of his Researches) is employed in an experimental discussion of this subject, which leads to no satisfactory

result. The controversy between Biot and Ampère*, on the nature of the elementary forces in electro-dynamic action, is another evidence that the discussion of this subject has not yet reached its termination. With regard to chemical polarity, I have already stated that this idea is as yet very far from being brought to an ultimate condition of definiteness; and the subject of chemical forces, (for the whole subject must be included in this idea of polarity,) which has already occasioned much perplexity and controversy, may easily occasion much more, before it is settled to the satisfaction of the philosophical world. The ideas of the classificatory sciences also have of late been undergoing much, and very instructive discussion, in the controversies respecting the relations and offices of the natural and artificial methods. And with regard to physiological ideas, it would hardly be too much to say, that the whole history of physiology up to the present time has consisted of the discussion of the fundamental ideas of the science, such as vital forces, nutrition, reproduction, and the like. We have had before us at some length, in the present work, a review of the opposite opinions which have been advanced on this subject; and have attempted in some degree to estimate the direction in which these ideas are permamently settling. But without attaching any importance to this attempt, the account there given may at least serve to show, how important a share in the past progress of this subject the discussion of its fundamental ideas has hitherto had.

4. There is one reflection which is very pointedly suggested by what has been said. The manner in which our scientific ideas acquire their distinct and ultimate form being such as has been described,—always involving much abstract reasoning and analysis of our conceptions, often much opposite argumentation and debate;—how

---

* *Hist. Ind. Sci.*, iii. 287.

unphilosophical is it to speak of abstraction and analysis, of dispute and controversy, as frivolous and unprofitable processes, by which true science can never be benefitted; and to put such employments in antithesis with the study of facts!

Yet some writers are accustomed to talk with contempt of all past controversies, and to wonder at the blindness of those who did not at first take the view which was established at last. Such persons forget that it was precisely the controversy, which established among speculative men that final doctrine which they themselves have quietly accepted. It is true, they have had no difficulty in thoroughly adopting the truth; but that has occurred because all dissentient doctrines have been suppressed and forgotten; and because systems, and books, and language itself, have been accommodated peculiarly to the expression of the accepted truth. To despise those who have, by their mental struggles and conflicts, brought the subject into a condition in which error is almost out of our reach, is to be ungrateful exactly in proportion to the amount of the benefit received. It is as if a child, when its teacher had with many trials and much trouble prepared a telescope so that the vision through it was distinct, should wonder at his stupidity in pushing the tube of the eye-glass out and in so often.

5. Again, some persons condemn all that we have here spoken of as the discussion of ideas, terming it *metaphysical*: and in this spirit, one writer* has spoken of the "metaphysical period" of each science, as preceding the period of "positive knowledge." But as we have seen, that process which is here termed "metaphysical," —the analysis of our conceptions and the exposure of their inconsistencies,—(accompanied with the study of facts,)—has always gone on most actively in the most

---

* M. Auguste Comte, *Cours de Philosophie Positive.*

prosperous periods of each science. There is, in Galileo, Kepler, Gassendi, and the other fathers of mechanical philosophy, as much of metaphysics as in their adversaries. The main difference is, that the metaphysics is of a better kind; it is more conformable to metaphysical truth. And the same is the case in other sciences. Nor can it be otherwise. For all truth, before it can be consistent with facts, must be consistent with itself: and although this rule is of undeniable authority, its application is often far from easy. The perplexities and ambiguities which arise from our having the same idea presented to us under different aspects, are often difficult to disentangle: and no common acuteness and steadiness of thought must be expended on the task. It would be easy to adduce, from the works of all great discoverers, passages more profoundly metaphysical than any which are to be found in the pages of barren *à priori* reasoners.

6. As we have said, these metaphysical discussions are not to be put in opposition to the study of facts; but are to be stimulated, nourished and directed by a constant recourse to experiment and observation. The cultivation of ideas is to be conducted as having for its object the connexion of facts; never to be pursued as.a mere exercise of the subtilty of the mind, striving to build up a world of its own, and neglecting that which exists about us. For although man may in this way please himself, and admire the creations of his own brain, he can never, by this course, hit upon the real scheme of nature. With his ideas unfolded by education, sharpened by controversy, rectified by metaphysics, he may *understand* the natural world, but he cannot *invent* it. At every step, he must try the value of the advances he has made in thought, by applying his thoughts to things. The Explication of Conceptions must.be carried on with a perpetual reference to the Colligation of Facts.

Having here treated of Education and Discussion as the methods by which the former of these two processes is to be promoted, we have now to explain the methods which science employs in order most successfully to execute the latter. But the Colligation of Facts, as already stated, may offer to us two steps of a very different kind,—the laws of Phenomena, and their Causes. We shall first describe some of the methods employed in obtaining truths of the former of these two kinds.

CHAPTER V.

ANALYSIS OF THE PROCESS OF INDUCTION.

SECT. I. *The Three Steps of Induction.*

1. WHEN facts have been decomposed and phenomena measured, the philosopher endeavours to combine them into general laws, by the aid of ideas and conceptions, these being illustrated and regulated by such means as we have spoken of in the last two chapters. In this task, of gathering laws of nature from observed facts, as we have already said*, the natural sagacity of gifted minds is the power by which the greater part of the successful results have been obtained; and this power will probably always be more efficacious than any Method can be. Still there are certain methods of procedure which may in such investigations give us no inconsiderable aid, and these I shall endeavour to expound.

2. For this purpose, I remark that the Colligation of ascertained facts into general propositions may be considered as containing three steps, which I shall term *the Selection of the Idea, the Construction of the Conception,* and *the Determination of the Magnitudes.* It will be recol-

* B. xi. c. 6.

lected that by the word *Idea*, (or Fundamental Idea,) used in a peculiar sense, I mean certain wide and general fields of intelligible relation, such as Space, Number, Cause, Likeness; while by *Conception* I denote more special modifications of these ideas, as a *circle*, a *square number*, a *uniform force*, a *like form* of flower. Now in order to establish any law by reference to facts, we must select the *true Idea* and the *true Conception*. For example; when Hipparchus found* that the distance of the bright star Spica Virginis from the equinoxial point had increased by two degrees in about two hundred years, and desired to reduce this change to a law, he had first to assign, if possible, the *idea* on which it depended;— whether it was regulated for instance, by *space*, or by *time*; whether it was determined by the positions of other stars at each moment, or went on progressively with the lapse of ages. And when there was found reason to select *time* as the regulative *idea* of this change, it was then to be determined how the change went on with the time;—whether uniformly, or in some other manner: the *conception*, or the rule of the progression, was to be rightly constructed. Finally, it being ascertained that the change did go on uniformly, the question then occurred what was its *amount:*—whether exactly a degree in a century, or more, or less, and how much: and thus the determination of the *magnitude* completed the discovery of the law of phenomena respecting this star.

3. Steps similar to these three may be discerned in all other discoveries of laws of nature. Thus, in investigating the laws of the motions of the sun, moon or planets, we find that these motions may be resolved, besides a uniform motion, into a series of partial motions, or Inequalities; and for each of these Inequalities, we have to learn upon what it directly depends, whether

* *Hist. Ind. Sci.*, i. 187.

upon the progress of time only, or upon some configuration of the heavenly bodies in space; then, we have to ascertain its law; and finally, we have to determine what is its amount.   In the case of such Inequalities, the fundamental element on which the Inequality depends, is called the *Argument*.   And when the Inequality has been fully reduced to known rules, and expressed in the form of a Table, the Argument is the fundamental series of numbers which stands in the margin of the Table, and by means of which we refer to the other numbers which express the Inequality.   Thus, in order to obtain from a Solar Table the Inequality of the sun's annual motion, the Argument is the number which expresses the day of the year; the Inequalities for each day being (in the Table) ranged in a line corresponding to the days.   Moreover, the Argument of an Inequality being assumed to be known, we must, in order to calculate the Table, that is, in order to exhibit the law of nature, know also the *Law* of the Inequality, and its *Amount*.   And the investigation of these three things, the Argument, the Law, and the Amount of the Inequality, represents the three steps above described, the Selection of the Idea, the Construction of the Conception, and the Determination of the Magnitude.

4. In a great body of cases, *mathematical* language and calculation are used to express the connexion between the general law and the special facts.   And when this is done, the three steps above described may be spoken of as the Selection of the *Independent Variable*, the Construction of the *Formula*, and the Determination of the *Coefficients*.   It may be worth our while to attend to an exemplification of this.   Suppose then, that, in such observations as we have just spoken of, namely, the shifting of a star from its place in the heavens by an unknown law, astronomers had, at the end of three successive years,

found that the star had removed by 3, by 8, and by 15 minutes from its original place. Suppose it to be ascertained also, by methods of which we shall hereafter treat, that this change depends upon the time; we must then take the *time*, (which we may denote by the symbol t,) for the *independent variable*. But though the star changes its place with the time, the change is not proportional to the time; for its motion which is only 3 minutes in the first year, is 5 minutes in the second year, and 7 in the third. But it is not difficult for a person a little versed in mathematics to perceive that the series 3, 8, 15, may be obtained by means of two terms, one of which is proportional to the time, and the other to the square of the time; that is, it is expressed by the *formula at + btt*. The question then occurs, what are the values of the *coefficients a* and *b*; and a little examination of the case shows us that a must be 2, and b, 1 : so that the formula is $2t + tt$. Indeed if we add together the series 2, 4, 6, which expresses the change proportional to the time, and 1, 4, 9, which is proportional to the square of the time, we obtain the series 3, 8, 15, which is the series of numbers given by observation. And thus the three steps which give us the Idea, the Conception, and the Magnitudes; or the Argument, the Law, and the Amount, of the change; give us the Independent Variable, the Formula, and the Coefficients, respectively.

We now proceed to offer some suggestions of methods by which each of these steps may be in some degree promoted.

SECT. II. *Of the Selection of the Fundamental Idea.*

5. When we turn our thoughts upon any assemblage of facts, with a view of collecting from them some connexion or law, the most important step, and at the same time that in which rules can least aid us, is the Selection of the Idea by which they are to be collected. So long

as this idea has not been detected, all seems to be hope-less confusion or insulated facts; when the connecting idea has been caught sight of, we constantly regard the facts with reference to their connexion, and wonder that it should be possible for any one to consider them in any other point of view.

Thus the different seasons, and the various aspects of the heavenly bodies, might at first appear to be direct manifestations from some superior power, which man could not even understand: but it was soon found that the ideas of time and space, of motion and recurrence, would give coherency to many of the phenomena. Yet this took place by successive steps. Eclipses, for a long period, seemed to follow no law; and being very remark-able events, continued to be deemed the indications of a supernatural will, after the common motions of the heavens were seen to be governed by relations of time and space. At length, however, the Chaldeans discovered that, after a period of eighteen years, similar sets of eclipses recur; and, thus selecting the idea of *time*, simply, as that to which these events were to be referred, they were able to reduce them to rule; and from that time eclipses were recognized as parts of a regular order of things. We may, in the same manner, consider any other course of events, and may inquire by what idea they are bound together. For example, if we take the weather, years peculiarly wet or dry, hot and cold, productive and unproductive, follow each other in a manner which, at first sight at least, seems utterly lawless and irregular. Now can we in any way discover some rule and order in these occurrences? Is there, for example, in these events, as in eclipses, a certain cycle of years, after which like seasons come round again? or does the weather depend upon the force of some extraneous body—for instance, the moon—and follow in some way her aspects? or

would the most proper way of investigating this subject
be to consider the effect of the moisture and heat of
various tracts of the earth's surface upon the ambient air?
It is at our choice to *try* these and other modes of obtaining a science of the weather: that is, we may refer the
phenomena to the idea of *time*, introducing the conception
of a cycle;—or to the idea of external *force*, by the conception of the moon's action;—or to the idea of *mutual
action*, introducing the conceptions of thermotical and
atmological agencies, operating between different regions
of earth, water, and air.

6. It may be asked, How are we to decide in such
alternatives? How are we to select the one right idea
out of several conceivable ones? To which we can only
reply, that this must be done by *trying* which will succeed.
If there really exist a cycle of the weather, as well as of
eclipses, it must be established by comparing the asserted
cycle with a good register of the seasons, of sufficient
extent. Or if the moon really influence the meteorological conditions of the air, the asserted influence must
be compared with the observed facts, and so accepted or
rejected. When Hipparchus had observed the increase
of longitude of the stars, the idea of a motion of the
celestial sphere suggested itself as the explanation of the
change; but this thought was verified only by observing
several stars. It was conceivable that each star should
have an independent motion, governed by time only, or
by other circumstances, instead of being regulated by its
place in the sphere; and this possibility could be rejected
by trial alone. In like manner, the original opinion of the
composition of bodies supposed the compounds to derive
their properties from the elements according to the law
of *likeness;* but this opinion was overturned by a thousand
facts; and thus the really applicable idea of chemical composition was introduced in modern times. In what has

already been said on the History of Ideas, we have seen how each science was in a state of confusion and darkness till the right idea was introduced.

7. No general method of evolving such ideas can be given. Such events appear to result from a peculiar sagacity and felicity of mind;—never without labour, never without preparation;—yet with no constant dependence upon preparation, or upon labour, or even entirely upon personal endowments. Newton explained the colours which refraction produces, by referring each colour to a peculiar *angle of refraction*, thus introducing the right idea. But when the same philosopher tried to explain the colours produced by diffraction, he erred, by attempting to apply the same idea, (the *course of a single ray*,) instead of applying the truer idea of the *interference of two rays*. Newton gave a wrong rule for the double refraction of Iceland spar, by making the refraction depend on the *edges* of the rhombohedron: Huyghens, more happy, introduced the idea of the *axis of symmetry* of the solid, and thus was able to give the true law of the phenomena.

8. Although the selected idea is proved to be the right one, only when the true law of nature is established by means of it, yet it often happens that there prevails a settled conviction respecting the relation which must afford the key to the phenomena, before the selection has been confirmed by the laws to which it leads. Even before the empirical laws of the tides were made out, it was not doubtful that these laws depended upon the places and motions of the sun and moon. We know that the crystalline form of a body must depend upon its chemical composition, though we are as yet unable to assign the law of this dependence.

Indeed in most cases of great discoveries, the right idea to which the facts were to be referred, was selected by many philosophers, before the decisive demonstration that it *was* the right idea, was given by the discoverer. Thus Newton showed that the motions of the planets might be explained by means of a central force in the sun: but though he established, he did not first select the idea involved in the conception of a central force. The idea had already been sufficiently pointed out, dimly by Kepler, more clearly by Borelli, Huyghens, Wren, and Hooke. Indeed this anticipation of the true idea is always a principal part of that which, in the History of the Sciences, we have termed the *Prelude* of a Discovery. The two steps, of *proposing* a philosophical problem, and of *solving* it, are, as we have elsewhere said, both important, and are often performed by different persons. The former step is, in fact, the Selection of the Idea. In explaining any change, we have to discover first the *Argument*, and then the *Law* of the change. The selection of the Argument is the step of which we here speak; and is that in which inventiveness of mind and justness of thought are mainly shown.

9. Although, as we have said, we can give few precise directions for this cardinal process, the Selection of the Idea, in speculating on phenomena, yet there is one Rule which may have its use: it is this:—*The idea and the facts must be homogeneous:* the elementary Conceptions, into which the facts have been decomposed, must be of the same nature as the Idea by which we attempt to collect them into laws. Thus, if facts have been observed and measured by reference to space, they must be bound together by the idea of space: if we would obtain a knowledge of mechanical forces in the solar system, we must observe mechanical phenomena. Kepler erred against this rule in his attempts at obtaining physical laws of the system; for the facts which he took were the *velocities*, not the *changes of velocity*, which are really the

mechanical facts.   Again, there has been a transgression of this Rule committed by all chemical philosophers who have attempted to assign the relative position of the elementary particles of bodies in their component molecules. For their purpose has been to discover the *relations* of the particles *in space;* and yet they have neglected the only facts in the constitution of bodies which have a reference to space—namely, *crystalline form*, and *optical properties*. No progress can be made in the theory of the elementary structure of bodies, without making these classes of facts the main basis of our speculations.

10. The only other Rule which I have to offer on this subject, is that which I have already given:—*the Idea must be tested by the facts*.   It must be tried by applying to the facts the conceptions which are derived from the idea, and not accepted till some of these succeed in giving the law of the phenomena.   The justice of the suggestion cannot be known otherwise than by making the trial. If we can discover a *true law* by employing any conceptions, the idea from which these conceptions are derived is the *right* one; nor can there be any proof of its rightness so complete and satisfactory, as that we are by it led to a solid and permanent truth.

This, however, can hardly be termed a Rule; for when we would know, to conjecture and to try the truth of our conjecture by a comparison with the facts, is the natural and obvious dictate of common sense.

Supposing the Idea which we adopt, or which we would try, to be now fixed upon, we still have before us the range of many Conceptions derived from it; many Formulæ may be devised depending on the same Independent Variable, and we must now consider how our selection among these is to be made.

CHAPTER VI.

GENERAL RULES FOR THE CONSTRUCTION OF THE CONCEPTION.

1. IN speaking of the discovery of laws of nature, those which depend upon *quantity*, as number, space, and the like, are most prominent and most easily conceived, and therefore in speaking of such researches we shall often use language which applies peculiarly to the cases in which quantities numerically measurable are concerned, leaving it for a subsequent task to extend our principles to ideas of other kinds.

Hence we may at present consider the Construction of a Conception which shall include and connect the facts, as being the construction of a Mathematical Formula, coinciding with the numerical expression of the facts; and we have to consider how this process can be facilitated, it being supposed that we have already before us the numerical measures given by observation.

2. We may remark, however, that the construction of the right Formula for any such case, and the determination of the Coefficients of such formula, which we have spoken of as two separate steps, are in practice almost necessarily simultaneous; for the near coincidence of the results of the theoretical rule with the observed facts confirms at the same time the Formula and its Coefficients. In this case also, the mode of arriving at truth is to try various hypotheses;—to modify the hypotheses so as to approximate to the facts, and to multiply the facts so as to test the hypotheses.

The Independent Variable, and the Formula which we would try, being once selected, mathematicians have devised certain special and technical processes by which the value of the coefficients may be determined. These we shall treat of in the next Chapter; but in the mean time we may note, in a more general manner, the mode in which, in physical researches, the proper formula may be obtained.

3. A person somewhat versed in mathematics, having before him a series of numbers, will generally be able to devise a formula which approaches near to those numbers. If, for instance, the series is constantly progressive, he will be able to see whether it more nearly resembles an arithmetical or a geometrical progression. For example, MM. Dulong and Petit, in their investigation of the law of cooling of bodies, obtained the following series of measures. A thermometer was placed in an inclosure of which the temperature was 0 degrees, and the rapidity of cooling of the thermometer was noted for many temperatures. It was found that

For the temperature	240	the rapidity of cooling was	10·69
,,	220	,,	8·81
,,	200	,,	7·40
,,	180	,,	6·10
,,	160	,,	4·89
,,	140	,,	3·88

and so on. Now this series of numbers manifestly increases with greater rapidity as we proceed from the lower to the higher parts of the scale. The numbers do not, however, form a geometrical series, as we may easily ascertain. But if we were to take the differences of the successive terms we should find them to be—

$$1·88, \ 1·41, \ 1·30, \ 1·21, \ 1·01, \ \&c.$$

and these numbers are very nearly the terms of a geometric series. For if we divide each term by the succeeding one, we find these numbers,

$$1·33, \ 1·09, \ 1·07, \ 1·20, \ 1·27,$$

in which there does not appear to be any constant tendency to diminish or increase. And we shall find that a geometrical series in which the ratio is 1·165, may be made to approach very near to this series, the deviations from it being only such as may be accounted for by conceiving them as errors of observation. In this manner a certain formula* is obtained, giving results which very nearly coincide with the observed facts, as may be seen in the margin.

The physical law expressed by the formula just spoken of is this :—that when a body is cooling in an empty inclosure at a constant temperature, the quickness of the cooling, for excesses of temperature in arithmetical progression, increases as the terms of a geometrical progression, diminished by a constant number.

3. In the actual investigation of Dulong and Petit, however, the formula was not obtained in precisely the manner just described. For the quickness of cooling depends upon two elements, the temperature of the hot body and the temperature of the inclosure; not merely upon the *excess* of one of these over the other. And it was found most convenient, first, to make such experiments as should exhibit the dependence of the velocity of cooling upon the temperature of the inclosure; which

* The formula is $v = 2{,}037\,(a^{t}-1)$ where v is the velocity of cooling, t the temperature of the thermometer expressed in degrees, and a is the quantity 1,0077.

The degree of coincidence is as follows :—

Excess of temperature of the thermometer, or values of t.		Observed values of v.		Calculated values of v.
240	. .	10·69	. .	10·68
220	. .	8·81	. .	8·89
200	. .	7·40	. .	7·34
180	. .	6·10	. .	6·03
160	. .	4·89	. .	4·87
140	. .	3·88	. .	3·89
120	. .	3·02	. .	3·05
100	. .	2·30	. .	2·33
80	. .	1·74	. .	1.72

dependence is contained in the following law:—The quickness of cooling of a thermometer in vacuo for a constant excess of temperature, increases in geometric progression, when the temperature of the inclosure increases in arithmetic progression. From this law the preceding one follows by necessary consequence*.

This example may serve to show the nature of the artifices which may be used for the construction of formulæ, when we have a constantly progressive series of numbers to represent. We must not only endeavour by trial to contrive a formula which will answer the conditions, but we must vary our experiments so as to determine, first one factor or portion of the formula, and then the other; and we must use the most probable hypothesis as means of suggestion for our formulæ.

4. In a *progressive* series of numbers, except the formula which we adopt be really that which expresses the law of nature; the deviations of the formula from the facts will generally become enormous, when the experiments are extended into new parts of the scale. True formulæ for a progressive series of results can hardly ever be obtained from a very limited range of experiments: just, as the attempt to guess the general course of a road or a river, by knowing two or three points of it in the neighbourhood of one another, would generally fail. In the investigation respecting the laws of the cooling of bodies just noticed, one great advantage of

---

* For if θ be the temperature of the inclosure, and t the excess of temperature of the hot body, it appears, by this law, that the radiation of heat is as a^{θ}. And hence the quickness of cooling, which is as the excess of radiation, is as $a^{\theta + t} - a^{\theta}$; that is, as $a^{\theta}(a^{t} - 1)$ which agrees with the formula given in the last note.

The whole of this series of researches of Dulong and Petit is full of the most beautiful and instructive artifices for the construction of the proper formulæ in physical research.

the course pursued by the experimenters was, that their experiments included so great a range of temperatures. The attempts to assign the law of elasticity of steam deduced from experiments made with moderate temperatures, were found to be enormously wrong, when very high temperatures were made the subject of experiment. It is easy to see that this must be so: an arithmetical and a geometrical series may nearly coincide for a few terms moderately near each other: but if we take remote corresponding terms in the two series, one of these will be very many times the other. And hence, from a narrow range of experiments, we may infer one of these series when we ought to infer the other; and thus obtain a law which is widely erroneous.

5. In astronomy, the serieses of observations which we have to study are, for the most part, not progressive, but *recurrent*. The numbers observed do not go on constantly increasing; but after increasing up to a certain amount they diminish; then, after a certain space, increase again; and so on, changing constantly through certain *cycles*. In cases in which the observed numbers are of this kind, the formula which expresses them must be a *circular function*, of some sort or other; involving, for instance, sines, tangents, and other forms of calculation, which have recurring values when the angle on which they depend goes on constantly increasing. The main business of formal astronomy consists in resolving the celestial phenomena into a series of *terms* of this kind, in detecting their *arguments*, and in determining their *coefficients*.

6. In constructing the formulæ by which laws of nature are expressed, although the first object is to assign the law of the phenomena, philosophers have, in almost all cases, not proceeded in a purely empirical manner, to connect the observed numbers by some expression of

calculation, but have been guided, in the selection of their formula, by some *hypothesis* respecting the mode of connexion of the facts. Thus the formula of Dulong and Petit above given was suggested by the theory of exchanges; the first attempts at the resolution of the heavenly motions into circular functions were clothed in the hypothesis of epicycles. And this was almost inevitable. "We must confess," says Copernicus[*], "that the celestial motions are circular, or compounded of several circles, since their inequalities observe a fixed law, and recur in value at certain intervals, which could not be except they were circular: for a circle alone can make that quantity which has occurred recur again." In like manner the first publication of the *law of the sines*, the true formula of optical refraction, was accompanied by Descartes with an hypothesis, in which an explanation of the law was pretended. In such cases, the mere comparison of observations may long fail in suggesting the true formulæ. The fringes of shadows and other diffracted colours were studied in vain by Newton, Grimaldi, Comparetti, the elder Herschel, and Mr. Brougham, so long as these inquirers attempted merely to trace the laws of the facts as they appeared in themselves; while Young, Fresnel, Fraunhofer, Schwerdt, and others, determined these laws in the most rigorous manner, when they applied to the observations the hypothesis of interferences.

7. But with all the aid that hypotheses and calculation can afford, the construction of true formulæ, in those cardinal discoveries by which the progress of science has mainly been caused, has been a matter of great labour and difficulty, and of good fortune added to sagacity. In the History of Science, we have seen how long and how hard Kepler laboured, before he converted the formula for the planetary motions, from an epicyclical combination,

[*] *De Rev.* l. i., c. 4.

to a simple ellipse. The same philosopher, labouring with equal zeal and perserverance to discover the formula of optical refraction, which now appears to us so simple, was utterly foiled. Malus sought in vain the formula determining the angle at which a transparent surface polarizes light : Sir D. Brewster*, with a happy sagacity, discovered the formula to be simply this, that the index of refraction is the tangent of the angle of polarization.

Though we cannot give rules which will be of much service when we have thus to divine the general form of the relation by which phenomena are connected, there are certain methods by which, in a narrower field, our investigations may be materially promoted ;—certain special methods of obtaining laws from observations. Of these we shall now proceed to treat,

---

Chapter VII.

SPECIAL METHODS OF INDUCTION APPLICABLE TO QUANTITY.

In cases where the phenomena admit of numerical measurement and expression, certain mathematical methods may be employed to facilitate and give accuracy to the determination of the formula by which the observations are connected into laws. Among the most usual and important of these Methods are the following :—

I. The Method of Curves.

II. The Method of Means.

III. The Method of Least Squares.

IV. The Method of Residues.

Sect. I. *The Method of Curves.*

1. The Method of Curves proceeds upon this basis ; that when one quantity undergoes a series of changes

* *Hist. Ind. Sci.*, ii. 377.

depending on the progress of another quantity, (as, for instance, the Deviation of the Moon from her equable place depends upon the progress of Time,) this dependence may be expressed by means of a *curve*. In the language of mathematicians, the variable quantity, whose changes we would consider, is made the *ordinate* of the curve, and the quantity on which the changes depend is made the *abscissa*. In this manner, the curve will exhibit in its form a series of undulations, rising and falling so as to correspond with the alternate increase and diminution of the quantity represented, at intervals of space which correspond to the intervals of time, or other quantity by which the changes are regulated. Thus, to take another example, if we set up, at equal intervals, a series of ordinates representing the height of all the successive high waters brought by the tides at a given place, for a year, the curve which connects the summits of all these ordinates will exhibit a series of undulations, ascending and descending once in about each fortnight; since, in that interval, we have, in succession, the high spring tides and the low neap tides. The curve thus drawn offers to the eye a picture of the order and magnitude of the changes to which the quantity under contemplation, (the height of high water,) is subject.

2. Now the peculiar facility and efficacy of the Method of Curves depends upon this circumstance;—that order and regularity are more readily and clearly recognized, when thus exhibited to the eye in a picture, than they are when presented to the mind in any other manner. To detect the relations of Number considered directly as Number, is not easy: and we might contemplate for a long time a Table of recorded Numbers without perceiving the order of their increase and diminution, even if the law were moderately simple; as any one may satisfy himself by looking at a Tide Table. But if these Numbers

are expressed by the magnitude of *Lines*, and if these Lines are arranged in regular order, the eye readily discovers the rule of their changes : it follows the curve which runs along their extremities, and takes note of the order in which its convexities and concavities succeed each other, if any order be readily discoverable.· The separate observations are in this manner compared and generalized and reduced to rule by the eye alone. And the eye, so employed, detects relations of order and succession with a peculiar celerity and evidence. If, for example, we thus arrange as ordinates the prices of corn in each year for a series of years, we shall see the order, rapidity, and amount of the increase and decrease of price, far more clearly than in any other manner. And if there were any recurrence of increase and decrease at stated intervals of years, we should in this manner perceive it. The eye, constantly active and busy, and employed in making into shapes the hints and traces of form which it contemplates, runs along the curve thus offered to it ; and as it travels backwards and forwards, is ever on the watch to detect some resemblance or contrast between one part and another. And these resemblances and contrasts, when discovered, are the images of laws of phenomena ; which are made manifest at once by this artifice, although the mind could not easily catch the indications of their existence, if they were not thus reflected to her in the clear mirror of space.

Thus when we have a series of good observations, and know the argument upon which their change of magnitude depends, the Method of Curves enables us to ascertain, almost at a glance, the law of the change ; and by further attention, may be made to give us a formula with great accuracy. The Method enables us to perceive among our observations an order which without the method is concealed in obscurity and perplexity.

3. But the Method of Curves not only enables us to obtain laws of nature from good observations, but also, in a great degree, from observations which are very *imperfect*. For the imperfection of observations may in part be corrected by this consideration;—that though they may appear irregular, the correct facts which they imperfectly represent, are really regular, And the Method of Curves enables us to remedy this apparent irregularity, at least in part. For when observations thus imperfect are laid down as ordinates, and their extremities connected by a line, we obtain, not a smooth and flowing curve, such as we should have if the observations contained only the rigorous results of regular laws; but a broken and irregular line, full of sudden and capricious twistings, and bearing on its face marks of irregularities dependent, not upon law, but upon chance. Yet these irregular and abrupt deviations in the curve are, in most cases, but small in extent, when compared with those bendings which denote the effects of regular law. And this circumstance is one of the great grounds of advantage in the Method of Curves. For when the observations thus laid down present to the eye such a broken and irregular line, we can still see, often with great ease and certainty, what twistings of the line are probably due to the irregular errors of observation; and can at once reject these, by drawing a more regular curve, cutting off all such small and irregular sinuosities, leaving some to the right and some to the left; and then proceeding as if this regular curve, and not the irregular one, expressed the observations. In this manner, we suppose the errors of observation to balance each other: some of our corrected measures being too great and others too small, but with no great preponderance either way. We draw our main regular curve, not *through* the points given by our observations, but *among* them : drawing it, as has been said by

one of the philosophers* who first systematically used this method, " with a bold but careful hand." The regular curve which we thus obtain, thus freed from the casual errors of observation, is that in which we endeavour to discover the laws of change and succession.

4. By this method, thus getting rid at once, in a great measure, of errors of observation, we obtain data which are *more true than the* individual *facts themselves.* The philosopher's business is to compare his hypotheses with facts, as we have often said. But if we make the comparison with separate special facts, we are liable to be perplexed or misled, to an unknown amount, by the errors of observation; which may cause the hypothetical and the observed result to agree, or to disagree, when otherwise they would not do so. If, however, we thus take the *whole mass of the facts,* and remove the errors of actual observation†, by making the curve which expresses the supposed observation regular and smooth, we have the separate facts corrected by their general tendency. We are put in possession, as we have said, of something more true than any fact by itself is.

One of the most admirable examples of the use of this Method of Curves is found in Sir John Herschel's *Investigation of the orbits of double stars*‡. The author there shows how far inferior the direct observations of the angle of position are, to the observations corrected by a curve in the manner above stated. "This curve once drawn," he says, " must represent, it is evident, the law of variation of the angle of position, with the time, not only for instants intermediate between the dates of observations, but even at the moments of observation themselves, much better than the individual *raw* observations can possibly (on an average) do. It is only requisite to try a case or two, to be satisfied that by substituting the curve for the points, we have made a nearer approach to nature, and in a great measure eliminated errors of observation." "In following the graphical process," he adds, "we have a conviction almost approaching to moral certainty that we cannot be greatly misled." Again, having thus corrected the raw observations, he makes another use of the graphical method, by trying whether an ellipse can be drawn "if not *through*, at least *among* the points, so as to approach tolerably near them all; and thus approaching to the orbit which is the subject of investigation."

5. The *obstacles* which principally impede the application of the method of curves are (I.) our *ignorance of the argument* of the changes, and (II.) the *complication of several laws* with one another.

(I.) If we do not know on what quantity those changes depend which we are studying, we may fail entirely in detecting the law of the changes, although we throw the observations into curves. For the true argument of the change should, in fact, be made the abscissa of the curve. If we were to express, by a series of ordinates, the *hour* of high water on successive days, we should not obtain, or should obtain very imperfectly, the law which these times follow; for the real argument of this change is not the *solar* hour, but the hour at which the *moon* passes the meridian. But if we are supposed to be aware that *this* is the argument, (which theory suggests and trial instantly confirms) we then do immediately obtain the primary rules of the time of high water, by throwing a series of observations into a curve, with the hour of the moon's transit for the abscissa.

In like manner, when we have obtained the first great or semi-menstrual inequality of the tides, if we endeavour to discover the laws of other inequalities by means of curves, we must take from theory the suggestion that the Arguments of such inequalities will probably be the *parallax* and the *declination* of the moon. This suggestion again is confirmed by trial; but if we were supposed to be entirely ignorant of the dependence of the changes of the tide on the distance and declination of the moon, the curves would exhibit unintelligible and seemingly capricious changes. For by the effect of the inequality arising from the parallax, the convexities of the curves which belong to the spring tides, are in some years made alternately greater and less all the year through; while in other years they are made all nearly equal. This difference does not betray its origin, till we refer it to the parallax; and the same difficulty in proceeding would arise if we were ignorant that the moon's declination is one of the arguments of tidal changes.

In like manner, if we try to reduce to law any meteorological changes, those of the height of the barometer for instance, we find that we can make little progress in the investigation, precisely because we do not know the Argument on which these changes depend. That there is a certain regular *diurnal* change of small amount we know; but when we have abstracted this inequality, (of which the Argument is the *time of day*,) we find far greater changes left behind, from day to day and from hour to hour; and we express these in curves, but we cannot reduce them to rule, because we cannot discover on what numerical quantity they depend. The assiduous study of barometrical observations, thrown into curves, may perhaps hereafter point out to us what are the relations of time and space by which these variations are determined; but in the mean time, this subject exemplifies to us our remark, that the method of curves is of comparatively small use, so long as we are in ignorance of the real Arguments of the Inequalities.

6. (II.) In the next place, I remark that a difficulty is thrown in the way of the method of curves by *the combination of several laws* one with another. It will readily be seen that such a cause will produce a complexity in the curves which exhibit the succession of facts. If, for example, we take the case of the tides, the height of high water increases and diminishes with the approach of the sun to, and its recess from, the syzygies of the moon. Again, this height increases and diminishes as the moon's parallax increases and diminishes; and again, the height diminishes when the declination increases, and *vicé versa;* and all these Arguments of change, the distance from syzygy, the parallax, the declination, complete their circuit and return into themselves in different periods. Hence the curve which represents the height of high water has not any periodical interval in which it completes its changes and commences a new cycle. The sinuosity which would arise from each inequality separately considered, interferes with, disguises, and conceals the others; and when we first cast our eyes on the curve of observation, it is very far from offering any obvious regularity in its form. And it is to be observed that we have not yet enumerated *all* the elements of this complexity: for there are changes of the tide depending upon the parallax and declination of the sun as well as of the moon. Again; besides these changes, of which the arguments are obvious, there are others, as those depending upon the barometer and the wind, which follow no known regular law, and which constantly affect and disturb the results produced by other laws.

In the tides, and in like manner in the motions of the moon, we have very eminent examples of the way in which the discovery of laws may be rendered difficult by the number of them which operate to affect the same quantity. In such cases, the inequalities are generally picked out in succession, nearly in the order of their magnitudes. In this way there were successively collected, from the study of the moon's motions by a series of astronomers, those Inequalities which we term the *Equation of the Centre*, the *Evection*, the *Variation*, and the *Annual Equation*. These Inequalities were not, in fact, obtained by the application of the Method of Curves; but the Method of Curves might have been applied to such a case with great advantage. The Method has been applied with great industry and with remarkable success to the investigation of the laws of the tides; and by the use of it, a series of Inequalities both of the Times and of the Heights of high water has been detected, which explain all the main features of the observed facts.

SECT. II. *The Method of Means.*

7. The Method of Curves, as we have endeavoured to explain above, frees us from the casual and extraneous irregularities which arise from the imperfection of observation; and thus lays bare the results of the laws which really operate, and enables us to proceed in search of those laws. But the Method of Curves is not the only one which effects such a purpose. The errors arising from detached observations may be got rid of, and the additional accuracy which multiplied observations give may be obtained, by operations upon the observed numbers without expressing them by spaces. The process of curves assumes that the errors of observation balance each other;—that the accidental excesses and defects are nearly equal in amount;—that the true quantities which would have been observed if all accidental causes of irregularity were removed, are obtained, exactly or nearly, by selecting quantities, upon the whole, equally distant from the extremes of great and small which our imperfect observations offer to us. But when, among a number of unequal quantities, we take a quantity equally distant from

the greater and the smaller, this quantity is termed the *Mean* of the unequal quantities. Hence the correction of our observations by the method of curves consists in taking the Mean of the observations.

8. Now without employing curves, we may proceed arithmetically to take the Mean of all the observed numbers of each class. Thus, if we wished to know the height of the spring tide at a given place, and if we found that four different spring tides were measured as being of the height of ten, thirteen, eleven, and fourteen feet, we should conclude that the true height of the tide was the *Mean* of these numbers,—namely, twelve feet; and we should suppose that the deviation from this height, in the individual cases, arose from the accidents of weather, the imperfections of observation, or the operation of other laws, besides the alternation of spring and neap tides.

This process of finding the Mean of an assemblage of observed numbers is much practised in discovering, and still more in confirming and correcting, laws of phenomena. We shall notice a few of its peculiarities.

9. It requires a knowledge of the *Argument* of the changes which we would study; for the numbers must be arranged in certain Classes, before we find the Mean of each Class; and the principle on which this arrangement depends is the Argument. This knowledge of the Argument is more indispensably necessary in the Method of Means than the Method of Curves; for when curves are drawn, the eye often spontaneously detects the law of recurrence in their sinuosities; but when we have collections of numbers, we must divide them into classes by a selection of our own. Thus, in order to discover the law which the heights of the tide follow, in the progress from spring to neap, we arrange the observed tides according to the *day of the moon's age;* and we then take the mean of all

those which thus happen at the same period of the moon's revolution. In this manner we obtain the law which we seek; and the process is very nearly the same in all other applications of this Method of Means. In all cases, we begin by assuming the Classes of measures which we wish to compare, the Law which we could confirm or correct, the Formula of which we would determine the coefficients.

10. The Argument being thus assumed, the Method of Means is very efficacious in ridding our inquiry of errors and irregularities which would impede and perplex it. Irregularities which are altogether accidental, or at least accidental with reference to some law which we have under consideration, compensate each other in a very remarkable way, when we take the means of many observations. If we have before us a collection of observed tides, some of them may be elevated, some depressed by the wind, some noted too high and some too low by the observer, some augmented and some diminished by uncontemplated changes in the moon's distance or motion: but in the course of a year or two at the longest, all these causes of irregularity balance each other; and the law of succession, which runs through the observations, comes out as precisely as if those disturbing influences did not exist. In any particular case, there appears to be no possible reason why the deviation should be in one way, or of one moderate amount, rather than another. But taking the mass of observations together, the deviations in opposite ways will be of equal amount, with a degree of exactness very striking. This is found to be the case in all inquiries where we have to deal with observed numbers upon a large scale. In the progress of the population of a country, for instance, what can appear more inconstant, in detail, than the causes which produce births and deaths? yet in each country, and even in each

province of a country, the proportions of the whole numbers of births and deaths remain nearly constant. What can be more seemingly beyond the reach of rule than the occasions which produce letters that cannot find their destination? yet it appears that the number of "dead letters" is nearly the same from year to year. And the same is the result when the deviations arise, not from mere accident, but from laws perfectly regular, though not contemplated in our investigation*. Thus the effects of the Moon's Parallax upon the Tides, sometimes operating one way and sometimes another, according to certain rules, are quite eliminated by taking the Means of a long series of observations; the excesses and defects neutralizing each other so far as concerns the effect upon any law of the tides which we would investigate.

11. In order to obtain very great accuracy, very large masses of observations are often employed by philosophers, and the accuracy of the result increases with the multitude of observations. The immense collections of astronomical observations which have in this manner been employed in order to form and correct the tables of the celestial motions are perhaps the most signal instances of the attempts to obtain accuracy by this accumulation of observations. Delambre's Tables of the Sun are founded upon nearly 3000 observations; Bürg's Tables of the Moon upon above 4000.

But there are other instances hardly less remarkable. Mr. Lubbock's first investigations of the laws of the tides of London[†], included above 13,000 observations, extending through nineteen years; it being considered that this large number was necessary to remove the effects of

---

* Provided the argument of the law which we neglect have no coincidence with the argument of the law which we would determine.

[†] *Phil. Tr.* 1831.

accidental causes*. And the attempts to discover the laws of change in the barometer have led to the performance of labours of equal amount: Laplace and Bouvard examined this question by means of observations made at the Observatory of Paris, four times every day for eight years.

12. We may remark one striking evidence of the accuracy thus obtained by employing large masses of observations. In this way we may often detect inequalities much smaller than the errors by which they are encumbered and concealed. Thus the diurnal oscillations of the barometer were discovered by the comparison of observations of many days, classified according to the hours of the day; and the result was a clear and incontestable proof of the existence of such oscillations, although the differences which these oscillations produce at different hours of the day are far smaller than the casual changes, hitherto reduced to no law, which go on from hour to hour and from day to day. The effect of law, operating incessantly and steadily, makes itself more and more felt as we give it a longer range; while the effect of accident, followed out in the same manner, is to annihilate itself, and to disappear altogether from the result.

SECT. III. *The Method of Least Squares.*

13. The Method of Least Squares is in fact a method of means, but with some peculiar characters. Its object is to determine the *best Mean* of a number of observed quantities; or the *most probable Law* derived from a number

---

* This period of nineteen years was also selected for a reason which is alluded to in a former note. (p. 553.) It was thought that this period secured the inquirer from the errors which might be produced by the partial coincidence of the arguments of different irregularities; for example, those due to the moon's parallax and to the moon's declination. It has since been found (*Phil. Tr.* 1838. *On the Determination of the Laws of the Tides from Short Series of Observations,*) that with regard to parallax at least, the Means of one year give sufficient accuracy.

of observations, of which some, or all, are allowed to be more or less imperfect. And the method proceeds upon this supposition;—that all errors are not *equally* probable, but that small errors are more probable than large ones. By reasoning mathematically upon this ground, we find that the best result is obtained (since we cannot obtain a result in which the errors vanish) by making, not the *Errors* themselves, but the *Sum of their Squares* of the *smallest* possible amount.

14. An example may illustrate this. Let a quantity which is known to increase uniformly, (as the distance of a star from the meridian at successive instants,) be measured at equal intervals of time, and be found to be successively 4, 12, 14. It is plain, upon the face of these observations that they are erroneous; for they ought to form an arithmetical progression, but they deviate widely from such a progression. But the question then occurs, what arithmetical progression do they *most probably* represent: for we may assume several arithmetical progressions which more or less approach the observed series; as for instance, these three; 4, 9, 14; 6, 10, 14; 5, 10, 15. Now in order to see the claims of each of these to the truth, we may tabulate them thus.

Observation

	4, 12, 14	Errors.	Sums of errors.	Sums of squares of errors.
Series (1)	4, 9, 14 ...	0, 3, 0 ...	3 ...	9
„ (2)	6, 10, 14 ...	2, 2, 0 ...	4 ..	8
„ (3)	5, 10, 15 ...	1, 2, 1 ...	4 ...	6

Here, although the first series gives the sum of the errors less than the others, the third series gives the sum of the squares of the errors least; and is therefore, by the proposition on which this Method depends, the *most probable* series of the three.

This Method, in more extensive and complex cases, is a

great aid to the calculator in his inferences from facts, and removes much that is arbitrary in the Method of Means.

SECT. IV. *The Method of Residues.*

15. By either of the preceding Methods we obtain, from observed facts, such laws as readily offer themselves; and by the laws thus discovered, the most prominent changes of the observed quantities are accounted for. But in many cases we have, as we have noticed already, *several* laws of nature operating at the same time, and combining their influences to modify those quantities which are the subjects of observation. In these cases we may, by successive applications of the Methods already pointed out, detect such laws one after another: but this successive process, though only a repetition of what we have already described, offers some peculiar features which make it convenient to consider it in a separate Section, as the Method of Residues.

16. When we have, in a series of changes of a variable quantity, discovered *one* Law which the changes follow, detected its argument, and determined its magnitude so as to explain most clearly the course of observed facts, we may still find that the observed changes are not fully accounted for. When we compare the results of our Law with the observations, there may be a difference, or as we may term it, a *Residue,* still unexplained. But this Residue being thus detached from the rest, may be examined and scrutinized in the same manner as the whole observed quantity was treated at first: and we may in this way detect in *it* also a Law of change. If we can do this, we must accommodate this new found Law as nearly as possible to the Residue to which it belongs; and this being done, the difference of our Rule and of the Residue itself, forms *a Second Residue.* This Second Residue we may again bring under our consideration; and may perhaps in *it* also discover some Law of change by

which its alterations may be in some measure accounted for. If this can be done, so as to account for a large portion of this Residue, the remaining unexplained part forms a *Third Residue;* and so on.

17. This course has really been followed in various inquiries, especially in those of Astronomy and Tidology. The *Equation of the Centre,* for the moon, was obtained out of the *Residue* of the Longitude, which remained when the *Mean Anomaly* was taken away. This Equation being applied and disposed of, the *Second Residue* thus obtained, gave to Ptolemy the *Evection.* The *Third Residue,* left by the Equation of the Centre and the Evection, supplied to Tycho the *Variation* and the *Annuul Equation.* And the Residue, remaining from these, has been exhausted by other equations, of various arguments, suggested by theory or by observation. In this case, the successive generations of astronomers have gone on, each in its turn executing some step in this Method of Residues. In the examination of the Tides, on the other hand, this method has been applied systematically and at once. The observations readily gave the *Semimenstrual Inequality;* the *Residue* of this supplied the corrections due to the Moon's *Parallax* and *Declination;* and when these were determined, the *remaining Residue* was explored for the law of the Solar Correction.

18. In a certain degree, the Method of Residues and the Method of Means are *opposite* to each other, For the Method of Residues extricates Laws from their combination, *bringing them into view in succession;* while the Method of Means discovers each Law, not by bringing the others into view, but by *destroying their effect* through an accumulation of observations. By the Method of Residues we should *first* extract the Law of the Parallax Correction of the Tides, and *then,* from the Residue left by this, obtain the Declination Correction. But we

might at once employ the Method of Means, and put together all the cases in which the Declination was the same; not allowing for the Parallax in each case, but taking for granted that the Parallaxes belonging to the same Declination would neutralize each other; as many falling above as below the mean parallax. In cases like this, where the Method of Means is not impeded by a partial coincidence of the Arguments of different unknown Inequalities, it may be employed with almost as much success as the Method of Residues. But still, when the Arguments of the Laws are clearly known, as in this instance, the Method of Residues is more clear and direct, and is the rather to be recommended.

19. If for example, we wish to learn whether the Height of the Barometer exerts any sensible influence on the Height of the Sea's Surface, it would appear that the most satisfactory mode of proceeding, must be to subtract, in the first place, what we know to be the effects of the Moon's Age, Parallax and Declination, and other ascertained causes of change; and to search in the *unexplained Residue* for the effects of barometrical pressure. The contrary course has, however, been adopted, and the effect of the barometer on the ocean has been investigated by the direct application of the Method of Means, classing the observed heights of the water according to the corresponding heights of the barometer without any previous reduction. In this manner, the suspicion that the tide of the sea is effected by the pressure of the atmosphere, has been confirmed. This investigation must be looked upon as a remarkable instance of the efficacy of the Method of Means, since the amount of the barometrical effect is much smaller than the other changes from among which it was by this process extricated. But an application of the Method of Residues would still be desirable on a subject of such extent and difficulty.

20. Sir John Herschel, in his *Discourse on the Study*

of Natural Philosophy (Articles 158—161), has pointed out the mode of making discoveries by studying Residual Phenomena; and has given several illustrations of the process. In some of these, he has also considered this method in a wider sense than we have done; treating it as not applicable to quantity only, but to properties and relations of different kinds.

We likewise shall proceed to offer a few remarks on Methods of Induction applicable to other relations than those of quantity.

CHAPTER VIII.

METHODS OF INDUCTION DEPENDING ON RESEMBLANCE.

SECT. I. *The Law of Continuity.*

1. THE Law of Continuity is applicable to quantity primarily, and therefore might be associated with the methods treated of in the last chapter: but inasmuch as its inferences are made by a transition from one degree to another among contiguous cases, it will be found to belong more properly to the Methods of Induction of which we have now to speak.

The *Law of Continuity* consists in this proposition,— That a quantity cannot pass from one amount to another by any change of conditions, without passing through all intermediate degrees of magnitude according to the intermediate conditions. And this law may often be employed to correct inaccurate inductions, and to reject distinctions which have no real foundation in nature. For example, the Aristotelians made a distinction between motions according to nature, as that of a body falling vertically downwards, and motions contrary to nature, as that of a body moving along a horizontal plane: the former, they held, became naturally quicker and quicker, the latter naturally slower and slower. But to this it might be replied, that a horizontal line may pass, by gradual motion, through various inclined positions, to a vertical position: and thus the retarded motion may pass into the accelerated; and hence there must be some inclined plane on which the motion downwards is naturally uniform: which is false, and therefore the distinction of such kinds of motion is unfounded. Again, the proof of the First Law of Motion depends upon the Law of Continuity: for since, by diminishing the resistance to a body moving on a horizontal plane, we diminish the retardation, and this without limit, the law of continuity will bring us at the same time to the case of no resistance and to the case of no retardation.

2. The Law of Continuity is asserted by Galileo in a particular application; and the assertion which it suggests is by him referred to Plato;—namely*, that a moveable body cannot pass from rest to a determinate degree of velocity without passing through all smaller degrees of velocity. This law, however, was first asserted in a more general and abstract form by Leibnitz†: and was employed by him to show that the laws of motion propounded by Descartes must be false. The Third Cartesian Law of Motion was this‡: that when one moving body meets another, if the first body have a less momentum than the second, it will be reflected with its whole motion: but if the first have a greater momentum than the second, it will lose a part of its motion, which it will transfer to the second. Now each of these cases leads, by the Law of Continuity, to the case in which the two bodies have *equal* momentums: but in this case, by the first part of

* *Dialog.* iii. 150. iv. 32   † *Opera*, i. 366.
‡ Cartes. *Prin.*, p. 35.

the law the body would *retain all* its motion ; and by the second part of the law it would *lose* a portion of it : hence the Cartesian Law is false.

3. I shall take another example of the application of this Law from Professor Playfair's Dissertation on the History of Mathematical and Physical Science*.   " The Academy of Sciences at Paris having (in 1724) proposed, as a Prize Question, the Investigation of the Laws of the Communication of Motion, John Bernoulli presented an Essay on the subject very ingenious and profound; in which, however, he denied the existence of hard bodies, because in the collision of such bodies, a finite change of motion must take place in an instant: an event which, on the principle just explained, he maintained to be impossible."   And this reasoning was justifiable : for we can form a *continuous* transition from cases in which the impact manifestly occupies a finite time, (as when we strike a large soft body) to cases in which it is apparently instantaneous.   Maclaurin and others are disposed, in order to avoid the conclusion of Bernoulli, to reject the Law of Continuity.   This, however, would not only be, as Playfair says, to deprive ourselves of an auxiliary, commonly useful though sometimes deceptive ; but what is much worse, to acquiesce in false propositions, from the want of clear and patient thinking.   For the Law of Continuity, when rightly interpreted, is *never* violated in actual fact.   There are not really any such bodies as have been termed *perfectly hard :* and if we approach towards such cases, we must learn the laws of motion which rule them by attending to the Law of Continuity, not by rejecting it.

4. Newton used the Law of Continuity to suggest, but not to prove, the doctrine of universal gravitation. Let, he said, a terrestrial body be carried as high as the

In the *Ency. Brit.*, p. 537.

moon: will it not still fall to the earth? and does not the moon fall by the same force*? Again: if any one says that there is a material ether which does not gravitate†, this kind of matter, by condensation, may be gradually transmuted to the density of the most intensely gravitating bodies: and these gravitating bodies, by taking the internal texture of the condensed ether, may cease to gravitate; and thus the weight of bodies depends, not on their quantity of matter, but on their texture; which doctrine Newton conceived he had disproved by experiment.

5. The evidence of the Law of Continuity resides in the universality of those ideas, which enter into our apprehension of Laws of Nature. When, of two quantities, one depends upon the other, the Law of Continuity necessarily governs this dependence. Every philosopher has the power of applying this law, in proportion as he has the faculty of apprehending the ideas which he employs in his induction, with the same clearness and steadiness which belong to the fundamental ideas of quantity, space and number. To those who possess this faculty, the Law is a Rule of very wide and decisive application. Its use, as has appeared in the above examples, is seen rather in the disproof of erroneous views, and in the correction of false propositions, than in the invention of new truths. It is a test of truth, rather than an instrument of discovery.

Methods, however, approaching very near to the Law of Continuity may be employed as positive means of obtaining new truths; and these I shall now describe.

SECT. II. *The Method of Gradation.*

6. To gather together the cases which resemble each other, and to separate those which are essentially distinct, has often been described as the main business of science; and may, in a certain loose and vague manner of speaking, pass for a description of some of the leading procedures in

---

* *Principia*, l. iii. prop. 6.   † *Ib.*, Cor. 2.

the acquirement of knowledge. The selection of instances which agree, and of instances which differ, in some prominent point or property, are important steps in the formation of science. But when classes of things and properties have been established in virtue of such comparisons, it may still be doubtful whether these classes are separated by distinctions of opposites, or by differences of degree. And to settle such questions, the *Method of Gradation* is employed; which consists in taking intermediate stages of the properties in question, so as to ascertain by experiment whether, in the transition from one class to another, we have to leap over a manifest gap, or to follow a continuous road.

7. Thus for instance, one of the early *Divisions* established by electrical philosophers was that of *Electrics* and *Conductors*. But this division Faraday has overturned as an essential opposition. He takes* a *Gradation* which carries him from Conductors to Non-conductors. Sulphur, or lac, he says, are held to be non-conductors, but are not rigorously so. Spermaceti is a bad conductor: ice or water better than spermaceti: metals so much better that they are put in a different class. But even in metals the transit of the electricity is not instantaneous: we have in them proof of a retardation of the electric current: "and what reason," Mr. Faraday asks, " why this retardation should not be of the same kind as that in spermaceti, or in lac, or sulphur? But as, in them, retardation is insulation, [and insulation is induction†] why should we refuse the same relation to the same exhibitions of force in the metals?"

The process employed by the same sagacious philosopher to show the *identity* of Voltaic and Franklinic

---

* *Researches*, 12th Series, Art. 1328.

† These words refer to another proposition, also established by the Method of Gradation.

electricity is another example of the same kind*. Machine [Franklinic] electricity was made to exhibit the same phenomena as Voltaic electricity, by causing the discharge to pass through a bad conductor, into a very extensive discharging train : and thus it was clearly shown that Franklinic electricity, not so conducted, differs from the other kinds, only in being in a state of successive tension and explosion instead of a state of continued current.

Again ; to show that the decomposition of bodies in the voltaic circuit was not due to the Attraction of the Poles†, Mr. Faraday devised a beautiful series of experiments, in which these supposed *Poles* were made to assume all possible electrical conditions:—in which the decomposition took place against air, which according to common language is not a conductor, nor is decomposed ; —against the metallic poles, which are excellent conductors but undecomposable : and hence he infers that the decomposition cannot justly be considered as due to the Attraction, or Attractive Powers, of the Poles.

8. The reader of the *Novum Organon* may perhaps, in looking at such examples of the Rule, be reminded of some of Bacon's classes of instances, as his *instantiæ absentiæ in proximo*, and his *instantiæ migrantes*. But we may remark that instances classed and treated as Bacon recommends in those parts of his work, could hardly lead to scientific truth. His processes are vitiated by his proposing to himself the *form* or *cause* of the property before him, as the object of his inquiry; instead of being content to obtain, in the first place, the *law of phenomena*. Thus his example‡ of a migrating instance is thus given. " Let the *nature inquired into* be that of whiteness; an instance migrating to the production of this property is glass, first whole, and then pulverized; or plain water, and

---

water agitated into a foam; for glass and water are transparent, and not white; but glass powder and foam are
white, and not transparent.   Hence we must inquire
what has happened to the glass or water in that migration.   For it is plain that the *form of whiteness* is conveyed and induced by the crushing of the glass and
shaking of the water."

9. We may easily give examples from other subjects
in which the method of gradation has been used to establish, or to endeavour to establish, very extensive propositions.   Thus Laplace's Nebular Hypothesis,—that systems like our solar system are formed by gradual condensation from diffused masses, such as the nebulæ among
the stars,—is founded by him upon an application of this
Method of Gradation.   We see, he conceives, among
these nebulæ, instances of all degrees of condensation,
from the most loosely diffused fluid, to that separation
and solidification of parts by which suns, and satellites,
and planets are formed: and thus we have before us
instances of systems in all their stages; as in a forest we
see trees in every period of growth.   How far the examples in this case satisfy the demands of the Method of
Gradation, it remains for astronomers and philosophers to
examine.

Again; this method was used with great success by
Maculloch and others to refute the opinion, put in currency by the Wernerian school of geologists, that the
rocks called *trap rocks* must be classed with those to
which a *sedimentary* origin is ascribed.   For it was shown
that a gradual *transition* might be traced from those
examples in which trap rocks most resembled stratified
rocks, to the lavas which have been recently ejected from
volcanoes: and that it was impossible to assign a different
origin to one portion, and to the other, of this kind of
mineral masses; and as the volcanic rocks were certainly

not sedimentary, it followed that the trap rocks were not of that nature.

Again; we have an attempt of a still larger kind made by Mr. Lyell, to apply this Method of Gradation so as to disprove all distinction between the causes by which geological phenomena have been produced, and the causes which are now acting at the earth's surface. He has collected a very remarkable series of changes which have taken place, and are still taking place, by the action of water, volcanoes, earthquakes, and other terrestrial operations; and he conceives he has shown in these a *gradation* which leads, with no wide chasm or violent leap, to the state of things of which geological researches have supplied the evidence.

10. Of the value of this Method in geological speculations, no doubt can be entertained. Yet it must still require a grave and profound consideration, in so vast an application of the Method as that attempted by Mr. Lyell, to determine what extent we may allow to the steps of our *gradation;* and to decide how far the changes which have taken place in distant parts of the series may exceed those of which we have historical knowledge, before they cease to be of the *same kind.* Those who, dwelling in a city, see, from time to time, one house built and another pulled down, may say that such *existing causes,* operating through past time, sufficiently explain the existing condition of the city. Yet we arrive at important political and historical truths, by considering the *origin* of a city as an event of a *different order* from those daily changes. The causes which are now working to produce geological results, may be supposed to have been, at some former epoch, so far exaggerated in their operation, that the changes should be paroxysms, not degrees;—that they should violate, not continue, the gradual series. And we have no kind of evidence whether the duration of our

historical times is sufficient to give us a just measure of the limits of such degrees;—whether the terms which we have under our notice enable us to ascertain the average rate of progression.

11. The result of such considerations seems to be this:—that we may apply the Method of Gradation in the investigation of geological causes, provided we leave the Limits of the Gradation undefined. But, then, this is equivalent to the admission of the opposite hypothesis: for a continuity of which the successive intervals are not limited, is not distinguishable from discontinuity. The geological sects of recent times have been distinguished as *uniformitarians* and *catastrophists*: the Method of Gradation seems to prove the doctrine of the uniformitarians; but then, at the same time that it does this, it breaks down the distinction between them and the catastrophists.

There are other exemplifications of the use of gradations in Science which well deserve notice: but some of them are of a kind somewhat different, and may be considered under a separate head.

SECT. III. *The Method of Natural Classification.*

12. The method of natural classification consists, as we have seen, in grouping together objects, not according to any selected properties, but according to their most important resemblances; and in combining such grouping with the assignation of certain marks of the classes thus formed. The examples of the successful application of this method are to be found in the Classificatory Sciences through their whole extent; as, for example, in framing the Genera of plants and animals. The same method, however, may often be extended to other sciences. Thus the classification of crystalline forms, according to their degree of symmetry, (which is really an important distinction,) as introduced by

Mohs and Weiss, was a great improvement upon Haüy's arbitrary division according to certain assumed primary forms. Sir David Brewster was led to the same distinction of crystals by the study of their optical properties; and the scientific value of the classification was thus strongly exhibited. Mr. Howard's classification of clouds appears to be founded in their real nature, since it enables him to express the laws of their changes and successions. As we have elsewhere said, the criterion of a true classification is, that it makes general propositions possible. One of the most prominent examples of the beneficial influence of a right classification, is to be seen in the impulse given to geology by the distinction of strata according to the organic fossils which they contain*: which, ever since its general adoption, has been a leading principle in the speculations of geologists.

13. The mode in which, in this and in other cases, the Method of Natural Classification directs the researches of the philosopher, is this:—his arrangement being adopted, at least as an instrument of inquiry and trial, he follows the course of the different members of the classification, according to the guidance which Nature herself offers; not prescribing beforehand the marks of each part, but distributing the facts according to the total resemblances, or according to those resemblances which he finds to be most important. Thus, in tracing the course of a series of strata from place to place, we identify each stratum, not by any single character, but by all taken together;— texture, colour, fossils, position, and any other circumstances which offer themselves. And if, by this means, we come to ambiguous cases, where different indications appear to point different ways, we decide so as best to preserve undamaged those general relations and truths which constitute the value of our system. Thus although

* *Hist. Ind. Sci.*, iii. 507.

we consider the organic fossils in each stratum as its most important characteristic, we are not prevented, by the disappearance of some fossils, or the addition of others, or their total absence, from identifying strata in distant countries, if the position and other circumstances authorize us to do so.   And by this Method of Classification, the doctrine of *Geological Equivalents** has been applied to a great part of Europe.

14. We may further observe, that the same method of natural classification which thus enables us to identify strata in remote situations, notwithstanding there may be great differences in their material and contents, also forbids us to assume the identity of the series of rocks which occur in different countries, when it has not been verified by such a continuous exploration of the component members of the series.   It would be in the highest degree unphilosophical to apply the special names of the English or German strata to the rocks of India, or America, or even of southern Europe, till it has appeared that in those countries the geological series of northern Europe really exists.   In each separate country, the divisions of the formations which compose the crust of the earth must be made out, by applying the Method of Natural Arrangement *to that particular case,* and not by arbitrarily extending to it the nomenclature belonging to another case.   It is only by such precautions, that we can ever succeed in obtaining geological propositions, at the same time true and comprehensive ; or can obtain any sound general views respecting the physical history of the earth.

15. The method of natural classification, which we thus recommend, falls in with those mental habits which we formerly described as resulting from the study of natural history.   The method was then termed the *Method of Type,* and was put in opposition to the *Method of Definition.*

* *Hist. Ind. Sci.,* iii. 532.

The Method of Natural Classification is directly opposed to the process in which we assume and apply *arbitrary* definitions; for in the former Method, we find our classes in nature, and do not make them by marks of our own imposition. Nor can any advantage to the progress of knowledge be procured, by laying down our characters when our arrangements are as yet quite loose and unformed. Nothing was gained by the attempts to *define* Metals by their weight, their hardness, their ductility, their colour; for to all these marks, as fast as they were proposed, exceptions were found, among bodies which still could not be excluded from the list of Metals. It was only when elementary substances were divided into *Natural Classes*, of which classes Metals were one, that a true view of their distinctive characters was obtained. Definitions in the outset of our examination of nature are almost always, not only useless, but prejudicial.

16. When we obtain a law of nature by induction from phenomena, it commonly happens, as we have already seen, that we introduce, at the same time, a Proposition and a Definition. In this case, the two are correlative, each giving a real value to the other. In such cases, also, the Definition, as well as the Proposition, may become the basis of rigorous reasoning, and may lead to a series of deductive truths. We have examples of such Definitions and Propositions in the laws of motion, and in many other cases.

17. When we have established Natural Classes of objects, we seek for Characters of our classes; and these Characters may, to a certain extent, be called the *Definitions* of our classes. This is to be understood, however, only in a limited sense: for these Definitions are not absolute and permanent. They are liable to be modified and superseded. If we find a case which manifestly belongs to our Natural Class, though violating our Definition, we do not shut out the case, but alter our defini-

tion. Thus, when we have made it part of our Definition of the *Rose* family, that they have *alternate stipulate leaves*, we do not, therefore, exclude from the family the genus *Lowæa*, which has *no stipulæ*. In Natural Classifications, our Definitions are to be considered as temporary and provisional only. When Mr. Lyell established the distinctions of the tertiary strata, which he termed *Eocene, Miocene,* and *Pliocene,* he took a numerical criterion (the proportion of recent species of shells contained in those strata) as the basis of his division. But now that those kinds of strata have become, by their application to a great variety of cases, a series of Natural Classes, we must, in our researches, keep in view the natural connexion of the formations themselves in different places; and must by no means allow ourselves to be governed by the numerical proportions which were originally contemplated; or even by any amended numerical criterion equally arbitrary; for however amended, Definitions in natural history are never immortal. The etymologies of *Pliocene* and *Miocene* may, hereafter, come to have merely an historical interest; and such a state of things will be no more inconvenient, provided the natural connexions of each class are retained, than it is to call a rock *oolite* or *porphyry*, when it has no roelike structure and no fiery spots.

The Methods of Induction which are treated of in this and the preceding chapter, and which are specially applicable to causes governed by relations of Quantity or of Resemblance, commonly lead us to *Laws of Phenomena* only. Inductions founded upon other ideas, those of Substance and Cause for example, appear to conduct us somewhat further into a knowledge of the essential nature and real connexions of things. But before we speak of these, we shall say a few words respecting the way in which inductive propositions, once obtained, may be verified and carried into effect by their application.

CHAPTER IX.

OF THE APPLICATION OF INDUCTIVE TRUTHS.

1. BY the application of inductive truths, we here mean, according to the arrangement given in page 484, those steps, which in the natural order of science, follow the discovery of each truth. These steps are, the *verification* of the discovery by additional experiments and reasonings, and its *extension* to new cases, not contemplated by the original discoverer. These processes occupy that period, which, in the history of each great discovery, we have termed the *Sequel* of the epoch; as the collection of facts, and the elucidation of conceptions, form its Prelude.

2. It is not necessary to dwell at length on the processes of the verification of discoveries. When the law of nature is once stated, it is far easier to devise and execute experiments which prove it, than it was to discern the evidence before. The truth becomes one of the standard doctrines of the science to which it belongs, and is verified by all who study or who teach the science experimentally. The leading doctrines of chemistry are constantly exemplified by each chemist in his *Laboratory*; and an amount of verification is thus obtained of which books give no adequate conception. In astronomy, we have a still stronger example of the process of verifying discoveries. Ever since the science assumed a systematic form, there have been *Observatories*, in which the consequences of the theory were habitually compared with the results of observation. And to facilitate this comparison, *Tables* of great extent have been calculated, with immense labour, from each theory, showing the place which the theory assigned to the heavenly bodies at successive times; and thus, as it were, challenging nature to deny

the truth of the discovery.   In this way, as I have elsewhere stated, the continued prevalence of an error in the
systematic parts of astronomy is impossible*.   An error,
if it arise, makes its way into the tables, into the ephe-
meris, into the observer's nightly list, or his sheet of re-
ductions ; the evidence of sense flies in its face in a thousand Observatories ; the discrepancy is traced to its source,
and soon disappears for ever.

3. In these last expressions, we suppose the theory,
not only to be tested, but also to be corrected when it is
found to be imperfect.   And this also is part of the business of the observing astronomer.   From his accumulated
observations, he deduces more exact values than had previously been obtained, of the *Coefficients* of these Inequalities of which the *Argument* is already known.   This
he is enabled to do by the methods explained in the fifth
chapter of this Book; the Method of Means, and especially the Method of Least Squares.   In other cases, he
finds, by the Method of Residues, some new Inequality ;
for if no change of the Coefficients will bring the Tables
and the observation to a coincidence, he knows that a
new Term is wanting in his formula.   He obtains, as far
as he can, the law of this unknown Term ; and when its
existence and its law have been fully established, there
remains the task of tracing it to its cause.

4. The condition of the science of Astronomy, with
regard to its security and prospect of progress, is one of
singular felicity.   It is a question well worth our consideration, as regarding the interests of science, whether,
in other branches of knowledge also, *a continued and connected system of observation and calculation,* imitating the
system employed by astronomers, might not be adopted.
But the discussion of this question would involve us in a
digression too wide for the present occasion.

* *Hist. Ind. Sci.,* ii. 287.

5. There is another mode of application of true theories after their discovery, of which we must also speak; I mean the process of showing that facts, not included in the original induction, and apparently of a different kind, are explained by reasonings founded upon the theory. The history of physical astronomy is full of such events. Thus after Bradley and Wargentin had observed a certain cycle among the perturbations of Jupiter's satellites, Laplace explained this cycle by the doctrine of universal gravitation*. The long inequality of Jupiter and Saturn, the diminution of the obliquity of the ecliptic, the acceleration of the moon's mean motion, were in like manner accounted for by Laplace. The coincidence of the nodes of the moon's equator with those of her orbit was proved to result from mechanical principles by Lagrange. The motions of the recently-discovered planets, and of comets, shown by various mathematicians to be in exact accordance with the theory, are verifications and extensions still more obvious.

6. In many of the cases just noticed, the consistency between the theory, and the consequences thus proved to result from it, is so far from being evident, that the most consummate command of all the powers and aids of mathematical reasoning is needed, to enable the philosopher to arrive at the result. In consequence of this circumstance, the labours just referred to, of Laplace, Lagrange, and others, have been the object of very great and very just admiration. Moreover, the necessary connexion of new facts, at first deemed inexplicable, with principles already known to be true;—a connexion utterly invisible at the outset, and yet at last established with the certainty of demonstration;—strikes us with the delight of a new discovery; and at first sight appears no less admirable than an original induction. Accordingly, men sometimes

* *Hist. Ind. Sci.*, ii. 220.

appear tempted to consider Laplace and other great
mathematicians as persons of a kindred genius to Newton.
We must not forget, however, that there is a great and
essential difference between inductive and deductive processes of the mind.   The discovery of a *new* theory,
which is true, is a step widely distinct from any mere
development of the consequences of a theory already
invented and established.

7. As an example, in another field, of the extension
of a discovery by applying it to the explanation of new
phenomena, we may adduce Wells's *Inquiry into the Cause
of Dew*.   For this investigation, although it has sometimes been praised as an original discovery, was, in fact,
only resolving the phenomenon into principles already
discovered.   The atmologists of the last century were
aware* that the vapour which exists in air in an invisible
state may be condensed into water by cold ; and they had
noticed that there is always a certain temperature, lower
than that of the atmosphere, to which if we depress
bodies, water forms upon them in fine drops.   This temperature is the limit of that which is necessary to constitute vapour, and is hence called the *constituent temperature.*
But these principles were not generally familiar in England till Dr. Wells introduced them into his *Essay on
Dew*, published in 1814; having indeed been in a great
measure led to them by his own experiments and reasonings.   His explanation of Dew,—that it arises from the
coldness of the bodies on which it settles,—was established
with great ingenuity; and is a very elegant confirmation
of the Theory of Constituent Temperature.

8. The example of all the best writers who have previously treated of the philosophy of sciences, from Bacon
to Herschel, draws our attention to those instances of the
application of scientific truths, which are subservient to

* *Hist. Ind. Sci.*, ii. 510.

the uses of practical life; to the support, the preservation, the pleasure of man. It is well known in how large a degree the furtherance of these objects constituted the merit of the *Novum Organon* in the eyes of its author; and the enthusiasm with which men regard these visible and tangible manifestations of the power and advantage which knowledge may bring, has gone on increasing up to our own day. Such useful inventions as we here refer to must always be objects of great philosophical, as well as practical interest; and it might be well worth our while, did our present limits allow, to discuss the bearing of such inventions upon the formation and progress of science. For the present, it must suffice to observe that those practical inventions which are of most importance in the Arts, are rarely or never of any material consequence to Science; for they are either mere practical processes, which the artist practises, but which the scientist cannot account for: or at most, they depend upon some of the inferior generalizations of science for their reason, and do not tend to confirm or illustrate the higher points at which theory has arrived. These considerations must be our apology for not entering into this discussion at the present advanced stage of our undertaking. As we have already said, knowledge is power; but its interest for us in the present work, is not that it is power, but that it is knowledge. The effect which the application of science to general practical uses has in *diffusing* a knowledge of theoretical principles, and thus in giving to men's minds an intellectual culture, is indeed well worthy our attention; but the consideration of this subject must be reserved for some future occasion.

We must now conclude our task by a few words on the subject of inductions involving Ideas ulterior to those already considered.

CHAPTER X.

OF THE INDUCTION OF CAUSES.

1. WE formerly* stated the objects of the researches of Science to be Laws of Phenomena and Causes; and showed the propriety and the necessity of not resting in the former object, but extending our inquiries to the latter also. Inductions, in which phenomena are connected by relations of Space, Time, Number and Resemblance, belong to the former class; and of the Methods applicable to such Inductions we have treated already. In proceeding to Inductions governed by any ulterior Ideas, we can no longer lay down any Special Methods by which our procedure may be directed. A few general remarks are all that we shall offer.

The principal Maxim in such cases of Induction is the obvious one:—that we must be careful to possess and to apply, with perfect clearness and precision, the Fundamental Idea on which the Induction depends.

We may illustrate this in a few cases.

2. *Induction of Substance.*—The Idea of Substance† involves this axiom, that the weight of the whole compound must be equal to the weights of the separate elements, whatever changes the composition or separation of the elements may have occasioned. The application of this Maxim we may term the *Method of the Balance.* We have seen‡ how the memorable revolution in Chemistry, the overthrow of phlogiston, and the establishment of the oxygen theory, was produced by the application of this Method. We have seen too§ that the same Idea leads us to this Maxim;—that *Imponderable Fluids* are not to be admitted as *chemical* elements of bodies.

* Book xi. c. 7.   † *Ib.* vi. c. 3.   ‡ *Ib.* vol. i. p. 397.
§ *Ib.* vol. i. p. 400.

Whether those which have been termed *Imponderable Fluids,*—the supposed fluids which produce the phenomena of Light, Heat, Electricity, Galvanism, Magnetism,—really exist or no, is a question, not merely of the *Laws*, but of the *Causes* of Phenomena. It is, as has already been shown, a question which we cannot help discussing, but which is at present involved in great obscurity. Nor does it appear at all likely that we shall obtain a true view of the cause of Light, Heat, and Electricity, till we have discovered precise and general laws connecting optical, thermotical, and electrical *phenomena* with those chemical doctrines to which the Idea of Substance is necessarily applied.

3. *Induction of Force.*—The inference of *Mechanical Forces* from phenomena has been so abundantly practised, that it is perfectly familiar among scientific inquirers. From the time of Newton, it has been the most common aim of mathematicians; and a persuasion has grown up among them, that mechanical forces, attraction and repulsion, are the only modes of action of the particles of bodies which we shall ultimately have to consider. I have attempted to show that this mode of conception is inadequate to the purposes of sound philosophy;—that the particles of crystals, and the elements of chemical compounds, must be supposed to be combined in some other way than by mere mechanical attraction and repulsion. Faraday has gone further in shaking the usual conceptions of the force exerted, in well-known cases. Among the most noted and conspicuous instances of attraction and repulsion exerted at a distance, were those which take place between electrized bodies. But the eminent electrician just mentioned has endeavoured to establish, by experiments of which it is very difficult to elude the weight, that the action in these cases does not take place at a distance, but is the result of a chain

of intermediate particles connected at every point by forces of another kind*.

4. *Induction of Polarity.*—The forces to which Mr. Faraday ascribes the action in these cases are *Polar Forces.* We have already endeavoured to explain the Idea of Polar Forces ; which implies† that at every point forces exactly equal act in opposite directions ; and thus, in the greater part of their course, neutralize and conceal each other; while at the extremities of the line, being by some cause liberated, they are manifested, still equal and opposite. And the criterion by which this polar character of forces is recognized, is implied in the reasoning of Faraday, on the question of one or two electricities, of which we formerly spoke‡. The maxim is this :—that in the action of polar forces, along with every manifestation of force or property, there exists a corresponding and simultaneous manifestation of an equal and opposite force or property.

5. As it was the habit of the last age to reduce all action to mechanical forces, the present race of physical speculators appears inclined to reduce all forces to polar forces. Mosotti has endeavoured to show that the positive and negative electricities pervade all bodies, and that gravity is only an apparent excess of one of the kinds over the other. As we have seen, Faraday has given strong experimental grounds for believing that the supposed remote actions of electrized bodies are really the effects of polar forces among contiguous particles. If this doctrine were established with regard to all electrical, magnetical and chemical forces, we might ask, whether, while all other forces are polar, gravity really affords a single exception to the universal rule ? Is not the universe pervaded by an omnipresent antagonism, a funda-

* *Researches*, 12th Series. † *Hist. Ind. Sci.*, i. 341.
  ‡ *Ib.* i. 335.

mental conjunction of contraries, everywhere opposite, nowhere independent? We are, as yet, far from the position in which Inductive Science can enable us to answer such inquiries.

6. *Induction of Ulterior Causes.*—The first Induction of a Cause does not close the business of scientific inquiry. Behind proximate causes, there are ulterior causes, perhaps a succession of such. Gravity is the cause of the motions of the planets; but what is the cause of gravity? This is a question which has occupied men's minds from the time of Newton to the present day. Earthquakes and volcanoes are the causes of many geological phenomena; but what is the cause of those subterraneous operations? This inquiry after ulterior causes is an inevitable result from the intellectual constitution of man. He discovers mechanical causes, but he cannot rest in them. He must needs ask, whence it is that matter has its universal power of attracting matter. He discovers polar forces: but even if these be universal, he still desires a further insight into the cause of this polarity. He sees, in organic structures, convincing marks of adaptation to an end: whence, he asks, is this adaptation? He traces in the history of the earth a chain of causes and effects operating through time: but what, he inquires, is the power which holds the end of this chain?

Thus we are referred back from step to step, in the order of causation, in the same manner as, in the palætiological sciences, we were referred back in the order of time. We make discovery after discovery in the various regions of science; each, it may be, satisfactory, and in itself complete, but none final. Something always remains undone. The last question answered, the answer suggests still another question. The strain of music from the lyre of Science flows on, rich and sweet, full and harmonious, but never reaches a close: no cadence is heard with which the intellectual ear can feel satisfied.

Of the Supreme Cause.—In the utterance of Science, no cadence is heard with which the human mind can feel satisfied. Yet we cannot but go on listening for and expecting a satisfactory close. The notion of a cadence appears to be essential to our relish of the music. The idea of some closing strain seems to lurk among our own thoughts, waiting to be articulated in the notes which flow from the knowledge of external nature. The idea of something ultimate in our philosophical researches, something in which the mind can acquiesce, and which will leave us no further questions to ask, of *whence*, and *why*, and *by what power*, seems as if it belonged to us;—as if we could not have it withheld from us by any imperfection or incompleteness in the actual performances of science. What is the meaning of this conviction? What is the reality thus anticipated? Whither does the development of this Idea conduct us?

We have already seen that a difficulty of the same kind, which arises in the contemplation of causes and effects considered as forming an historical series, drives us to the assumption of a First Cause, as an Axiom to which our Idea of Causation in time necessarily leads. And as we were thus guided to a First Cause in order of Succession, the same kind of necessity directs us to a Supreme Cause in order of Causation.

On this most weighty subject it is difficult to speak fitly; and the present is not the proper occasion, even for most of that which may be said. But there are one or two remarks which flow from the general train of the contemplations we have been engaged in, and with which this Work must conclude.

We have seen how different are the kinds of cause to which we are led by scientific researches. *Mechanical*

Forces are insufficient without *Chemical Affinities;* Chemical agencies fail us, and we are compelled to have recourse to *Vital Powers;* Vital Powers cannot be merely physical, and we must believe in something hyperphysical, something of the nature of a *Soul.* Not only do biological inquiries lead us to assume an animal soul, but they drive us much further; they bring before us *Perception,* and *Will* evoked by Perception. Still more, these inquiries disclose to us *Ideas* as the necessary forms of Perception, in the actions of which we ourselves are conscious. We are aware, we cannot help being aware, of our Ideas and our Volitions as belonging to *us,* and thus we pass from *things* to *persons;* we have the idea of *Personality* awakened. And the idea of Design and *Purpose,* of which we are conscious in our own minds, we find reflected back to us, with a distinctness which we cannot overlook, in all the arrangements which constitute the frame of organized beings.

We cannot but reflect how widely diverse are the kinds of principles thus set before us;—by what vast strides we mount from the lower to the higher, as we proceed through that series of causes which the range of the sciences thus brings under our notice. Yet we know how narrow is the range of these sciences when compared with the whole extent of human knowledge. We cannot doubt that on many other subjects, besides those included in physical speculation, man has made out solid and satisfactory trains of connexion;—has discovered clear and indisputable evidence of causation. It is manifest, therefore, that, if we are to attempt to ascend to the Supreme Cause—if we are to try to frame an idea of the Cause of all these subordinate causes;— we must conceive it as more different from any of them, than the most diverse are from each other;—more elevated above the highest, than the highest is above the lowest.

But further;—though the Supreme Cause must thus be inconceivably different from all subordinate causes, and immeasurably elevated above them all, it must still include in itself all that is essential to each of them, by virtue of that very circumstance that it is the Cause of their Causality. Time and Space,—Infinite Time and Infinite Space,—must be among its attributes; for we cannot but conceive Infinite Time and Space as attributes of the Infinite Cause of the Universe. Force and Matter must depend upon it for their efficacy; for we cannot conceive the activity of Force, or the resistance of Matter, to be independent powers. But these are its lower attributes. The Vital Powers, the Animal Soul, which are the Causes of the actions of living things, are only the Effects of the Supreme Cause of Life. And this Cause, even in the lowest forms of organized bodies, and still more in those which stand higher in the scale, involves a reference to Ends and Purposes, in short, to manifest Final Causes. Since this is so, and since, even when we contemplate ourselves in a view studiously narrowed, we still find that we have Ideas, and Will and Personality, it would render our philosophy utterly incoherent and inconsistent with itself, to suppose that Personality, and Ideas, and Will, and Purpose, do not belong to the Supreme Cause from which we derive all that we have and all that we are.

But we may go a step further;—though, in. our present field of speculation, we confine ourselves to knowledge founded on the facts which the external world presents to us, we cannot forget, in speaking of such a theme as that to which we have thus been led, that these are but a small, and the least significant portion of the facts which bear upon it. We cannot fail to recollect that there are facts belonging to the world within us, which more readily and strongly direct our thoughts to the Supreme Cause of all things. We can plainly discern that we have Ideas elevated above the region of mechanical causation, of animal existence, even of mere choice and will, which still have a clear and definite significance, a permanent and indestructible validity. We perceive as a fact, that we have a Conscience, judging of Right and Wrong; that we have Ideas of Moral Good and Evil; that we are compelled to conceive the organization of the moral world, as well as of the vital frame, to be directed to an end and governed by a purpose. And since the Supreme Cause is the cause of these facts, the Origin of these Ideas, we cannot refuse to recognize Him as not only the Maker, but the Governor of the World; as not only a Creative, but a Providential Power; as not only a Universal Father, but an Ultimate Judge.

We have already passed beyond the boundary of those speculations which we proposed to ourselves as the basis of our conclusions. Yet we may be allowed to add one other reflection. If we find in ourselves Ideas of Good and Evil, manifestly bestowed upon us to be the guides of our conduct, which guides we yet find it impossible consistently to obey;—if we find ourselves directed, even by our natural light, to aim at a perfection of our moral nature from which we are constantly deviating through weakness and perverseness; —if, when we thus lapse and err, we can find, in the region of human philosophy, no power which can efface our aberrations, or reconcile our actual with our ideal being, or give us any steady hope and trust with regard to our actions, after we have thus discovered their incongruity with their genuine standard;—if we discern that this is our condition, how can we fail to see that it is in the highest degree consistent with all the indications supplied by such a philosophy as that of which we have been attempting to lay the foundations, that the Supreme

Cause, through whom man exists as a moral being of vast capacities and infinite hopes, should have Himself provided a teaching for our ignorance, a propitiation for our sin, a support for our weakness, a purification and sanctification of our nature?

And thus, in concluding our long survey of the grounds and structure of science, and of the lessons which the study of it teaches us, we find ourselves brought to a point of view in which we can cordially sympathize, and more than sympathize, with all the loftiest expressions of admiration and reverence and hope and trust, which have been uttered by those who in former times have spoken of the elevated thoughts to which the contemplation of the nature and progress of human knowledge gives rise. We can not only hold with Galen, and Harvey, and all the great physiologists, that the organs of animals give evidence of a purpose;— not only assert with Cuvier that this conviction of a purpose can alone enable us to understand every part of every living thing;—not only say with Newton that "every true step made in philosophy brings us nearer to the First Cause, and is on that account highly to be valued;"—and that "the business of natural philosophy is to deduce causes from effects, till we come to the very First Cause, which certainly is not mechanical:" —but we can go much further, and declare, still with Newton, that "this beautiful system could have its origin no other way than by the purpose and command of an intelligent and powerful Being, who governs all things, not as the soul of the world, but as the Lord of the Universe; who is not only God, but Lord and Governor."

When we have advanced so far, there yet remains one step. We may recollect the prayer of one, the master in this school of the philosophy of science: "This also we humbly and earnestly beg;—that human things may not prejudice such as are divine;—neither that from the unlocking of the gates of sense, and the kindling of a greater natural light, anything may arise of incredulity or intellectual night towards divine mysteries; but rather that by our minds thoroughly purged and cleansed from fancy and vanity, and yet subject and perfectly given up to the divine oracles, there may be given unto faith the things that are faith's."   When we are thus prepared for a higher teaching, we may be ready to listen to a greater than Bacon, when he says to those who have sought their God in the material universe, "Whom ye ignorantly worship, him declare I unto you."   And when we recollect how utterly inadequate all human language has been shown to be, to express the nature of that Supreme Cause of the Natural, and Rational, and Moral, and Spiritual world, to which our Philosophy points with trembling finger and shaded eyes, we may receive, with the less wonder but with the more reverence, the declaration which has been vouchsafed to us:

ΕΝ ΑΡΧΗ ΗΝ 'Ο ΛΟΓΟΣ, ΚΑΙ 'Ο ΛΟΓΟΣ ΗΝ ΠΡΟΣ ΤΟΝ ΘΕΟΝ, ΚΑΙ ΘΕΟΣ ΗΝ 'Ο ΛΟΓΟΣ.

THE END.

LONDON. HARRISON AND CO., PRINTERS, ST. MARTIN'S LANE.

9 781108 064033